2-19-58

THE WORLD'S SUGAR

Publications of the Institute

CURRENT SERIES

STUDIES ON FOOD, AGRICULTURE, AND WORLD WAR II
STUDIES IN COMMODITY ECONOMICS AND AGRICULTURAL POLICY
STUDIES IN TROPICAL DEVELOPMENT
MISCELLANEOUS PUBLICATIONS

DISCONTINUED SERIES

COMMODITY POLICY STUDIES   GRAIN ECONOMICS SERIES
FATS AND OILS STUDIES   WAR-PEACE PAMPHLETS
WHEAT STUDIES

A complete list of publications
of the Food Research Institute
will be furnished on request.

A Publication of the
FOOD RESEARCH INSTITUTE
STANFORD UNIVERSITY

One of a group of studies on
FOOD, AGRICULTURE, AND WORLD WAR II

THE WORLD'S SUGAR

Progress and Policy

By
VLADIMIR P. TIMOSHENKO
and
BORIS C. SWERLING

Stanford University Press

STANFORD, CALIFORNIA
1957

FOOD RESEARCH INSTITUTE

Established at Stanford University, Stanford, California, in 1921, jointly by Carnegie Corporation of New York and the Trustees of the Leland Stanford Junior University, for research in the production, distribution, and consumption of food.

STANFORD UNIVERSITY PRESS, STANFORD, CALIFORNIA
LONDON: OXFORD UNIVERSITY PRESS

Library of Congress Catalog Card Number: 57-11674

DIRECTOR'S PREFACE

1017755

This book is the twelfth to appear in the Food Research Institute's projected series of some twenty volumes designed to illuminate the complex aspects of food, agriculture, and World War II. It is one of a group of studies dealing with world-wide commodity developments before, during, and after the war. Coffee, tea, and cocoa were considered in an earlier volume of this group; the world fertilizer economy in another. Monographs on aspects of the world economics of grain, the fats and oils, and the livestock industries are in preparation.

With the passage of time the wartime distortions of the international sugar situation in production, trade, prices, and consumption, and the efforts of governments to cope with them, have tended to diminish in contemporary interest. The authors, members of the Institute's staff, have accordingly chosen not to construct their study as a detailed analysis and record of wartime problems and events in the world of sugar, but to treat the war period as an episode, though a major one, in the long-range evolution of the international sugar market. Major politico-economic problems of sugar currently remain unresolved. Their treatment in perspective in this book promotes understanding, itself a step toward mitigation.

Grateful acknowledgment is made to the Rockefeller Foundation for a grant of funds which facilitated preparation and publication of this work. The Foundation is in no way responsible for the treatment of the subject. The final responsibility, in general and in detail, rests with the authors themselves.

M. K. Bennett
Director

Stanford, California
November 1957

vii

AUTHORS' PREFACE

There is no dearth of published material on sugar. Statistical and scientific data flow forth in a steady stream of trade periodicals, technical journals, and governmental compilations. For some individual sugar areas, careful regional reports are also available as a result of official inquiries or individual scholarship. Detailed descriptions of technology are to be found in a number of standard manuals; policy is administered with the aid of a multitude of executive decrees.

Borrowing heavily from this wealth of sources, the present study seeks to contribute something significant by way of economic interpretation, objective analysis, and broad perspective. It focuses primarily on problems of the international sugar economy. When dealing with national policies, it aspires to a well-balanced appraisal despite the controversial political environment in which this commodity is immersed. Technological questions are of considerable interest, but only in so far as they have general economic relevance. While the book appears within a general series on Food, Agriculture, and World War II, the war years are viewed in the context of a longer historical development.

Both conception and execution of the study are open to some serious objections. A global approach, though it offers interesting interregional comparisons and useful insights, cannot probe deeply into the economy of individual sugar territories. Fully adequate treatment of the problems of the international sugar market would call for a separate chapter on the countries of the Far East, but the study goes to press without one. Omission of a distinct chapter on Latin America is less serious and in any case is compensated for somewhat by the attention accorded to Cuba. While conclusions have been drawn from the technological history of the industry with considerable care, there is a risk that sugar technicians may take exception to a few matters of detail. It can only be hoped that the merits of the study more than offset its deficiencies.

The manuscript owes a great debt to many helping hands. Every member of the staff of the Food Research Institute has made some particular contribution. Special thanks should go to our colleagues Vernon D. Wickizer and Karl Brandt for a critical reading of the entire study; to Miss Rosamond Peirce for statistical assistance; and to Mrs. Catherine Whittemore, Mrs. Patricia Theimer, and P. Stanley

King for skillful services graciously rendered. Mrs. Betty Sutton has borne the greater part of the secretarial burden. Requests for vital pieces of information have been generously handled in Havana, Washington, New York, London, Paris, Brussels, Taipei, and Honolulu. It is with the kind permission of the American Economic Association that verbatim extracts have been taken from two articles originally appearing in the *American Economic Review*.

Collaboration has been so close and for so protracted a period that the parentage of particular portions of the book cannot always be clearly established. It can be said, however, that chapters 9 and 10, as well as discussion of Continental Europe and the USSR in other portions of the book, are the work of the senior author. Chapter 8 has been contributed by R. J. Hammond, and is a product of his scholarship, his intimate knowledge of the British Ministry of Food, and his friendship.

STANFORD UNIVERSITY
July 1956

V. P. T.
B. C. S.

CONTENTS

PART I. INTRODUCTION

PART II. SUGAR CROPS AND AGRICULTURAL DEVELOPMENT

xi

PART III. SUGAR POLICY IN PEACE AND WAR

PART I
INTRODUCTION

CHAPTER 1

INTERCOMMODITY COMPETITION
AND RELATIONSHIPS

Emergency wartime policies, like persistent commodity patterns in time of peace, call attention to the complicated links by which one commodity is connected to others. Some links represent the tight bonds of fixed relationships that seriously limit the scope for human decision; others open up a wide range of alternatives for discretion and social choice. A few, so unobtrusive as to be almost invisible, provide the medium by which pressure at a particular point is transmitted to unexpected places.

In these respects sugar is not unique, but it does afford striking illustrations both of the rigid and of the pliable. For one thing, the white granular end product can be obtained from the sugar beet, a root crop of the temperate zones, or from sugar cane, a giant tropical grass. Competition is fundamentally between the entirely different systems of agricultural economy that these two crops support. While the processing sector is more characterized by fixed relationships between input and output than the agricultural, a variety of end products is obtained in different societies by stopping the process short at subordinate stages. In consumption, as sweetener, as blender, and as preservative, the fortunes of sugar are associated with a wide variety of commodities that includes the grains, fats and oils, fruit and confectionery, beverages and dairy products. Questions of wartime food management, peacetime policy, and technological advance can be more profitably discussed if this intricate commodity network is recognized at the outset.

FACTORY AND FARM

The process of manufacturing sugar is similar though not identical for beets and cane. In each case, sugar syrup must be extracted from the raw product, color and impurities removed, water boiled off at restrained temperatures under vacuum, crystallization induced, and sugar crystals separated by centrifugal action from the final juice, leaving uncrystallizable sugar and residual impurities in the molasses (1, 2, 3). Either process may yield a completely pure, white sugar, or a "raw" sugar that will lose less than 10 per cent of its weight when reprocessed in a separate bone-char refinery. Only in the initial step, extraction of the juice, do the two processes markedly differ. The beet roots, once screened to remove dirt and trash, and washed, are sliced into small chiplike "cossettes." From the great surface area thereby exposed, sugar is removed by a diffusion process, i.e., the denser liquid within the root's cells passes by osmosis through the wall membrane into the thinner surrounding liquid, until the densities of internal liquid and outside water are equalized. While diffusion is technically possible also for cane, it is not practical. Instead, shredders and crushers rupture the cane stalk, and juice is squeezed out by triangular sets of horizontal mills operating at slow speeds and under high pressures, with the addition of varying amounts of maceration water.

What is of particular interest in contrasting the beet-sugar factory and the cane-sugar mill is the nature of the agricultural economy in their immediate vicinity. Pressed beet pulp, equivalent to 50 per cent of the raw crop by weight and containing 8 to 10 per cent dry matter, is valuable as a by-product feedstuff. In Europe, the pulp remaining from a ton of beets is considered to have the feed value of 25 kilograms of barley (*4*, p. 119). Within limits, beet molasses can be mixed with dried pulp and incorporated in the feeding program. In addition, the beet tops and crowns that are removed during the harvest contribute to the protein content of the animal diet. Acreages planted to sugar beets accordingly provide both a cash income from sugar extracted and also incidental products, yields of which compare favorably with alternative feed crops. In Germany, sugar-beet by-products are estimated to have the same feed value as grass and two-thirds the value of fodder beets, i.e., mangels, that could be obtained from the same acreage (*5*, p. 160). In feed-deficit Europe, there is even some feeding of sugar beets whole, notably in Denmark and Germany. In prewar Germany, dried cossettes were occasionally fed to animals without extracting sugar, while low export prices and marketing restrictions forced Czechoslovakia to feed denatured sugar to animals for a time in the 1930's. Some of these practices were exceptional but the broad historical pattern is clear. Introduction of the sugar beet is associated with the transformation of European agriculture from the old three-field system to more advanced crop rotations. Almost everywhere it is typified by a highly diversified farm operation in which cereals are grown as well as root crops, and livestock is produced.

In place of rotation, diversification, and a crop-and-livestock economy, major cane areas are typically highly specilized monocultures externally oriented both for markets and for inshipments of the basic food supply. The mere fact that cane is a perennial, available for successive croppings, limits competition from other crops for use of the land. The bagasse, the fibrous residue remaining after the cane has been milled, is generally required to fire the furnaces in fuel-deficit tropical areas. Unlike beet pulp, it is not directly usable for feed, though Hawaiian producers have recently accomplished the separation of edible pith from inedible fibers (*6*, pp. 8–9). Cane tops and leaves are a far smaller portion of the total plant than beet top and crown, and have far less feed value. Though draft animals may forage on these field by-products, cane leaves are in fact described as "trash" and, instead of being carefully harvested like beet leaves in Europe, are frequently burned off the field to facilitate the cutting of cane. Animals may be foddered on young cane, but the operation is not economical; in Australia, where this use is reported separately, less than 2 per cent of the cane acreage goes for this purpose (*7*, p. 986). To be sure, cane molasses outrates beet molasses for animal feeding. But since it must be combined with some bulkier material it finds a better outlet in the mixed-feeds industry of metropolitan countries than in tropical agriculture. Increasingly, it is also being sold direct to American farmers in liquid form for increasing the palatability of hay and other coarse roughage. Indeed, 6½ gallons (11.7 pounds per gallon) are considered to have the approximate feed value of a bushel of corn in the United States (*33*, p. 1), and a still more satisfactory product is provided as molasses with urea, high in nitrogen content.

To speak of cane monoculture is no exaggeration of the degree to which this crop can dominate the economy of a region. Particularly in certain offshore territories of metropolitan countries, sugar overshadows all other commodities, in export trade and in local agriculture alike. The portion of arable land devoted to cane can run extremely high. The figure is almost nine-tenths in British Mauritius, and over two-thirds in Barbados; in French Martinique and Réunion it is not far below one-half. In Hawaii and Puerto Rico, the percentages are 72 and 44 respectively (*8*). By some, so high a degree of specialization has been considered a symptom of commercial colonialism in such independent nations as Cuba, where an estimated three-quarters of the cultivated land is planted to cane. The same attitude has been common in the Philippines and Indonesia, though in these cases the share of sugar in the total local agriculture is minor. In fact, it is more appropriate to look for an economic than a colonial explanation for monoculture. No dependent territory is more heavily committed to cane production than are sections of Australia's northeast coast, where cane occupies as much as 99 per cent of all cultivated land (*9*, p. 40). Tucuman in Argentina, the sugar parishes of Louisiana, the northeastern states of Brazil, and Natal in South Africa are cane-dominated segments of national economic systems.

Relations between crops, in both beet and cane regions, are naturally more complex than our description of the typical case might suggest. Into the standard rotation system based on root crops and cereals, the grower on irrigated land of the western United States may also incorporate alfalfa as a forage crop and field beans or potatoes for market (*10*, pp. 23–42). In Europe, beets must compete with grain and mangels for their place in the rotation (*11*, pp. 238–44), and consequently larger beet acreages may be planted in the spring if winter wheat and barley suffer crop damage. It is on the irrigated lands of California that the range of choice is widest in response to market incentives of the moment. Truck and canning crops, especially lettuce and tomatoes, and even cotton, are direct competitors for water and land in that state, the standard rotation with cereals being usually uneconomical (*12*, p. 22). When United States price support programs come to be implemented by crop restrictions, the shift can be from beets to potatoes in the Red River Valley area of Minnesota, from wheat to beets in the Pacific Northwest, from cotton to beets in California.

Historically, the introduction of sugar cane outside the Orient has been associated more with reclaiming virgin land or turning cattle ranges to field crops than with bidding the land away from other cultivated plants. However, the same increase in competition from overseas grains that promoted expansion of beets at the expense of cereals in nineteenth-century Europe promoted cane in place of wheat in the upper Nile (*13*, p. 15). Over long periods, land shifts slowly between cane and pineapples in Hawaii, out of bananas and into cane or from cane to citrus in Jamaica (*14*, p. 115), from cane to cotton in the coastal Canete Valley of central Peru (*15*, p. 122), from cane to rice in British Guiana.

Only in the densely populated Orient does cane come into close annual competition with other crops, notably rice. In Formosa, wherever water is in ample supply, there is a rational response to changes in the relative prices of cane and rice. This contest was tempered before the war in the large irrigation project in the Kanan Taishu area. Water was apportioned so as to enforce a rigid 3-year rotation including cane, rice, and an upland crop, commonly peanuts or sweet potatoes (*27*, pp. 57–62). In the lowlands of east and central Java before the war, an 18-month cane crop in rotation with rice and an upland crop (here usually corn or cassava) was equally standard practice (*16*, pp. 10–11). Cane is also rotated with rice in some parts of the Philippines (*17*, p. 138), while flax, vegetables, peanuts, or sweet potatoes may be interplanted with sugar cane to increase the yield of Formosan fields (*18*, p. 9). The major Western analogue of Oriental rotation is in Louisiana, where two or three cane crops have been followed by corn and legumes as a matter of scientific agricultural practice. There is also an Eastern parallel to the advance of sugar crops in the United States when wheat or cotton acreages are cut back: in Pakistan, cane plantings increased somewhat when jute acreages were restricted (*19*, p. 9).

New exigencies arising as a result of war forced severe readjustments in certain cane areas. Endowed with superabundant sugar supplies in conquered territories, the Japanese pushed cotton and local food crops almost to the exclusion of sugar cane in the Philippines and Indonesia and, in Formosa, ultimately reversed a policy that had favored sugar exports over rice (*20*, p. 10). Shipping shortages of necessity required conversion of some cane land to local foodstuffs elsewhere: manioc and sweet potatoes in Mauritius, peanuts and rice in Cuba (*21*, p. 31). The official requirement that local foodstuffs be cropped along with cane has been continued in Barbados, but mechanization of agriculture has helped free acreage formerly required to feed draft animals and there has been no sacrifice of sugar output. In the early postwar years, India's "Grow More Food" campaign classified cane as an industrial crop and promoted other crops at its expense. The war itself presented a temporary opportunity to persons disturbed by certain unattractive features of the cane monoculture. Francis B. Sayre, the United States High Commissioner, considered wartime damage to Philippine sugar factories a blessing in disguise, and urged that sugar be neglected in postwar reconstruction of the national economy (*22*, pp. 207–08). "Diversification" has frequently been suggested as the answer to Cuba's economic problems (cf. *23*, p. 7). What is impressive about those experiences is the tenacity with which cane has tended to reassert itself, even in such a place as Indonesia, where the local environment was not particularly hospitable.

Throughout the world, the extraction of sugar from beets today is accomplished by means of heavy capital equipment. To be sure, during the Second World War beets were extended to new regions of the USSR beyond the range of existing factories, and improvised extraction in the form of syrup had to suffice. While those were exceptional circumstances, primitive processing is not uncommon for cane. In some areas, people simply chew the cane stalk. There are many regions where the juice is squeezed out by inefficient rollers, and the juice boiled to a hard mass of varying sweetness and stickiness, containing more or less sugar, residual moisture, and impurities. Production of these noncentrifugal grades exceeds centrifugal sugar by a wide margin in India, Pakistan, Colombia, mainland China, and Burma, and there is also an important output in Brazil, Mexico, and Venezuela, under a host of local names: *muscovado, chancaca* (Peru), *piloncillo, panela* (Colombia), *panocha* (Philippines), and *rapadura*. Most important of all is the *gur* of India, which comprises three-quarters of a national sugar production second only to Cuba's. The Philippines in the 1920's and Brazil in the 1930's represent regions where cane was steadily diverted from inferior processing to modern mills, but white sugar and gur to some degree compete regularly for India's cane on a price basis. High sugar prices tend to pull cane away from gur, resulting in lower gur consumption and high gur prices, with a likelihood that the pull will be reversed the following season. This competition has been complicated in the postwar period by government policy, which has typically set a price ceiling on white sugar and on cane destined for the factory but not on the gur industry. Finally, as an alternative to the crystalline form, sugar may be used as a liquid. Some 10 per cent of Egypt's crop is converted to a syrup known as "honey sugar"; while a portion of the Barbados crop ends up as so-called "fancy molasses" of high purity and characteristic flavor.

The manufacture of beet sugar and cane sugar results in incidental by-products adaptable to a wide range of uses. Some, though involving large capital investment for individual productive units, are of quite secondary importance in aggregate. A waxy substance, present in cane in concentrations of little more than one-tenth of one per cent, is extracted in Australia and in Cuba for polishes, copying carbon, etc., uses in which its properties are similar to Brazil's carnauba wax. Monosodium glutamate, a flavor intensifier, is a minor product processed from beet molasses in Ohio and California. Though sugar cane may contain as much fiber as it does sucrose, bagasse is left over for bulk industrial use only when other local fuels exist or boiler efficiency is exceptional. In Peru and elsewhere, bagasse is in fact converted into paper products: newsprint, in place of the wood-pulp product; corrugated cardboard and commercial papers that displace other paper or cotton containers. Cuba, Hawaii, Australia, and Louisiana manufacture from bagasse various wallboards and insulating sheets that serve instead of lumber and alternative building materials. The carbohydrates in bagasse can also be broken down chemically for the production of furfural, an industrial chemical used in making nylons, which consequently provides a minor link with the textile industry.

By far the major industrial by-product is molasses, the tonnage of which is about one-quarter that of the sugar recovered from beets or cane. If price incentives are sufficient, as they are in the United States, there will be secondary extraction of sugar from beet molasses (*2*, p. 438). In addition to the considerable portion of molasses that ends up in the form of animal products, some goes into edible uses and the food-processing industries, whether as rum and other spirituous beverages, as carbon dioxide, or as yeast or vinegar. More is converted by fermentation and distillation into ethyl alcohol—an industrial chemical widely used for solvents and plastics; for aldehydes, to be converted into such products as acetic anhydride for the rayon textile industry; for synthetic rubber; for antifreeze; and for anhydrous or fuel alcohol (*24*). Not merely the utilization but also the production of industrial alcohol adds major commodities to the list of those significantly linked to the sugar industry, for alcohol can be obtained from such alternative carbohydrates as the grains and potatoes, from reprocessing beverage alcohols and wines, as well as synthetically from petroleum products. In some uses, beet and cane molasses compete directly with hydrol, a by-product of the wet corn-milling industry, and with citrus molasses, a by-product of canning operations.

The competitive position of molasses in industry varies dramatically from country to country and as between war and peace. Rum has long been manufactured in the West Indies but the aggregate production figures are not impressive by the standards of today. In recent years, production of dried food yeast has been undertaken in Jamaica to help offset the serious protein and B-vitamin deficiencies in tropical diets, but the experiment failed because entrenched food habits could not be overcome. Petroleum-deficit countries may favor subsidized fuel alcohol over imported fuels. Thus Brazil enforces the addition of stated proportions of anhydrous alcohol to gasoline, and indeed seeks to regulate domestic sugar prices by regulating the flow of cane as between alcohol and sugar. In France, beet growers share with grape and apple interests an outright subsidy to the agricultural sector that takes the form of government purchases of industrial alcohol processed directly from sugar beets, indirectly from beet molasses, or from wine and cider. In Germany, a similar role is reserved for potato growers. Most of the French alcohol is normally marketed as fuel under compulsory admixture regulations similar to those in Brazil. When market weakness forces Cuba to restrict its harvest of standing cane, considerable tonnages may be processed for industry as invert syrup, a high-grade molasses from which no sucrose has been extracted, while frost-damaged cane may be similarly salvaged in Louisiana or Florida. Several of these programs seek merely to dispose of sugar surpluses, but sugar-crop acreages may in fact be expanded mainly for the industrial alcohol they will yield. Italy increased beet production in the 1930's to counteract petroleum sanctions at the time of the Ethiopian war, and Czechoslovakia did the same during World War II to help meet Germany's need for synthetic rubber. Alcohol under such circumstances becomes as much a joint product as a by-product of the sugar industry.

As a raw material for industrial chemicals in the United States economy, cane molasses has within the past two decades been subject to rapid shifts in fortune as a result of changing technology (24). Though industrial alcohol formerly went overwhelmingly into solvents, more than half the output is now absorbed by the aldehydes. This link with rayon and the synthetic fibers, as well as lesser ties with the rapidly growing plastics industry, has given a definite upward trend to industrial alcohol production, but molasses has benefited little from this rising trend. Until quite recently, more than 70 per cent of ethyl alcohol was based on fermentation of molasses. Now a synthetic alcohol from petroleum products has gradually taken over more than half the market, and industrial alcohol is no longer the major outlet for molasses. Even enormous alcohol requirements for synthetic rubber production brought only sporadic benefit, creating a highly inflated demand during World War II and again during 1950–51. The synthetic alcohol industry has enjoyed less volatility by virtue of long-term supply contracts with petroleum producers. Competitive displacement from the industrial market, as compared with an expanding market for animal products in the American diet, has pushed molasses increasingly into feed use in the last decade, to the point that in some postwar years more molasses has been fed than has gone to all other uses combined.

At no time does the full range of commodity interrelationships become more evident than during actual hostilities, when the resources of an economy are under severest pressure. Food and industrial uses come into direct competition since cane can go directly into high-grade molasses, with higher consequent recovery of industrial alcohol, or into sugar, with only incidental production of by-product blackstrap (*25*, pp. 28–33; *26*).

The war superimposed upon standard peacetime outlets for industrial alcohol the heavy requirements of smokeless powder and of antifreeze for military vehicles. Alcohol production in the United States had to be quadrupled between 1940 and 1944. By the latter date, however, the entirely new synthetic rubber industry was absorbing fully one-half the supply. As an alternative to molasses, grains provided a major raw material which could yield alcohol by fermentation and distillation. But the Japanese conquest of Southeast Asia, which was responsible for denying the Allies its normal sources of natural rubber, had also created a serious shortage of vegetable oilseeds. Though the United States entered the war with large stocks of corn and wheat, much of the corn was dissipated during 1943 in feeding heavy hogs, partly for meat but secondarily for lard, while stocks of wheat also became inadequate to satisfy all claims made upon them. To economize on the use of alcohol-processing facilities, the manufacture of whisky was for a time prohibited. So intense became the demand for industrial alcohol that various regional interests strongly urged the utilization of their particular surplus carbohydrates, whether inedible sweet potatoes in Florida or sawmill wastes in the Pacific Northwest. Petroleum products, which had been heavily counted on as the raw material for synthetic rubber manufacture, came under comparable pressure from military requirements, particularly for high-octane aviation gasoline. The emerging feed shortage put molasses in heavy demand from the agricultural sector as well.

Disposition of Cuba's sugar cane in wartime reflected not merely these variable external requirements, but also the local need for fuel alcohol to compensate for a shortage of landed petroleum products on the island, while prohibition of whisky manufacture in the United States diverted increasing quantities of Cuban molasses into a highly profitable rum business. The tight situation in the industrial-chemical field was one factor that contributed to the introduction of sugar rationing in the United States in 1942, while sacrifice of a potential 900,000 tons of sugar to invert syrup for the synthetic rubber program contributed to the severe sugar shortage of 1944 and 1945. Such were the close ties between grain, rubber, petroleum, vegetable oils, animal fats, whisky, and sugar. The size of the American sugar ration responded to such superficially unrelated factors as the amount of petroleum required for high-octane gasoline, the volume of grain stocks, the rate of sinkings of molasses-carrying ocean tankers, and the size of the animal population. Intelligent alteration of normal peacetime relations between inputs and outputs was the essence of wartime adaptation.

SUGAR AND OTHER FOODS

Despite the very different production organizations they support, pure beet sugar and pure cane sugar are chemically indistinguishable and perfect substitutes one for the other as foodstuffs. Their primary function is as sweetener, and in that role they gradually push aside a variety of primitive sweeteners as levels of living advance. Honey and maple sugar still play a minor role, while in the East the inflorescence of the coconut palm yields sticky juice that affords a sugar (*16*, p. 65). In Mediterranean regions, where natural fruits, dates, and raisins are available in abundance and consumption of wine is high, correspondingly less sugar may be used. Availability of such modern processing facilities as wet milling of corn has put corn sugar (dextrose) and corn syrup (glucose) into serious competition with sucrose for some purposes. Until after the Civil War, refined sugar was a luxury commodity even in the United States; many farms grew a sugar sorghum for a home-processed syrup, and indeed the practice has not altogether disappeared in the southeastern states. In times of stringency, peoples tend to revert to some of the more primitive substitutes, including the Japanese shift to sweet-potato starch during World War II.

In the Middle Ages, sugar was a high luxury in Western Europe, used mainly in exotic sweets and pharmaceutical preparations. While sugar plays a few dramatic roles in modern medicine, as an extender for blood plasma and for intravenous feeding, its chief uses involve combination with other foodstuffs. Indeed, its place as an article of mass consumption in the modern diet begins in association with the beverages and the growing popularity of tea, coffee, and cocoa in the seventeenth century. Confectionery has been of continuing importance, with tree nuts, fruits, chocolate, and desiccated coconut typical concomitants. For beverages and for confection, sugar continues to make its contribution mainly as sweetener. But the cheapening of sugar has made other of its properties increasingly important. As blender, it combines with flour in bread, cake, and pastry, with dairy products in ice cream, and with fats in both. As preservative of fruits and vegetables, it is a key ingredient of canned, bottled, and frozen foodstuffs, to the point where "preserves" have replaced brine preparations in the family pantry. Mediterranean areas, with the possibility of converting their surplus fruit into storable form, may therefore become heavy sucrose users despite the large availability of natural sugars (*28*, p. 47). Soft beverages have come in as an important form of consumption, contending for a place in competition with natural fruit juices, fermented fruit beverages, and beer.

Particularly in recent decades, rising consumption levels have accordingly meant a relative decline in the direct use of sugar in the household, and its close association with modern improvements in commercial food processing, preserving, canning, dehydrating, and freezing. Indeed, the entire increase in United States sugar consumption since 1929 has been as a raw material for food manufacturing. Interesting repercussions can be identified, on industrial organization and on the problem of social control. New patterns of vertical integration become appropriate. Caribbean sugar companies become investment outlets not only for metropolitan refiners or shipping and trading firms, but also for chocolate manufacturers and soft-beverage companies, while canning interests may protect their source of supply by purchase of sugar enterprises. During the war, some American manufacturers of soft beverages sought to control syrup manufacturing plants beyond the United States border, just as attempts to evade rationing in postwar England included an unusually heavy import of fondant and other fat-sugar mixtures. Since even less satisfactory forms could sometimes serve, some Caribbean sugar was diverted to local manufacture of low-quality hard candy for export, during the period of greatest wartime stringency. Where the food-manufacturing sector absorbs a high proportion of the national sugar supply, severe rationing dislocates a major industry at the same time that it forces the final consumer to tighten his belt.

This role as raw material for the food manufacturing industry has incidentally also altered the form in which sugar leaves the refinery, and the relations among sweeteners. Though raw and inferior grades of sugar have little industrial use, the fact that manufacturers frequently must mix sugar with water has given rise to the production of liquid sugar in the United States, a peculiar reversion from granulated article to syrup. To that extent, jute, cotton, and paper containers give way to tanks, tank cars, and piping. On the West Coast, tank cars originally designed for carrying petroleum now haul sugar, while it is not unusual in some areas for stainless-steel milk trucks to carry liquid sugar on their return haul (*29*, p. 12). The switch to liquid, moreover, has been to the competitive disadvantage of beet sugar. Liquid cane sugar can be obtained by stopping the concentration of the mother liquor short of crystallization, whereas liquid beet sugar involves the more expensive process of granulation and remelting. Industrial use has also given a particular opportunity to corn sugar and corn syrup, which perform particularly well in bread baking, confectionery, ice cream and sherbets, and for bringing out natural fruit flavors in the canning industry (*30*, pp. 2–3).

While its major role, whether in the food industry or within the household, is that of a joint ingredient, sugar has come to occupy an important position quantitatively in the modern diet. In high-income countries, sugar may now provide 15 per cent and more of the total calories consumed. Over long periods, therefore, per capita consumption of sugar tends to move inversely with utilization of such starchy staples as potatoes and bread, which it in a sense displaces. At the same time, it is a pure carbohydrate with none of the nutritional advantages of the protective foods, no minerals, no vitamins, no protein. Quantitative importance with nutritional deficiency gives rise to two conflicting attitudes. On the one hand, a nutrition-minded public is regularly reminded of the virtues of better balanced or lower caloric foodstuffs. In the early days of wartime sugar shortage in the United States, attitudes of this sort were deliberately fostered by governmental agencies. But wartime food policy in the United Kingdom more commonly classifies sugar with fats, meat, and bacon—commodities of high food value and relatively small bulk, which make less heavy demands on shipping space than do the customary peacetime cargoes of food grains and feedstuffs (*31*, p. 21).

This wide net of commodity linkages creates major complications when a wartime ministry of foods is substituted for the sensitive transmission mechanism that is the price system. To be sure, the contests between beet and cane as systems of production, or between sugar and cereal in the diet, take place slowly, almost imperceptibly, over long periods of time. But in wartime the competition for land, for materials, for shipping space, even for the consumer's dollar, is unusually intense. At least the political environment in which the food administrator operates alerts him to ramifications of every decision, if only through the complaints of affected groups. Indeed the price of sugar in wartime Britain had on occasion to be reduced, to the discomfort of food officialdom (*32*, pp. 182–93), in order to stabilize the cost-of-living index in the face of rising prices for clothing. The administrator, happily, does not always have his way. But wartime food management can be intelligently directed only if there is full knowledge of intercommodity relationships: the scope for choice, the feasible improvisation, and the limits of discretion.

CITATIONS

1 G. L. Spencer and G. P. Meade, *Cane-Sugar Handbook* (New York, 1945).

2 R. A. McGinnis, ed., *Beet-Sugar Technology* (New York, 1951).

3 Andrew van Hook, *Sugar: Its Production, Technology and Uses* (New York, 1949).

4 J. Dubourg, *Sucrerie de Betteraves* (Paris, 1952).

5 Hans Lüdecke, *Zuckerrübenbau* (Hamburg, 1953).

6 Hawaiian Sugar Planters' Association, *Annual Report of the President, 1952* (Honolulu).

7 Australia, Commonwealth Bur. Census and Stat., *Year Book of the Commonwealth of Australia, 1953.*

8 Food and Agriculture Organization of the UN (FAO), *Yearbook of Food and Agricultural Statistics 1953: Part 1, Production* (Rome, 1954).

9 Queensland Cane Growers' Association *et al.*, *The Australian Sugar Year Book 1951* (Brisbane).

10 J. D. Black and C. T. Corson, *Sugar: Produce or Import?* (Carnegie Endowment for International Peace, Agricultural Series, No. 6, Berkeley, Calif., 1947).

11 E. Woermann, "Betriebswirtschaftliche Fragen des deutschen Zuckerrübenbaus," *Agrarwirtschaft* (Hanover), September 1952.

12 C. H. Wadleigh, "Expansion of Research on Sugar Beets . . . ," *The California Sugar Beet 1953* (Stockton).

13 Charles Issawi, *Egypt: An Economic and Social Analysis* (London, 1947).

14 Caribbean Commission, *Monthly Information Bulletin* (Port - of - Spain), December 1953.

15 F. D. Barlowe, Jr., *Cotton in South America* (Memphis, 1952).

16 J. E. Metcalf, *The Agricultural Economy of Indonesia* (U.S. Dept. Agr., Agricultural Monograph 15, 1952).

17 D. H. Grist, *Rice* (London, 1953).

18 Taiwan, Provincial Agr. College and Joint Commis. on Rural Reconstruction, "Inter-Regional Competition between Rice and Sugarcane . . ." (by S. C. Hsieh, Taipei, 1953, mimeo.).

19 F. O. Licht's *Sugar Information Service* (Ratzeburg), March 1, 1954.

20 Mutual Security Administration, Mission to China, *Economic Development of Formosa 1951–1952* (Taipei, 1953).

21 E. B. Wilson, *Sugar and Its Wartime Controls 1941–1947* (New York, n.d.).

22 E. H. Jacoby, *Agrarian Unrest in Southeast Asia* (New York, 1949).

23 International Bank for Reconstruction and Development, *Report on Cuba* (Washington, D.C., 1951).

24 U.S. Dept. Agr., Production and Marketing Administration (PMA), Sugar Branch, *Marketing Industrial Molasses* (by B. K. Doyle, Agr. Inf. Bull. No. 82, 1951).

25 P. G. Berdeshevsky, "Molasses: A Carbohydrate at War," *Sugar*, July 1944.

26 U.S., Civilian Prod. Admin., Bur. of Demobilization, *Alcohol Policies of the War Production Board and Predecessor Agencies May 1940 to January 1945* (by Virginia Turrell, Historical Reports on War Administration, WPB, Special Study No. 16, 1946).

27 Saburi Ebi, "Sugar Industry of Java and Formosa—A Comparative Study" (Econ. Coop. Admin., Mission to China, Tokyo, 1947, mimeo.).

28 FAO, *Sugar*, Commodity Series, Bull. No. 22 (Rome, 1952).

29 U.S. Dept. Agr., PMA, *Marketing Liquid Sugar* (by F. J. Poats, Marketing Research Rept. No. 52, 1953).

30 U.S. Dept. Agr., PMA, *Competitive Relationships Between Sugar and Corn Sweeteners* (Agr. Inf. Bull. No. 48, 1951).

31 E. H. Whetham, *British Farming 1939–49* (London, 1952).

32 R. J. Hammond, *Food: The Growth of Policy* (History of the Second World War, U.K. Civil Series, H.M.S.O., 1951).

33 U.S. Dept. Agr., PMA, Sugar Branch, *Feeding Molasses to Livestock* (1953).

CHAPTER 2

SUPPLY: AN OVER-ALL VIEW

Sugar remains one of the leading staples in international commerce, despite the fact that it is produced in a host of countries throughout temperate and tropical regions. Including shipments to metropolitan countries from offshore possessions or dependencies, some 12 million metric tons,[1] or over one-third of world production, moved across national frontiers on the eve of World War II (*1, 2*). This great traffic has been characterized by considerable irregularity both in direction and in volume. Major shifts in peacetime commercial policies have had the effect of abruptly opening or closing available markets. A commodity so committed to overseas trade was bound to be peculiarly vulnerable to the changing naval fortunes of war, while the sugar-beet system of Continental Europe suffered repeated dislocation as the result of territorial realignments. The effects of policy and war on national supplies have been further complicated by major developments on the technological front, which have affected beet and cane at different dates and in unequal degree.

The nineteenth century was peculiarly associated with the ascendancy of beet sugar. Abolition of the slave trade after 1833 disrupted the social system upon which the cane-sugar exports of the West Indies had been organized. The era of duty-free import of sugar into England, which lasted from 1874 to 1901, completed the disintegration of the Jamaican plantations (*3*, p. 162) and opened wide the key British market to beet sugar from Europe. Major technological improvements in processing sugar syrups—multiple-effect evaporators, vacuum pans, centrifugal machines—represented the successful application of French and German science to the beet-sugar industry, just as the selection of commercial varieties of beets was an outstanding achievement of applied agricultural science. Under the stimulus of government subsidization, the share of beet in world

---

[1] Sugar statistics are variously reported, in English short tons of 2,000 pounds each, metric tons of 2,204.6 pounds, English long tons of 2,240 pounds, and Spanish long tons of 2,271.6 pounds. For the most part, the conventional national unit will be used when national issues are under discussion, and the metric system for international comparisons.

sugar production had reached 15 per cent by 1850 and climbed to two-thirds by 1900 (*4*, p. 21). Fundamentally, beet sugar was for domestic consumption, but beet supplied one-fourth of world exports as late as 1914 (*5*, p. 145). At that date, England had no local production and depended on imports from the Continent for the bulk of her sugar supplies.

This ascendancy, however, was being undermined even before the First World War devastated Europe's beet-sugar economy. Subsidization of beet-sugar exports had become so expensive as to be a burden to the public treasuries in Germany, Austria-Hungary, and Russia. By agreeing to embargo imports of subsidized sugar, England paved the way for the Brussels Sugar Convention of 1902, which largely removed the artificial competition faced by tropical sources of supply. Moreover, the transmission of improved factory equipment to cane areas, especially Louisiana, Cuba, Hawaii, and Java, had begun before 1900; the successful application of steam to milling operations had already paved the way for the modern cane factory; and cane agriculture was also coming in for scientific attention, notably in Java and Hawaii. Indeed Cuban production passed one million tons in the early 1890's, before revolt against Spain caused a severe setback (*6*, p. 262). But outstanding achievements in cane processing came following 1900, after the Spanish-American War (1898–1902) opened the way for a heavy export of American capital to formerly Spanish territories—Cuba, Puerto Rico, and the Philippines. Although the development of beet-sugar production on the United States mainland after 1890 represented a major overseas offshoot of the European beet system, by 1913 the world was again producing more sugar from cane than from beets.

THE INTERWAR PERIOD

World War I severely disrupted the European beet economy. Much of the fighting on the Western Front surged over the French beet departments of Aisne, Pas-de-Calais, Nord, Somme, and Oise. Production fell lower still during the years of economic disorganization immediately following the war. New territorial boundaries further unsettled national supply systems. As against a European beet-sugar production of 8.3 million metric tons in 1913/14, the total in 1919/20 was only 2.6 million (*7*, p. 106). The extent of decline in Germany, France, and the USSR is indicated in Chart 1D. Some beet-sugar regions formerly controlled by Germany, Russia, and Austria-Hungary were taken over by Poland and Czechoslovakia, which now became sugar-surplus areas. Not until 1928/29 did European production surpass its prewar level.

The gap in European supplies, and rising levels of sugar consumption in the United States and Japan, permitted such cane areas as Cuba, Java, and the Dominican Republic to utilize their modernized milling facilities in production for export. But the basis of their prosperity was beginning to slip away in Europe as well as overseas. Great Britain entered the ranks of beet producers by introducing a direct subsidy on home-grown beets effective October 1, 1924. Recovery of beet production on the Continent (excluding the USSR) had progressed so far by the middle 1920's that Czechoslovakia found it necessary to reduce its output for lack of export markets. Intensified protection of beet sugar arising out of the Great Depression implied further difficulties for exporting nations.

Expansion of protected supplies of cane sugar proceeded more rapidly still. Higher output in certain Latin-American countries, notably Argentina, Mexico, and Brazil, was almost exclusively for local consumption and had little effect on international markets. But the rise of Indian production of white sugar under protectionist policies introduced in 1930 and 1931 represented a direct displacement of imports from Java; Japan similarly succeeded in establishing Formosa as a base of sugar self-sufficiency by the end of the 1920's, also at the expense of the trade with Java. These market adversities in the Far East unfortunately coincided with major achievements in agricultural productivity, for Java had just succeeded in developing new varieties of cane capable of exceptionally high yields.

The policy environment in the United Kingdom and in the United States was highly complicated. To be reconciled were the somewhat conflicting interests of domestic beet producers, metropolitan cane refiners, offshore cane producers, and foreign suppliers, not to mention the consumer and the public treasury.

Although Britain maintained free trade in sugar from 1874 to 1901, and only a light revenue tariff immediately after that date, a substantial wartime tariff was continued after 1918 for both revenue and protectionist purposes. Imperial preferences were introduced in 1919, the beet subsidy in 1924. Home production of beet sugar exceeded one-half million tons by 1934–35. Imports had been 80 per cent beet sugar before World War I, but were 92 per cent cane sugar by 1930. By 1937, of Britain's sugar consumption 60 per cent was Empire-grown, compared with a trifling 4 per cent in 1913 and

CHART 1 A–C.—WORLD SUGAR PRODUCTION, 1913–37*

(*Million metric tons, raw value*)

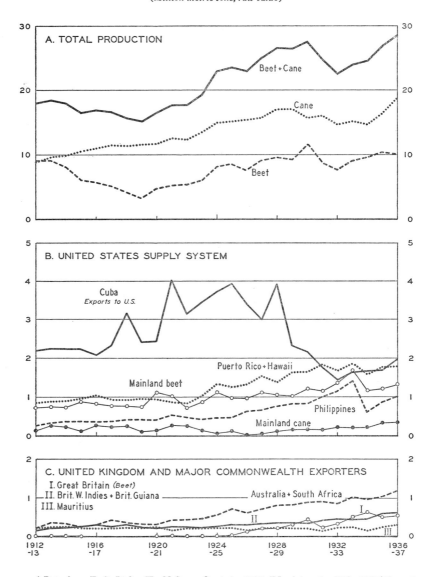

* Data from F. O. Licht, *World Sugar Statistics 1937* (Magdeburg); *ibid.*, *1937/38*; and *Anuario Azucarero de Cuba 1949* (Havana), p. 111.

CHART 1 D–F.—WORLD SUGAR PRODUCTION, 1913–37*

(*Million metric tons, raw value*)

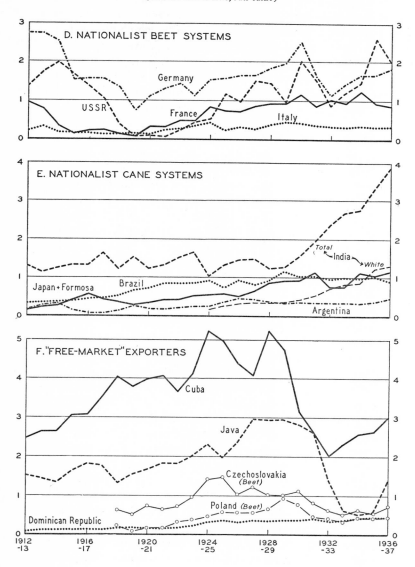

D. NATIONALIST BEET SYSTEMS

Germany

USSR

France

Italy

E. NATIONALIST CANE SYSTEMS

Total
India
White

Japan + Formosa

Brazil

Argentina

F. "FREE-MARKET" EXPORTERS

Cuba

Java

Czechoslovakia
(*Beet*)

Poland (*Beet*)

Dominican Republic

1912
-13
1916
-17
1920
-21
1924
-25
1928
-29
1932
-33
1936
-37

* Sources as for Chart 1 A–C.

24 per cent in the middle 1920's. Two Dominions, Australia and South Africa, as well as colonial territories such as the British West Indies and Mauritius, shared in this advance. The sugar output of the British Empire (excluding that of India, which doubled) in 1937 was four to five times as high as in 1913. Moreover, while imports of beet sugar had typically been in refined form, the tariff schedule in 1928 was framed to promote imports as raw. That measure was a particular blow to Czech beet-sugar exports, but provided the basis for an important re-export business by British cane-sugar refiners.

The main developments in American sugar supply have been the notable growth of insular cane-sugar production (in Hawaii and Puerto Rico, as well as the Philippines), the rise of a domestic beet-sugar industry since 1890 to a level exceeded only by prewar Germany and the USSR, the recovery of Louisiana cane production from the effects of cane disease in the 1920's, and the expansion of Florida cane production. In addition, preferential treatment of imports from Cuba since the Reciprocity Treaty of 1903 had brought the almost complete displacement of full-duty sugars by 1912. Until 1934 the tariff remained the major instrument of American sugar policy, providing the basis for slow growth in Hawaii and the beet areas. But tariff increases in 1921 and 1922 fostered more rapid expansion in Puerto Rico and particularly in the Philippines. When the high Smoot-Hawley tariff was introduced in 1930, the chief beneficiaries again turned out to be these offshore producers who now displaced about one million tons of Cuban sugar and offered an increasing threat to mainland sugar interests as well.

This combination of sugar policies in the United States and the United Kingdom, which together remained the destination for 60 per cent of the overseas trade in sugar, as well as measures being taken in Europe and the Far East, placed exporters to nonpreferential markets in an increasingly precarious position. Total Cuban exports reached nearly 5 million tons in 1925, as against half that amount in 1914, but progressive displacement from the United States market cut exports below 2 million tons in the early 1930's, despite relatively stable sales to England and France (*8*, p. 117). Java, lacking either a preferential market abroad or a sizable domestic outlet, became a still more striking victim of the imperialization of sugar trade. Protectionist policies in India and Japan offset such advantages in Java's Far Eastern sales as location, low cost, and a white sugar processed without bone char and therefore inoffensive to the religious scruples of India's Hindus. The Japanese "incident" on the Chinese mainland was shortly to reduce the import capacity of the last major Asiatic outlet, though domination of Manchuria provided the basis for a Japanese re-export trade in refined sugar. Java was compelled to cut production from 3 million tons in 1930/31 to just over 500,000 in 1935/36. Only lesser exporters, such as the Dominican Republic and Peru, were able to maintain sales at a fairly stable level.

INTERNATIONAL CONTROL OF SUGAR

Marketing difficulties and unprofitable prices paved the way for several attempts at concerted action among exporting countries after 1927. Eventually all "free market" exporters,[2] except Brazil and the Dominican Republic, adhered to the ("Chadbourne") International Sugar Agreement, signed at Brussels on May 9, 1931 (9, pp. 40–50). Through painful reduction of output, exceeding 50 per cent in the case of Cuba, Czechoslovakia, and Poland, and more than 80 per cent in Java, the Agreement succeeded in working off more than 3 million tons of excess stocks that had piled up in exporter hands. But the expected improvement in prices failed to materialize. Though the Chadbourne group cut production by more than 6 million tons, expansion in protected areas had completely nullified their sacrifice by 1935/36: Chadbourne countries contributed less than 25 per cent of world supplies in 1933/34 as compared with 45 per cent in 1929/30. It became clear that exporter countries by themselves could not solve the problems of the sugar industry. Nationalistic policies elsewhere were stimulating domestic sources, protecting local producers against the world depression, and deliberately breaking the link between the world price and local prices. The times were running counter to economic specialization in highly efficient exporter areas.

Whereas the Brussels Sugar Convention earlier in the century had come into being largely because the burden of export subsidies was becoming excessive in the exporting countries, overexpansion in the 1930's was a primary concern of importing countries. The threat to United States mainland producers from expansion in offshore areas (and the outbreak of a general strike and revolution in Cuba in 1933) paved the way for the Jones-Costigan Act of May 9, 1934, which established a sugar-quota system for the entire United States market. The Cuban position as United States supplier was partially restored, offshore domestic production cut back, and a definite limit placed

---

[2] For world sugar trade, the so-called "free market" refers broadly to that portion of total import requirements not reserved for particular categories of suppliers. Even in the "free market," import duties and export restrictions prevail.

on mainland production of cane and beet sugar. Moreover, the Philippines Independence Act of March 24, 1934, by granting those islands an assured export quota to the United States of 850,000 long tons, removed the immediate danger of further pressure on Java's shrunken markets in the Far East. Similarly, the situation in the British Commonwealth came to be stabilized: by the Sugar Industry (Re-organisation) Act of 1936, which set a limit on the volume of subsidized domestic beet sugar; by a system of certification of colonial imports that reduced the incentive to expand; and by independent crop-limitation programs introduced in South Africa in 1936 and in Australia as early as 1930.

The logical corollary of these domestic restrictions was an international arrangement that delimited the free market by enforcing the status quo in importing countries, and that apportioned available outlets among exporters. Such were the purposes of the International Sugar Agreement signed at London on May 6, 1937, by 21 exporting and importing countries (*9*, pp. 51–56). While United States' formal commitments were trivial, the sugar-quota system introduced in 1934 and renewed in 1937 was consistent with the spirit of the Agreement. Moreover, so long as the United States honored the Independence Act quota, the Philippines agreed to export only to American territory (except for a small share in any over-all increase in quotas). As for the British supply system, the Agreement reinforced the limit on beet-sugar production legislated in 1936, set basic export quotas for the British Colonial Empire and Dominions at about existing levels, and essentially reserved the remainder of United Kingdom consumption to be supplied by free-market exporters. Two countries, while not signatories of the Agreement, did make certain committing statements of a minor sort. Canada indicated her intention not to stimulate domestic production by fiscal action, while Japan did not expect "such an increase of net exports of sugar as might generally neutralize the effects of the Agreement," and agreed to "respect the spirit of the Agreement as far as possible." Under Article 4, all signatory countries agreed not to permit domestic prices to rise as the world price rose, in order to encourage consumption and deter further expansion of domestic sugar production.

International sugar controls have been largely free from many faults common to commodity agreements. Whereas cartels are often criticized for reducing the volume of trade and maintaining submarginal producers in operation, the 1937 Agreement sought to defend low-cost exporters against the excesses of nationalistic policies. Quotas have served less to freeze obsolete trade channels than to promote international exchange. International agreements have been the child, not the father, of governmental interference in the sugar industry.

Control has not been exercised aggressively, nor have price objectives been excessive. Consumer interests have been respected, importing countries represented. The very expansibility of sugar production, and the dependence of many sugar areas on large volume to reduce unit costs, serve also as protection against an excessive world price. Under the Chadbourne Agreement, at Java's insistence, a price rise was to call forth an increase in export quotas automatically. The presence of low-cost producers within an international commodity agreement may thus assure moderation in much the same way that potential disintegration (should prices go too high and output fall too low) serves to mitigate the severity of a private cartel. Since so many countries must participate in an effective agreement, the aggregate of quotas is more likely to be too high than too low. This, too, is all to the good so far as the consumer is concerned. Legitimate criticism would be aimed less in the direction of excessive restraint than of laxity. For sugar, it is especially true that "what tends to emerge from international commodity conferences, if anything, is a mere bridge between incompatible national policies, with limited adjustments to their inconsistencies" (10, p. 49).

WORLD WAR II AND ITS AFTERMATH

The International Sugar Agreement had been operating barely a year when the specter of approaching war dominated all other factors in the world sugar market. Even before Munich, the British government had begun to stockpile certain essential foodstuffs, though sugar stockpiling was conceived and executed on a modest scale. Market prices had strengthened only moderately before Britain requested higher export quotas under the Agreement and won the right to admit exceptional shipments from the exporting Dominions (see chap. 8). The third year of the Agreement, beginning September 1, 1939, almost exactly coincided with the first year of hostilities in Europe.

While the system of sugar supply that developed in the interwar period was nationalistic to a high degree, it did not conform to autarchic principles as neatly as did the European beet economy (Chart 2). Long ocean voyages were required to serve the eastern seaboard of the United States with sugar from Puerto Rico (or Cuba),

CHART 2.—NATIONAL SUGAR PRODUCTION, CONSUMPTION, AND TRADE, AVERAGE 1935–39*

(Million short tons, raw value)

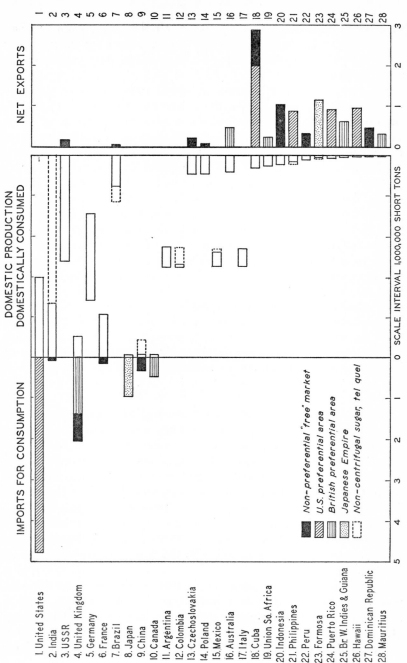

IMPORTS FOR CONSUMPTION

DOMESTIC PRODUCTION
DOMESTICALLY CONSUMED

NET EXPORTS

1. United States
2. India
3. USSR
4. United Kingdom
5. Germany
6. France
7. Brazil
8. Japan
9. China
10. Canada
11. Argentina
12. Colombia
13. Czechoslovakia
14. Poland
15. Mexico
16. Australia
17. Italy
18. Cuba
19. Union So. Africa
20. Indonesia
21. Philippines
22. Peru
23. Formosa
24. Puerto Rico
25. Br. W. Indies & Guiana
26. Hawaii
27. Dominican Republic
28. Mauritius

Non-preferential "free" market
U.S. preferential area
British preferential area
Japanese Empire
Non-centrifugal sugar, tel quel

SCALE INTERVAL 1,000,000 SHORT TONS

* Data from U.S. Dept. Agr., For. Agr. Serv., *Foreign Crops and Markets*, May 10, 1954, pp. 429–31; U.S. Dept. Agr., Off. For. Agr. Rel., *Foreign Agriculture Circular FS 2/52*, Aug. 4, 1952; U.S. Dept. Agr., Bur. Agr. Econ., *World Sugar Situation*, August 1946; *Statistical Bulletin of the International Sugar Council* (London). Exports for USSR, Poland, and Czechoslovakia are averages for crop years 1935/36 to 1937/38.

or the Pacific Coast with shipments from Hawaii. Supplying the United Kingdom simultaneously from the British West Indies, from Mauritius, from the Union of South Africa, and above all from Australia involved stupendous problems of wartime logistics. Canada similarly was supplied from British overseas areas under preferential arrangements. Japan produced little sugar on the home islands, but relied on keeping open the ocean supply lines to Formosa and the South Pacific Islands. Even France was in a more exposed position than net trade figures might imply, since North African colonies imported considerable amounts of Caribbean sugar after it had been refined in France. Moreover, the economic life of such preferential sugar suppliers as Mauritius, Hawaii, Barbados, and Martinique, as well as of independent countries like Cuba and the Dominican Republic, was heavily committed to overseas trade, and these islands were accordingly highly vulnerable to the naval misfortunes of war.

The sharp rise of sugar prices immediately after World War II broke out proved short-lived. In London, private trading gave place to complete government control. By contrast with 1914, Britain was no longer heavily dependent on free-market supplies, particularly enemy-controlled supplies. More than two-thirds of her consumption was now supplied from domestic, Dominion, or colonial sources, and early steps were taken to bulk-purchase the unsold portions of the South African, Australian, and Mauritian crops. Though American refiners employed Cuban raw sugar in taking over the British re-export trade with Norway, Finland, Greece, and Switzerland, Caribbean cane sugar went begging for markets in 1940, particularly after the fall of France made Continental Europe and its beet-sugar economy a closed German preserve.

During 1941 the situation in the West steadily deteriorated. Ships became too scarce to be spared for the long ocean voyage from Australia to England. By February, shipments from the Philippines to the United States were subject to interruptions, due in part to the difficulty of chartering the Japanese ships that were the normal carriers. Passage of the Lend-Lease Act in March permitted the United Kingdom to shift away from sources as distant as Oceania, the Indian Ocean, and South Africa, but put increasing pressure on dollar sources in the Caribbean area. In June the Germans moved on the Russian front, the Ukraine was soon occupied, and the USSR's normal production of some $2\frac{1}{2}$ million tons was decimated. The three-fifths of the Soviet population remaining outside the conquered territories had also to count on the Western Hemisphere for additional supplies.

Pearl Harbor was the culmination of this process of deterioration in the West. The Japanese, by overrunning Southeast Asia after December 1941, gained control of sugar in the Dutch East Indies and the Philippine Islands. Amply supplied from imperial sources (especially Formosa), Japan quickly cut the cane-sugar output in these islands to negligible amounts. For almost three years the world sugar economy was split into three entirely self-contained compartments: the Japanese, in which potential supplies far exceeded requirements; the German, in which supplies were adequate to forestall any severe stringency; and the Allied, in which a severely restricted supply area had to serve the needs of a vastly enlarged consuming territory. At that, it was well into 1944 before lack of physical supplies, rather than the shipping shortage, became the key sugar problem for the Western nations.

Allied military successes at first accentuated rather than eased the strain on Western Hemisphere—especially Caribbean—supplies. North Africa, long a sugar-deficit area, was the first region to be liberated from Axis control. Victory in Western Europe and the economic dislocations of the immediate postwar period meant lost beet harvests in Europe or, even where beet harvests did not suffer, breakdown of processing operations due to insufficient transport, equipment, fuel, and materials. Similarly in the Far East, even liberation of the Philippines added nothing to available supplies but instead created a temporary need for imports. But disorganization of beet production did not come simultaneously in all areas, as it had at the end of World War I, since this time restoration had begun in France and the USSR before the final defeat of Germany (Chart 3B).

The years of greatest sugar stringency were 1945 and 1946. With the 1945 European beet crop (excluding that of the USSR) cut to about half the prewar output, a heavy local deficit was created. Cuba's 1945 crop was somewhat reduced by drought, and even more potential sugar had been lost by diversion to industrial alcohol during 1944. Overseas supplies from Mauritius, Puerto Rico, Australia, and Hawaii were languishing as a combined result of the shipping shortage, a sustained lack of fertilizers, and a scarcity of hands. Nor did the sugar surpluses of Formosa reappear on their prewar scale after the fall of Japan. Only in Cuba could there be expansion somewhat commensurate with the contemporary needs. Available sugar was divided among competing claimants by international allocation, but the severe shortage required an extremely low priority for the needs of the defeated Axis powers as well as stricter rationing in the United

CHART 3 A–C.—SUGAR PRODUCTION: DISORGANIZATION AND RESTORATION, 1937–53*

(*Million metric tons, raw value*)

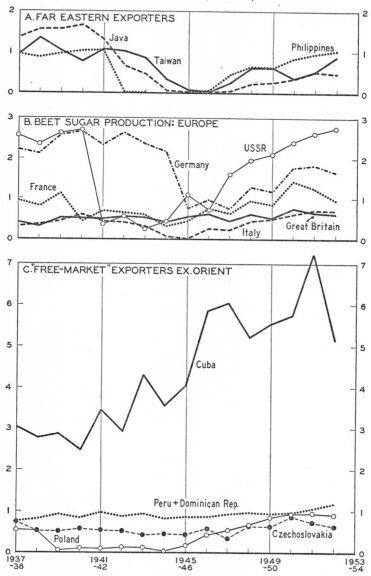

* Data from F. O. Licht's *International Sugar Statistical Year and Address Book 1950*, Vol. I (Ratzeburg); and F. O. Licht's *Sugar Information Service* (Ratzeburg), Dec. 31, 1953.

CHART 4 A–C.—SUGAR TRADE: DISLOCATION AND RESTORATION, 1937–53*

(*Million metric tons, raw value*)

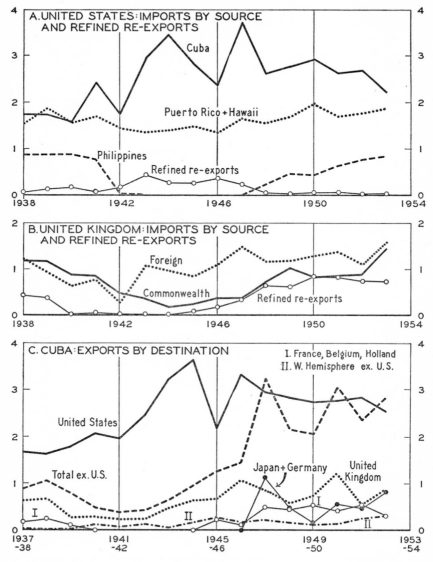

* Data from U.S. Dept. Agr., Prod. and Mkt. Admin., Sugar Branch, *Sugar Statistics,* Vol. I (1953); U.S. Dept. Agr., Comm. Stab. Serv., Sugar Div., *Sugar Reports No. 22,* February 1954, p. 23; Commonwealth Econ. Comm. [Gr. Brit.], *Plantation Crops; Statistical Bulletin of the International Sugar Council* (London).

States during 1945 and 1946 than had been necessary in wartime.

Gradually world stocks were replenished and a semblance of normality was restored. Though the European beet crop in 1947/48 turned out poorly, the following year surpassed the 1938/39 figure. Under the Marshall Plan, European output was shortly to exceed prewar levels by a higher margin in beet sugar than in any other major agricultural commodity. By 1948/49, shipments from Mauritius and Australia to England were not far below the prewar rate. United States imports from Puerto Rico and Hawaii recovered even earlier, though full restoration of the Philippine trade had not taken place as late as 1953.

So much had the supply situation eased by 1950 that a preliminary draft of a new international sugar agreement was prepared for the meeting of the continuing International Sugar Council, scheduled for July 20 (*11*, p. 6). In the meantime, the Korean war temporarily reversed the market, encouraged additional plantings in Cuba, and resulted in a 1952 Cuban crop of 7¼ million metric tons. The way was paved for re-enactment of the drama of the interwar period: Cuban restriction, followed by a new international agreement.

Even though world commodities show great powers of recuperation from the shock of war, certain changes prove irreversible. Out of the Napoleonic Wars, at a time when the tropics were the world's exclusive source of sugar, had come the stimulus for commercial manufacture of sugar from European-grown beets. World War I helped to shift the balance in favor of cane, for it gave a strong, if temporary, incentive to expansion in Cuba and Java, and alerted the United Kingdom to the risks of depending upon the Continent for the major part of a vital foodstuff.

The repercussions of World War II were more widely diffused. For one thing, the beet-sugar surpluses of Poland, East Germany, and Czechoslovakia passed under Soviet domination almost as quickly as they were released from Nazi control. Production expanded impressively for local consumption in Latin America, which remained outside the theater of military operations but enjoyed the stimulus of high export earnings. Cut off from outlets across the Atlantic, Peru came to depend almost entirely on export markets nearby. Most important of all, however, were developments in the Far East. Whereas all cane regions had been spared the devastations of war a generation earlier, World War II meant occupation for Indonesia and the Philippines, and devastation for their sugar industries. The Allied victory also separated Formosa from Japan, and brought further disorganization to the island when Chinese Nationalist personnel migrated from the mainland. Communist control of mainland China completed the detachment of a major deficit region from the world market and paved the way for intra-Soviet arrangements between eastern Europe and the Orient.

As late as 1953, the effects of war continued to plague the sugar industry of the Far East. Rebellion, social conflicts, and ascendant nationalism hampered re-establishment of plantation production in Indonesia. Cane sugar has managed at best a checkered career in competition with rice in postwar Formosa, although a combination of exceptional circumstances raised exports in 1953 to the prewar level. Assured of a place in the United States market, the Philippines enjoyed the most sustained recovery, but only by shifting the center of the industry from Luzon to outlying Negros.

TRADE: SOME CROSS-CURRENTS

While international sugar trade is primarily a commerce in raw cane sugar, a traffic of some importance takes place also in a wide range of associated commodities and services. For many years, Central Europe was almost the exclusive source of beet seed for all countries, but wars, nationalism, and special local problems have largely brought the trade to a halt. At the same time, the international exchange of "seed cane" (i.e., cuttings from the cane stalk) and cane "fuzz" (i.e., the multitude of tiny seeds resulting from successful pollination), as well as of cane seedlings, has become increasingly important. That trade is of trivial volume, but it has provided the means for spreading the results of a varietal revolution throughout the world. Indeed, the "Co." varieties of cane developed at the Coimbatore Sugar Cane Breeding Station in India have been more important to the sugar industry of Natal, and the "POJ" varieties developed at the Proefstation Oost Java have contributed more to sugar yields in the West Indies, than in the land where they were developed. Sugar beets have occasionally moved across national boundaries for processing—from Canada to United States factories, or between adjacent countries in Western Europe—and Puerto Rico earlier in the century processed some cane grown in the Dominican Republic, but such a traffic is quite exceptional.

The product of the cane mill or beet factory may be shipped in various forms. While beet sugar is generally exported after refining, it remains an open question as to which form, raw or refined, is the more economical for cane-sugar trade (cf. *12*, pp. 110–13). The solution is left to politics, and not to economics. The commercial policy of metropolitan consuming nations penalizes refined imports, while England, Holland, Belgium, and France regularly do a considerable re-export business on the basis of imported raw sugar. A significant volume of sugar also moves as a component in manufactured foodstuffs—chocolate exported by Switzerland and Holland, canned fruit from Australia, biscuits and preserves from Great Britain, frozen fruit and soft drinks from the United States. So far as bulk cargoes are concerned, the trade in cane molasses aggregates several hundred million gallons, mainly destined for the United States or the United Kingdom. Beet molasses, or industrial alcohol derived from it, also enters world trade, not merely within Europe but also periodically from Europe to the United States. Rum is a standard export of cane-producing regions, while the United Kingdom imports limited quantities of baled bagasse for manufacturing insulating board (*13*, pp. 341–42).

For centuries, sugar has also been identified with an international movement of persons. Cane was as closely associated with the slave trade in the West Indies as was cotton in the southeastern United States. The problem of providing an ample supply of field labor in tropical regions has in the past led to migration of East Indians to British Guiana and the British West Indies, or of Chinese, Japanese, and Filipino laborers to Hawaii, while the Florida cane fields continue to require transient laborers from the British West Indies. Migrant workers from the Philippines and Mexico have similarly played an important part in the United States beet fields. In Europe, large numbers of Poles have commonly assisted in the harvest of the German beet crop, and Belgians find employment in the French beet fields. A comparable seasonal migration of agricultural workers required for field-crop production on the United States mainland has jeopardized the smooth operation of the cane harvest in Puerto Rico in recent years (*14*, p. 257).

But the transient employed in the sugar industry is not always a casual field worker or an unskilled laborer. "Sugar tramp" is the label applied to the itinerant engineers who have plied their trade in various lands. Japanese invasion during World War II meant first imprisonment and later international dispersal for the outstanding scientific personnel that had staffed Java's prewar sugar industry. Particularly in the case of cane agronomists, Indonesia's loss will bring compensating gains for the West Indies and the other regions in which these scientists are now applying their talents.

The export of raw sugar from highly specialized cane mono-cultures in tropical regions has its counterpart in a heavy flow of imported supplies in the reverse direction. Indeed, a multilateral exchange among England, the British North American colonies, the West Indies, and Africa, involving sugar, slaves, rum, and provisions, played a crucial role in the commercial development of New England in the seventeenth and eighteenth centuries (*15*). Mauritius, Cuba, and Barbados remain dependent on imports for a substantial portion of their basic food supply. Cuba ranks as a leading market for rice, wheat and wheat flour, and lard exported from the United States. Bagging of the world's sugar crop has represented an important market for the burlap and jute of India-Pakistan, a market that is currently being undermined by the introduction of bulk shipment of raw sugar. Moreover, the technological changes in field and factory have resulted in a large trade in fertilizers and mill machinery from industrial countries to the cane regions. About 95 per cent of the sugar machinery manufactured in the United Kingdom is for export to cane-producing areas within and beyond the Common-wealth (*16*, p. 46), while overseas exports from Continental Europe are also considerable. As motor transport and field mechanization spread, imported motor vehicles replace locally manufactured ox-carts, the trade in tractors and agricultural machinery expands, and imports of petroleum products displace locally grown feedstuffs.

There are, finally, important trade connections between the cane regions and other specialized trading nations. Among the countries with net imports of sugar running from 75,000 to 250,000 tons a year will be found the animal-product economies of Uruguay and New Zealand; Ceylon, with its exports of tropical tree products; the rubber economy of Malaya; and Chile, with its concentration on copper. Even larger net imports are absorbed by Switzerland and Canada. All of these countries are heavily committed to the international trading system; all have hitherto depended overwhelmingly on imports for their sugar supply. Postwar Japan, with its small home-island production and large prewar consumption, and lacking its for-mer sugar empire, tops the list of nations whose continuing capacity to import sugar depends on the sustained health of world trade. While the interwar period witnessed the mushrooming of protected cane-sugar production under the aegis of the major metropolitan im-porters, there has been a tendency since the war for local aspirations to encourage domestic sugar production in smaller importing coun-tries such as Ceylon and Iran, Chile and Uruguay. The retreat from the world market continues. Even such an agency as the Food and Agriculture Organization of the United Nations condones this process (cf. *17*, p. 44) at the same time that it concerns itself with the problems of surplus agricultural commodities, sugar included.

CITATIONS

1 U.S. Dept. Agr., For. Agr. Serv., *Foreign Agriculture Circular*, Sept. 14, 1954.

2 *Foreign Agriculture Circular*, June 25, 1954.

3 P. D. Curtin, "The British Sugar Duties and West Indian Prosperity," *Journal of Economic History*, Spring, 1954.

4 H. C. Prinsen Geerligs, *The World's Cane Sugar Industry, Past and Present* (Manchester, 1912).

5 Lois B. Bacon and F. C. Schloemer, *World Trade in Agricultural Products: Its Growth, Its Crisis, and the New Trade Policies* (International Institute of Agriculture, Rome, 1940).

6 Ramiro Guerra y Sánchez, *Azúcar y Población en las Antillas* (3d ed., Havana, 1944).

7 F. O. Licht, *World Sugar Statistics, 1937* (Magdeburg).

8 *Anuario Azucarero de Cuba 1951* (Havana).

9 B. C. Swerling, *International Control of Sugar 1918–41* (Food Research Institute Commodity Policy Studies No. 7, Stanford, Calif., 1949).

10 J. S. Davis, *International Commodity Agreements: Hope, Illusion, or Menace?* (New York, 1947).

11 United Nations, Interim Coord. Com. for Internat. Commod. Arrangements, *Review of International Commodity Problems 1950* (New York, 1951).

12 Gr. Brit., Col. Off., *The Processing of Colonial Raw Materials* (by Charlotte Leubuscher, 1951).

13 A. C. Barnes, *Agriculture of the Sugar-Cane* (London, 1953).

14 Caribbean Commission, *Monthly Information Bulletin* (Port-of-Spain), June 1954.

15 H. A. Innis, *The Cod Fisheries, The History of an International Economy* (rev. ed., Toronto, 1954).

16 *Sugar*, April 1952.

17 *Sugar*, April 1951.

PART II
SUGAR CROPS AND AGRICULTURAL DEVELOPMENT

CHAPTER 3

SUGAR YIELDS: CONTRASTS AND PROGRESS

The contribution of agriculture to economic development and growth has been insufficiently studied and is little understood. Despite some ambitious attempts at comprehensive analysis (1, 2), the facts at present are evidently liable to widely divergent interpretations. There are some economists for whom "it is evident that there is an association between agriculture and economic stagnation" (3, p. 55). Others, sharing Ricardo's low opinion of landlords (4, p. 225), allege a resistance to change on the part of landed gentry who squander the agricultural surplus on high living and spend "relatively insignificant amounts . . . on improvements of agricultural estates" (5, p. 73). A few scholars, to be sure, have attempted to review the wide range of institutional devices by which the surplus product of agriculture has in fact been siphoned off to support programs of accelerated industrialization in Japan (6), the Soviet Union (7), and elsewhere (8). There is also a growing recognition that investment in rural improvements—even simple ones like higher-quality seed, fertilizers, and agricultural extension services, without heavy call on funds or prior reorganization of the agricultural economy—holds out to the economically less advanced countries their best avenue to prompt increases in productivity (9, p. 53; 31, pp. 187–89). Yet the historical connections between the modernization of agriculture and the industrial revolution seem not to be sufficiently widely appreciated.

The relations between agricultural and economic development can profitably be analyzed through the study of specific agricultural staples. Examination of significant developments over time, as well as comparative analysis of concurrent economic organization found in different regions, are both amenable to a commodity approach. A number of pertinent questions can be posed with some hope of useful empirical answers: matters of technological leadership and cultural lag, interregional transmission and disparity, social stimulus and inertia. Which kinds of institutions facilitate, and which deter, technological change? What sources of "technical assistance" can be called upon to overcome scientific bottlenecks? How much momentum can be generated in auxiliary fields, as well as in the area of initial improvement? What are the instrumentalities for transmitting known techniques from the leading to the less progressive producers and from the more advanced to the less advanced countries? History abounds with answers to just such questions as these, for a large number of individual commodities. At a time when the questions have attained political and even military significance, we do well to analyze the pertinent commodity data.

SUGAR CROPS: A CASE STUDY

For such a study, sugar enjoys certain unique advantages. The data are abundant for centuries back. With sugar crops produced over a wide range of regions in both temperate and tropical zones, the opportunities for comparative studies of agricultural systems are excellent. Less obvious is the intimate link between sugar-crop farming, scientific practices, and industrial manufacture. This is particularly true of sugar beets. The low sugar content of the eighteenth-century parent beet had to be painstakingly raised by experimental processes before commercial operation became feasible at all. Beets played an important role in modernizing European agriculture from their very introduction. Here was a row crop for clean cultivation, fitting readily into rotation systems; a root crop to aerate and improve the condition of the soil; a voracious consumer of barnyard manure and commercial fertilizer; a crop rich in by-product feed to support an animal population that itself promoted an intensive and diversified agriculture. At times, the crop has certainly been introduced locally as much for these effects on agricultural practices as for the production of sugar as such.

For some regions, the sugar beet's combination of physical requirements and economic characteristics has given it a pivotal role. A good example is provided by the irrigated valleys of the semiarid western United States, where beet production is feasible in spite of low rainfall, high elevations, and mildly alkaline soils (*10*, pp. 3–5). A commercial crop by its very nature and one with a peculiarly assured market, the sugar beet can provide a cash offset to the fixed charges of irrigated agriculture and for that reason has played a considerable role in the crop pattern of areas opened up by large-scale reclamation projects in the American West.

While cane does not coordinate so well with animal production and diversified agriculture, for both cane and beets the separation between agriculture and manufacturing is far less distinct than one expects for major food crops. Factories are of necessity spotted throughout the countryside to process bulky semiperishable roots or stalks into raw or refined sugar. In many cases, the factory's interest in a high-quality raw material leads it to engage in agricultural operations, experimentally if not commercially. To the cultivated field it brings its own high state of technology and strong capital position. Elsewhere growers, as part proprietors in the factory, transmit to their agricultural practices the factory's concern for a crop rich in sugar. Whether or not factory and farm are linked by ties of ownership, mere rural location brings a close association between processor and grower in the particular factory district. Processors find it to their own interest to provide agricultural advisory services, to supply high-grade selected seed, to publicize new methods, and to assist growers with their field problems. In the view of Sir John Russell: "The sugar beet is much more than a crop: it has introduced a new system of farming and a new sort of business into the countryside" (cf. *12*, p. 14).

Quite aside from the peculiar technological role of the factory, both crops provide inherent incentives to high standards of agricultural practice. With yields measurable in such terms as sugar content and purity of juice, growers are alerted to the advantages of applying the latest improvements, especially if the price of sugar crops is related to sugar content rather than to crop tonnage alone. Considerations such as these help to keep a fairly narrow gap between the more progressive and the less progressive beet farmers. Nowhere do we find sizable pockets of growers turning out a decidedly inferior article; and a purely subsistence agriculture is quite out of the question for this crop.

Not only do sugar factories incidentally raise agricultural productivity, but they are particularly likely to provide that "dynamic stimulus" commonly associated with manufacturing enterprise. The natural decentralization of processing facilities reaches out for a wide range of supporting skills and services throughout the countryside. In addition, mill and factory are heavily dependent on efficient transport services. The crops accordingly contribute to (and benefit from) expansion and modernization of transport facilities. Where railroads already exist in reasonably good condition, an important bulk cargo is provided for financing heavy fixed charges, though the network may be overburdened with seasonal peak loads.

While a study of sugar beets throws light on the process of technological change in advanced countries, cane is more interesting for what can be learned about the processes of transmission between

TABLE 1.—AVERAGE YIELDS, BEET-SUGAR REGIONS, 1949–51*

(Metric tons)

Region	Yield per hectare		Extraction rate
	Beet sugar (raw value)	Sugar beets	Raw sugar per 100 tons of beets
California, U.S.A.	6.56	42.29	15.5
Netherlands	5.78	41.45	14.0
Belgium	5.61	39.23	14.2
Denmark	5.28	33.98	15.5
United States	5.05	33.31	15.2
Germany	4.11	27.53	15.0
United Kingdom	3.91	28.37	13.7
France	3.86	29.46	13.0
Italy	3.65	26.68	13.7
Europe ex-USSR	3.64	24.88	14.6
Michigan, U.S.A.	3.59	23.35	15.4
Czechoslovakia	3.55	21.41	16.6
Poland	3.37	20.55	16.4
Hungary	2.27	15.34	14.8
Spain	2.06	15.93	12.9
USSR	1.85	13.71	13.5

* Calculated from data in F. O. Licht's *International Sugar Statistical Year and Address Book 1951/52* (Ratzeburg, 1952), pp. 197–98; U.S. Dept. Agr., Bur. Agr. Econ., *The World Sugar Situation*, October 1951, pp. 25–26; and *ibid.*, January 1953, pp. 17–18.

modernized and less advanced agricultural economies. Cane sugar was one of the first tropical products to be adapted to large-scale capitalistic organization; it has more recently become an important medium for injecting scientific agricultural practice into the tropical world. Spectacular achievements in the beet-sugar industry during the nineteenth century originally appeared disastrous to the cane sector, but much of the factory technology developed for beets was subsequently also adapted to cane. Especially since 1900, cane agriculture has made its own unique contributions to the efficiency of sugar production.

INTERNATIONAL COMPARISON OF SUGAR YIELDS

Data on sugar yield per acre are a particularly good indicator of improved economic performance. For one thing, statistics are, on the whole, more reliable than those for other crops planted so widely around the globe. It is, moreover, possible to distinguish between (a) tons of perishable crops per acre, the result of a strictly agricultural operation, and (b) per-acre yields of final sugar, which

TABLE 2.—REPRESENTATIVE YIELDS, CANE-SUGAR REGIONS, 1950–51
(Metric tons)

Region	Yield per hectare harvested		"Over-all recovery"
	Cane sugar (raw value)	Sugar cane	Raw sugar per 100 tons of cane
Hawaii	20.5	168.5	12.2
Java (1934–38)	16.6	137.8	12.0
Peru (1951–52)	16.2	143.8	11.2
Barbados	11.2	100.1	11.1
Formosa (1936–38)	9.6	76.0	12.5
Java	9.0	89.6	10.7
Australia	8.5	65.1	13.1
Puerto Rico	7.8	66.2	11.8
Mauritius	7.1	58.1	12.3
Union of South Africa	7.0	58.3	12.0
Florida, U.S.A.	6.5	70.8	9.3
Formosa	6.0	48.3	12.5
Jamaica	5.9	53.6	11.0
Cuba (1949–50)	4.8	35.7	13.5
Louisiana, U.S.A.	3.7	43.6	8.6
Argentina	2.5	28.9	8.8

* Calculated from data in FAO, *Yearbook of Food and Agricultural Statistics 1952*, Vol. 1, *Production* (Rome, 1953); *ibid., 1953*; U.S. Dept. Agr., Bur. Agr. Econ., *World Sugar Situation* (January 1953), p. 19; Hawaiian Sugar Planters' Association, *Sugar Manual* (Honolulu, 1954); U.S. Mutual Sec. Admin., Mission to China, *A Survey of the Taiwan Sugar Industry, Part III Economics* (Consultant Rept. No. 1-A, Taipei, 1952), p. III-5. See footnote p. 56 for discussion of problems of noncomparability.

respond both to tonnage and quality of the crop *and* to efficiency of extraction within the factory. Though he lacks the technical background in plant pathology, soil chemistry, genetics, entomology, and industrial engineering essential for a full understanding of differences in yields, the economist can make a significant contribution to their systematic analysis (cf. *26*, pp. 40–58; *27*, pp. 1–5).

Even a superficial reading of recent performance in selected countries (Tables 1 and 2) supports several interesting inferences, with which subsequent discussion will deal in more detail. In beet and cane regions alike, high sugar yields per acre are associated much more closely with high crop tonnage (attained in the Netherlands and Belgium for beets, in Hawaii and prewar Indonesia for cane) than with high extraction rates (in which respect Poland and Czechoslovakia, if their reported statistics can be relied upon, do best among beet-growing countries, and Cuba and Australia among cane producers). But only the very highest sugar-beet tonnages per acre exceed the cane tonnages of low-ranking Cuba and Argentina. Despite higher sucrose content and correspondingly superior extraction rates for beets, even outstanding beet-sugar yields per acre are not impressive by cane standards. Though the present tables do not so indicate, yields of beets grown without irrigation are generally more variable from year to year than yields of unirrigated sugar cane in major producing regions.

The range of yields within the two production systems is also of interest. Beet sugar is produced, for the most part, in regions that are broadly similar in their level of agricultural science and industrial technology. Despite a considerable range of climatic conditions, from Mediterranean to continental, the lowest-yield countries produce at least one-quarter as much beets and sugar per acre as do the highest; extraction rates vary within a rather narrow range. Not so with cane. Hawaii produces 8 times as much sugar per acre as Argentina, and more than 10 times as much cane per acre as India (which is not listed, for lack of reliable statistics). The range in extraction rates, though far less spectacular, places Cuba and Australia well above Argentina (or Mexico). Moreover, there is a wide gap between the rates of a few forerunners and the average performance. While part of the differential in cane yields results from pushing the crop to the climatic limits of its natural habitat, the range is an indicator of different intensities of agricultural operation, different degrees of application of modern science to rural economies, different social arrangements, and different stages of industrial advance. The highest yields attained in Hawaii may in fact not be entirely economical; the prewar performance in Indonesia may have rested on a social organization that has disappeared forever; and there are good reasons why some of the laggards can never expect to emulate those in the front rank. Nevertheless, the fact that cane practice falls so short of the potential under existing technology in many parts of the world invites explanation.

Cropping systems.—The differences in yields are accounted for in part by a peculiar combination of natural circumstances and cultural practices. Although the sugar beet is a biennial plant that requires 2 seasons to produce seed, when used for the manufacture of sugar the crop must be harvested during the first year of growth, about 8 months after planting. Yields are accordingly a proper indicator of productivity of land per year; they would have to be adjusted upward if one were concerned with output per month of growth or if full account were to be taken of the incidental benefit to succeeding crops from the intensive fertilization and cultivation of beets and the improved condition in which deep-growing beets leave the soil.

The length of the growing season for cane conforms to no such single pattern. It ranges from less than 10 months in Louisiana, where spring and winter frosts set rigid limits, up to almost 2 full years in Hawaii and Peru. Eighteen months is not uncommon in Indonesia, Formosa, and South Africa; 12 to 14 in the Caribbean area and in Queensland; while most other regions harvest after a comparable lapse of time. Since the total acreage under cane can regularly be as much as double the harvested area, and even more than double if land in cane is being rapidly extended, yields per harvested acre do not reflect land productivity at all well. The relatively low position of Louisiana, Argentina, and Cuba is considerably improved, and the lead enjoyed by Hawaii and Peru much reduced, if yields are computed on the basis of total acreage in cane, while the South African performance appears more modest still.

Moreover, cane is a perennial plant that permits the harvesting of successive "ratoon" crops without the expense of replanting. Practices in this respect also vary through a considerable range. Production of plantation sugar in Indonesia and Formosa has typically been without benefit of any ratoons at all, while in Cuba standard practice calls for about seven cuttings and the figure can go higher. India generally cuts one to two ratoons (17, p. 8); Natal and Louisiana, two; Hawaii, two to four; Demerara, three (19, p. 39); and three to five are rather typical for the Caribbean area. Ratoon crops generally ripen somewhat earlier than plant cane, but harvested tonnages can be expected to fall off each successive year. The decision to replant will accordingly involve economic considerations. If sugar prices are low or field labor in short supply, the considerable savings in planting expenses associated with ratooning will tend to become the dominating factor. High land rents would operate in the reverse direction, contributing to a higher percentage of "plant" cane and correspondingly higher average yields. Countries with a 24-month growth cycle maintain a ratooning pattern with a regularity that overrides transitory market conditions, since any decision commits their land for a longer period.

Sugar yields are also peculiarly affected by the length of the harvest and processing season, whether beet "campaign" or cane *zafra*. Cane, once cut, must quickly surrender its sugar to the mill or inversion will convert the sucrose to uncrystallizable form, but in cool climates beets afford at least the possibility of storage for some weeks. Since the cane-grinding season accordingly coincides with the field harvest, while the beet factory has more leeway, it might appear that the beet-sugar factory could plan to operate for more months each year than the typical cane-sugar mill. Actually, the standard beet campaign lasts for only about 100 days in most beet areas, while the cane harvest easily lasts 5 months in the Caribbean, 9 months in South Africa, and practically all year round in Peru and Hawaii.

Under north European conditions, there is a strong incentive not to begin the beet harvest until well into the fall, since both tonnage and sucrose rise steadily unless jeopardized by fresh rainfall or an early frost. Moreover, variable yields must be anticipated, and capital is abundant relative to land under Western European conditions. Consequently, the capacity of the processing unit tends to be large as compared with the agricultural output in any given season. Storage of beets, while possible, involves a slow wastage of sugar content, and excess factory capacity allows protracted storage to be avoided except in years of unusually high yields. In the USSR, where capital tends to be used more sparingly, the factory season can be drawn out as much as 150 days before it is considered necessary to build new capacity, but sugar recoveries suffer accordingly.

While sucrose content of cane rises slowly to a seasonal peak, it also falls relatively slowly thereafter. Moreover, heavy equipment in modern mills can be operated economically only if overhead costs are spread over a reasonably long processing season. Grinding accordingly commences well before and continues for some time after maximum sucrose content has been reached. As an extreme case, one factory in Mysore (India) is kept in operation practically the year round, at the expense of a very low sugar recovery (25, p. 50). But under irrigated conditions in Hawaii and Peru, where there is ample sunshine for ripening to proceed and a careful planting schedule is maintained, cane can be ground for almost 12 months a year with little sacrifice of available sucrose.

Natural hazards and cultural practices.—The range of natural conditions suitable either for sugar beets or for sugar cane is extremely wide. Beets are grown successfully in both maritime and continental climates of the temperate zone, while their more recent introduction in Turkey, Iran, or the Imperial Valley of California represents an extension to the subtropics. The fact that cane is essentially a tropical plant by no means assures uniformity of climatic environment. Frost must be reckoned with in Argentina, India, Taiwan, and Queensland, which are partly outside the tropics, and in Louisiana and Natal, which are entirely so. At issue are perennially low yields, as well as high variability from year to year. Low yields in Argentina and Louisiana result from a growing season too short to permit the crop to reach full maturity. Cane yields in subtropical United Provinces, Bihar, Punjab, and Bengal, where three-quarters of India's sugar crop is grown, amount to only 12 to 15 tons per acre, as against 35 to 40 tons in Bombay and Madras. Drought is an important hazard in Natal, where the crops were particularly hard hit between 1950 and 1952, the more so because the growing period overlaps two calendar years. The strictly tropical regions are liable to heavy storm and wind damage: Mauritius suffered a severe cyclone in 1944; Philippine recovery was set back by a typhoon in 1950; and hurricanes are a seasonal threat in the Caribbean. By comparison, Peru, Hawaii, and Java enjoy more equable conditions.

The two components upon which commercial recovery primarily depends, crop tonnage and sucrose content per ton, do not necessarily increase and decrease together. Both crops need large amounts of water for high tonnages, and a hot dry spell for best ripening of juice and correspondingly high sucrose. Thus the maritime beet regions of Western Europe tend to have relatively high tonnage but relatively low sucrose as compared with regions farther east. Variation in the seasonal pattern of rainfall is also important, however. The Far Eastern maritime region of the USSR has too little of its moisture in the spring and too much in August-September. Cane has an almost insatiable thirst for water and hence will grow well in the hot, humid tropics, but sugar yields suffer unless there is a pronounced dry period for maturation (cf. *22*, p. 34). Since sucrose content is usually higher in beets, year-to-year variations in sugar yield are more closely associated with fluctuations in tonnage than with variations in sucrose, despite the tendency for the two to move in inverse directions, whereas cane areas are particularly sensitive to changes in sucrose content. In the Caribbean, even hurricanes may damage real property more than standing cane, for the heavy rainfall will be absorbed for plant growth while cane's capacity to recover from wind damage is quite considerable, particularly as compared with alternative local crops.

To some extent natural difficulties can be counteracted by capital investment or improved cultural practices. Even frost damage, a serious hazard in marginal cane areas, is amenable to some restraints.

Louisiana has found it advantageous to plant cane in the fall; if frost comes, the undamaged root structure will sustain more rapid growth and a longer growing period than if planting takes place in the spring. Similarly, Natal has come to blanketing the soil with cane trash, as a protection against dry winds. But control of water offers even greater possibilities. Probably not more than 10 per cent of the world's cane acreage is under irrigation (*14*), but the high yields enjoyed in Peru and Hawaii would be unthinkable without it, or the lowest yields in Natal with it. The South African terrain, however, affords little opportunity for the controlled application of water. For most beet areas, natural rainfall is sufficiently reliable that year-to-year variations in yield are accepted and the expense of supplying water avoided. Irrigation of sugar beets in dry continental areas, where natural rainfall is inadequate, results in an excellent crop, as experience in California illustrates. The advance of beets into subtropical regions in Turkey, Iran, and Central Asia has generally relied on irrigation. In some respects, complete reliance on irrigation is ideal for both crops: sufficient water can be supplied for optimum tonnage, while moisture can be withheld once the time comes for allowing sucrose content to rise.

Devastating attacks of pests and diseases have occasionally proved a blessing in disguise. Major developments in the study of cane pathology and the breeding of new high-yielding varieties originated in the 1880's, when disease wiped out Java's production (*15*, p. 29). Similarly, the outbreak of rind disease in the West Indies in the 1890's led to the disappearance of the Otaheite variety, which had been grown for three hundred years, and cleared the way for a succession of improved varieties. While the major achievements of cane breeding and selection have come in the twentieth century, beet advances in the direction of high tonnage and high sucrose were spectacular long before 1900. But the need for developing varieties resistant to curly top in the American Far West and to leaf spot in the Middle West revived an interest in breeding that has borne fruit in unexpected directions (see chapter 5), and there have been recent European achievements of a comparable sort. A number of cane areas have had dramatic success in overcoming local menaces. Hawaii has pioneered with biological controls of insects since 1904, culminating in the introduction of *bufo marinum,* an insectivorous frog. In Queensland, where the white grub was causing serious crop losses as recently as 1934, modern chemicals in the form of benzene hexachloride have come to the rescue, while Natal has used chemical immersion of seed cane to prevent crop losses from pineapple disease which formerly struck the planted cane setts (*13*, p. 27). By contrast, rampant disease holds back cane tonnages in India, and even the cane-breeding stations, which have successfully propagated many high-yield varieties, appear to have paid insufficient attention to disease-resisting capacity (*17*, p. 10).

New varieties and disease-control methods are internationally transmissible, and irrigation can be supplied by capital investment at key points, but the maximum yields that result from proper application of fertilizer require minute and detailed attention to the characteristics of individual plots of ground. For the beet areas, a high level of cultural practices is usual; fertilizer application is heavy, and includes the plowing under of barnyard manure. But even today less than 20 per cent of the world's cane acreage is properly fertilized (*14*). Filter press cake and, occasionally, molasses must serve in place of manure for building up the humus content. Cane trash can also be incorporated into the soil, except in places where the fields are burned off to facilitate the harvest. In India, the Hindus object to the use of human excreta and they burn the cow dung that a large livestock population supplies, so that by far the greater part of the cane goes unfertilized (*16*, p. 108). But no tropical crop (rice possibly excepted) has as long a history of scientific agriculture as sugar cane. Commercial fertilizers have been used in Hawaii since 1879. Agricultural experiment stations devoted to plant selection and the improvement of cultural practices were set up in Java as early as 1885, in Hawaii by 1895, and in Queensland by 1900 (*20*, p. 711). Yields in Java represented a particularly labor-intensive form of agriculture, with care given to individual plants resembling more closely that of the horticultural garden than of the cultivated field.

So far as scientific research and instances of high performance are concerned, India has by no means been laggard. In prize competitions, as many as 75–80 tons of cane per acre have been obtained in Uttar Pradesh with appropriate nitrogen, irrigation, and hoeing, and up to 100 tons in tropical Bombay. But the Uttar Pradesh average remains around 10 tons. Canes developed at the Coimbatore Sugarcane Breeding Institute, capable of averaging 25 to 30 tons with proper care even under the hard climatic conditions of northern India, accomplish little when introduced in the field (*21*, pp. 19–20). Elsewhere in sugar regions the gap between cane research and field practice is less conspicuous. A good part of the explanation lies in the fact that India's gur industry has lacked direct contact with modern technology, while the white sugar industry, with its interest in the productivity of surrounding cane lands, is only two and a half decades old.

The peculiar nature of sugar yields to some extent places the interests of factory and grower in direct conflict. High crop tonnage with low sucrose makes for a more expensive factory operation per ton of sugar produced. But the beet grower, who is interested in the feed use of tops and by-product pulp, is not necessarily deterred by low sucrose. Indeed there have long been available three types of beet seed, offering high sugar but low tonnage, high tonnage but low sugar, and moderate levels of both. In cane, tonnage of raw material processed by the factory will be increased with no compensating return if careless harvest labor or imperfect mechanical harvesters result in "trashy" cane, while plant breeding to raise sugar yields per acre may at times profitably sacrifice high sucrose to high tonnage.

Sugar recovered per ton of cane and beets does not exclusively reflect factory performance, nor field tonnage the contribution of the agricultural sector. Obviously the percentage of sucrose present in the crop is itself an important consideration. But nonsucrose components can also facilitate or deter the processing operation. This is particularly true of cane. High fiber content makes milling more difficult, with the consequence that more sucrose will be left in the bagasse. Besides, the more invert sugars and impurities present in the juice, the greater the waste of uncrystallizable sucrose in the final molasses. The diffusion process is efficient enough in separating sucrose from beet pulp that in the beet economy the term "extraction" refers to the actual production of final sugar per ton of beets sliced. For cane, however, "extraction" is a more limited term, applied to the milling (i.e., grinding) operation alone. The extraction rate, a measure of cane-milling efficiency, relates the amount of sucrose in cane juice actually extracted *to* the sucrose content of the cane before milling. "Recovery" relates the amount of final sugar produced *to* the amount of sucrose in the extracted cane juice, while "over-all recovery" becomes a comprehensive measure equivalent to the "extraction" rate in beet-sugar processing. High "extraction" need not imply high "recovery," since the former responds to variations in fiber content and the latter to percentage of nonsucrose elements in the juice. It is typical of primitive cane processing that extraction of juice is inefficient, much sugar being left in the bagasse, while recovery is quite high, since only the water is boiled off and most impurities are consumed along with sugar. Thus 10 tons of cane affords roughly one ton of gur (purity 60 to 80 per cent) in the Indian village industry, and about the same weight of white sugar in a full-scale Indian factory (*25*, p. 42). Output figures for noncentrifugal sugar accordingly reflect fairly well the amount of centrifugal sugar that could have been recovered from the cane utilized, but they over-state the amount of sugar (sucrose) actually available for current consumption.

The amount of final sugar recovered from sugar crops is highly responsive to the speed with which the harvesting, transportation, handling, and manufacturing activities are completed. Efficient co-ordination of field and factory operations is particularly important for cane, which begins to deteriorate rapidly within a few hours after harvest. In years of short crops, the chances are that factory per-formance will suffer as a result of irregularities in flow and occasional stoppages. In an exceptionally good season, factories may have to begin early and prolong activities unduly, so that both underripe and overripe cane (or long-stored beets) are handled, if necessary, at rates in excess of rated processing capacity (*25*, p. 27). Similarly, if agricultural capacity is expanding more rapidly than milling, the harvest will be prolonged beyond the time of maximum sucrose con-tent. This was true, for example, in the USSR during the late 1930's, and recently has been the case in Natal, where, in addition, transpor-tation of the large crop overburdened rail facilities, with consequent delays between harvesting and processing (*13*, p. 27). In these re-spects, however, the cane monoculture has decided advantages, for the entire social and transport organization can be concentrated during the harvest season on the single task of handling the crop. The high present yields in Peru and Hawaii are as much an achieve-ment in cane transportation as they are in agriculture or in manu-facturing, while the breakdown of social organization seems to have been responsible in considerable part for the reduced sugar yields of postwar Java.

Statistical data for cane sugar production are sufficiently com-prehensive (Table 3) that it is possible to illustrate these general principles from the milling results of representative regions. Hawaii, which is second among listed areas in over-all recovery, clearly is not outstanding in any of the natural qualities of its cane. Instead, it has enjoyed price incentives and a market outlet that have generally encouraged it to wring the last ounce of sugar from available cane.

TABLE 3.—INDICATORS OF FACTORY EFFICIENCY AND RELATED DATA,
SELECTED CANE REGIONS, 1951*

Region	Over-all sugar recovery and losses[a]			Characteristics of raw material		
	Recovered as sugar	Lost in:		Raw cane		Juice
		Bagasse	Boiling house[b]	Sucrose	Fiber	Purity
	(*Per cent sucrose in cane*)			(*Per cent cane*)		(*Per cent[c]*)
Taiwan	90.5	3.2	6.3	13.7	13.0	87.5
Hawaii	87.7	4.0	8.3	12.7	12.3	87.6
Puerto Rico[d] ...	87.1	5.7	7.2	13.3	14.1	86.4
Barbados	86.5	6.6	7.9	12.6	14.9	....
Cuba[e]	84.9	5.6	9.5	14.1	11.5	83.3
Queensland	84.7	4.8	10.5	15.6	13.3	88.4
Jamaica	84.5	5.7	9.7	12.7	14.7	85.2
Trinidad[f]	84.0	6.8	9.2	12.0	16.8	83.8
Mauritius	84.0	5.4	10.7	13.0	11.8	87.4
Natal	82.5	7.0	11.3	13.3	16.3	87.6
British Guiana ..	80.8	8.6	10.6	10.6	14.8	82.5
India[g]	80.4	8.5	11.1	12.5	16.3	81.1

* Data from South African Sugar Journal, *South African Sugar Year Book 1949–50* (Durban), pp. 126–27, and *ibid.*, *1950–51*, pp. 127–28; *International Sugar Journal* (London), December 1953, pp. 338–39; Hawaiian Sugar Planters' Association, Exp. Sta., *1953 Factory Report* (Special Release 96, Honolulu, 1954), pp. 4–5; Cuba, *Final Laboratory Reports, 1952 Crop* (Havana, mimeo.), and *ibid.*, *1953 Crop*; Taiwan Sugar Corp., Chief Engineering Dept., *Annual Manufacturing Data of Taiwan Factories 1951–1952* (Taipei).

[a] Figures do not always add up to exactly 100 per cent in original sources.

[b] Including undetermined losses.

[c] Sucrose as per cent of total dry solids contained in first expressed juice.

[d] Crop year 1949/50.

[e] Average of crop years 1952 and 1953, for reporting mills.

[f] Crop year 1952.

[g] Crop year 1947/48.

The same incentives were important in developing the Taiwan industry during the period of Japanese control, though market prospects for the island's industry have been less favorable since the war. For the combination of high sucrose with relatively low fiber content, and high purity, Queensland is the leading region, but Cuba's natural advantages for production of a one-year crop are also evident. The industry in Natal loses considerable sugar as a result of high fiber content, but the notable case of low sucrose is in British Guiana, where seasonal distribution of rainfall affords too short a dry period for ripening. The excellent results obtained in Puerto Rico, despite a raw material of only moderately good quality, reflect the availability of American capital for large-scale processing units and the relatively high price assured in the United States market. The record suggests that, while India's cane agriculture is laggard, her white sugar industry is much more on a par with other countries. If account is taken of relative fiber content and purity of juice in the two areas, Indian factories do at least as well as those in Mauritius.

Though Table 3 gives no indication, the quality of cane in some regions can vary considerably from year to year. Variations in sucrose content are particularly common. In recent years, a range between 11.1 and 12.7 per cent has not been unusual for Trinidad, between 13.3 and 14.2 for Natal, or between 14.9 and 15.6 for Queensland. Such variability makes a considerable difference to the sugar actually produced in a specific year, but does not destroy the broad international comparisons here considered. So far as factory losses are concerned, the results from year to year are quite stable.

INTENSIVE VS. EXTENSIVE EXPANSION

The peculiar determinants of sugar yields complicate comparisons over time quite as much as international differences at a given moment. The history of production in each region is a composite of changes in acreage, crop tonnage per acre, and recovery of sugar from available tonnages. Performance in these respects has not necessarily been moving in the same direction, let alone at the same rate. In the twentieth century, the record for beets and for cane is strikingly different.

Beet regions.—For beet sugar, it is the impressive achievement in the nineteenth century that is notable. Careful selection and breeding from a common ancestor, the White Silesian beet, assured a high tonnage crop, rich in sugar content, and the level of processing technology was also high. By contrast, the rate of advance in the twentieth century has been extremely modest (Table 4). To a considerable degree, expansion in output since 1900 has come merely from higher acreage. For all of Europe (including the USSR), sugar yields per acre were actually lower around the middle of this century than they had been fifty years earlier; a slight increase in extraction rate was more than offset by lower beet tonnage. While part of the explanation lies in the difficulty of re-establishing a high level of agricultural practice following the devastation of two world wars, the postwar period 1949–51 saw a fully restored beet economy. So far as agricultural performance was concerned, the USSR had yet to recover the levels attained before World War I, but even in front-ranking Holland and Belgium the advance over 1900 was slight.

These results reflect several divergent tendencies. Newer pro-

Table 4.—Trends in Beet-Sugar Yields, 1900–51*

(Metric tons)

Region	1900–02	1910–12	1927–29	1934–36	1949–51
RAW SUGAR PER HECTARE					
Netherlands	4.61	5.04	5.29	6.24	5.78
Belgium	4.58	4.30	4.35	5.30	5.61
United States	2.44[b]	3.03	3.84	3.79	5.05
Germany	4.20	4.30	4.31	4.79	4.11[c]
Great Britain	—	1.76[a]	2.95	3.87	3.91
France	3.54	3.24	3.74	4.18	3.86
Europe ex-USSR	—	—	3.73	4.22	3.64
USSR	1.90	2.68	1.81	1.68	1.85
SUGAR BEETS PER HECTARE					
Netherlands	32.33	32.87	32.87	39.42	41.45
Belgium	33.20	28.86	29.39	34.89	39.23
United States	—	25.09[a]	24.34	23.75	33.31
France	27.99	23.79	26.76	29.47	29.46
Great Britain	—	17.15[a]	19.28	24.37	28.37
Germany	29.81	27.10	26.80	30.39	27.53[c]
Europe ex-USSR	—	—	23.86	26.53	24.88
USSR	14.16	18.17	11.42	11.53[d]	13.71
SUGAR EXTRACTION PER 100 TONS OF BEETS					
United States	12.2[b]	13.5	16.4	16.3	15.2
Germany	14.1	15.8	16.1	16.5	15.0[c]
Europe ex-USSR	—	—	15.6	15.9	14.6
Belgium	13.9	14.9	14.8	15.2	14.2
Netherlands	14.3	15.3	16.1	15.8	14.0
Great Britain	—	10.3[a]	15.2	15.8	13.7
USSR	13.4	14.7	15.9	14.8	13.5
France	12.7	13.5	14.0	14.2	13.0

* Calculated from data in F. O. Licht, *World Sugar Statistics, 1937*, and *World Sugar Statistics, 1936/37, 1935/36 and Estimates 1937/38*. For United States from U.S. Dept. Agr., Prod. and Mkt. Admin., Sugar Branch, *Statistical Summary of Operations of the U.S. Beet Sugar Industry for the 45-Year Period 1901–45*; and Table 1. Crop years beginning in years designated.

[a] 1912–14.
[b] 1901–03.
[c] Western Germany.
[d] By 1937–38, USSR yields were averaging about 16 tons per hectare.

ducing regions, such as Italy and the United Kingdom, and newer regions in the older producing countries, were frequently at a disadvantage with respect to either climate or soil. Extension of Soviet production beyond the Ukraine, to be sure, included a small amount of irrigated culture with correspondingly high yields, but the geographic shift was more often to regions of harsh climatic conditions. Even in high-yielding European countries, repeated planting increased losses from pest infestation, notably nematodes, while the restrictions on output generally in effect during the decade 1928–38 provided little positive inducement to raise yields.

For the United States, sugar-beet production is essentially a twentieth-century enterprise, and the table accordingly reflects the gradual improvement in grower practices that is so important for this crop. But there have also been severe setbacks as the result of disease, largely overcome since 1925 by the breeding of disease-resistant varieties. As might be expected, the American emphasis is more on productivity per man-hour and less on yield per acre than in Europe. The United States has consequently pioneered in the application of farm machinery to beet cultivating and harvesting, though with some sacrifice both of beet tonnage and of extraction rate. If sugar yields per acre in the United States nevertheless compare far less unfavorably with European results than, say, yields of wheat, an important part of the explanation lies in a major geographical shift. The center of the industry has moved from Ohio and Michigan to irrigated valleys in the western states, where yields are excellent.

While increases in beet tonnage, in sugar content of beet, and in factory extraction of sucrose advanced together before 1900, practically the entire improvement in European yields since that date has come from developments on the factory side. Gains in extraction have come partly by consolidating smaller producing units, partly by building factories of more efficient size in newer areas (e.g., in the United Kingdom) or during postwar reconstruction, and partly from improved technological processes within the factory. With the possible exception of the introduction of continuous diffusion in place of the batch method, none of these processing developments have been of major importance. Even the capacity of factories has not increased much in 50 years. Units were being constructed in California in 1898 and 1899 (*18*, pp. 2–3), with daily slicing capacities in the general range (2,000–3,000 tons per day) of the larger factories built in the last two decades.

Cane regions.—By contrast with the comparative stability in the beet industry, the interwar years saw major achievements in cane regions. Expansion in acreage planted to sugar cane was substantial in Mexico, Cuba, and the British Dominions, but, more important, productive efficiency was upgraded almost everywhere. The least advanced areas were turning to manufacture of a high-quality centrifugal product in place of various inferior types of sugars obtained by relatively primitive processing. This transition implied introduction of the highly capitalized central factory that goes along with grinding and extracting a superior sugar. Simultaneously, improved systems of transportation had to be organized, if the mill was to be served with a bulky and perishable raw material in sufficient quantity and from a sufficiently wide area. Spectacular success in developing new varieties was quickly reflected in field results almost everywhere. Improvement in extraction of sugar available in the original cane occurred at a slower pace, as did the spread of irrigation, fertilizer application, and improved cultural practices, but the combined impact was considerable.

At least some of these developments were taking place in all major cane areas, but at very unequal speeds. Typically, advance came first in the mill, and agriculture benefited only subsequently. In the Philippines, production of low-grade muscovado and panocha sugars exceeded that of centrifugal as recently as 1921 (*23*, p. 37), but almost all sugar was centrifugal by the early 1930's. Over 80 per cent of Brazil's sugar was being manufactured in modern mills (*usinas*) in 1947/48, as against about 40 per cent in 1925/26, though output from the multitude of small-scale *engenhos* has fallen little in absolute terms (*24*, p. 56). India also initiated an important white-sugar industry. But in none of these countries was anything like the same degree of advance registered for cane agriculture. Factory advances have also been important in Cuba, Puerto Rico, and the Dominican Republic in this century, while processing was substantially improved in the British West Indies during and since the interwar period. In Argentina, the recovery rate was reported at 6.8 per cent in 1920, but 9.8 per cent in the exceptionally good year 1940 (*29*, p. 34). If some regions benefited most from applied engineering, as typified by the modern mill, others were more properly identified with the successes of applied agricultural science, while the agricultural and manufacturing sectors sometimes advanced together. In South Africa since the 1930's, fibrous Uba cane has been displaced by new varieties that not only yield higher tonnages but also are less intractable to the milling process.

While yield statistics, even at best, are too unreliable to permit specific year-to-year changes to be interpreted closely,[1] the long-run

---

[1] Small year-to-year changes in yields, as well as international differences of the order of 10 to 15 per cent, cannot be interpreted too rigidly, for a number of reasons. Cane cut for replanting is frequently ignored, since it typically absorbs

patterns reflected in Charts 5 A–C warrant considerable confidence. If the doubling of yields in 10 years may be considered a spectacular achievement in agriculture (28, p. 18), the results attained by a modernizing cane industry in Taiwan between 1920/21 and 1931/32 border on the astonishing. Recovery rates rose by two-thirds, cane yields tripled, and sugar yields rose from less than one to more than 4 short tons per acre. World War II temporarily destroyed these gains almost completely, and even 7 years after it was over, the best prewar performance had not been repeated. In Queensland, though total sugar production rose more than fivefold over the period charted, yield trends are less impressive. If improvement has been at a slower rate, it has also been better sustained. Sugar yielded less than one ton per acre as late as 1882/83, though it had risen to nearly 4 tons by 1939/40. Recovery rates, generally over 14 per cent in recent years, stood below 10 per cent in 1901/02; improvement at this point, based both on higher-sucrose cane and on improved milling efficiency, in effect contributes 40 per cent of the expansion of Queensland output in the twentieth century. (Very recently, growth of cane production has overtaxed the processing capacity of mills, with adverse effects on recovery rates.)

By contrast, yields in Cuba are in the doldrums, despite high milling efficiency. Cane tonnages have declined during the period charted, and even high recovery rates permit sugar yields per acre barely to hold their own. Increased Cuban production was accordingly on an extensive basis entirely, reflecting a doubling in acreage. Even a marginal region like Louisiana can occasionally surpass Cuba in cane tonnage per acre, though low and variable sucrose content remains a handicap to the Louisiana cane-sugar industry. In Florida, for which no data are plotted, cane tonnages exceeded Cuba's from the beginning of the industry in the late 1920's (23, p. 35). The

---

only about 5 per cent of the crop, but can exceed 10 per cent in Louisiana. In some regions, e.g. Queensland, a small acreage is cut for green fodder, while in others, some cane is eaten directly as a confection. Unless relevant acreages are carefully distinguished, cane yields per acre may accordingly be underestimated somewhat. Sugar recovery rates are properly applied only to cane harvested for crushing, but the latter will not be identical at the factory and in the field, particularly with primitive methods of hauling. Sugar itself may be reported as raw or refined, or merely *tel quel*, so that discrepancies can creep in at this point as well. Special problems also arise in interpreting yields when a local industry is shifting from the older non-centrifugal to the modern centrifugal basis, though that is not a consideration for the countries and during the period charted. In the case of Cuba, however, cane processed into high-test molasses in occasional years is not adequately distinguished from cane processed into sugar.

CHART 5 A.—YIELDS OF RAW CANE SUGAR PER HARVESTED ACRE,
SELECTED REGIONS, 1919–53*

(*Short tons*)

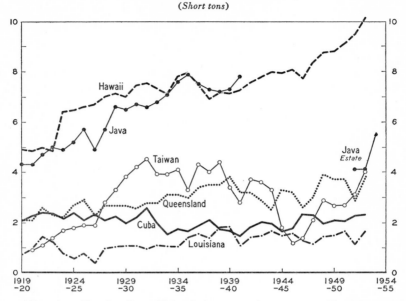

* Data from Hawaiian Sugar Planters' Association, *Sugar Manual* (Honolulu, 1954), pp. 16, 34, 36; Queensland Cane Growers' Association *et al.*, *Australian Sugar Year Book 1951* (Brisbane), p. 169; Australia, Commonwealth Bur. Census and Stat., *Primary Industries Part I—Rural Industries 1952–53* (Bull. No. 47), p. 48; International Institute of Agriculture, *International Yearbook of Agricultural Statistics* (Rome), various issues; U.S. Mutual Sec. Admin., Mission to China, *A Survey of the Taiwan Sugar Industry, Part III Economics* (Consultant Rept. No. 1, Taipei, 1950), p. 5; Republic of China, Min. Econ. Affairs, *Taiwan Sugar* (1955), p. 5; H. C. and R. J. Prinsen Geerligs, *Cane Sugar Production 1912–37* (London, 1938), p. 82; and Republic of Indonesia, Kantor Pusat Statistik, *Statistik Perkebunan Tanam²an Perdagangan 1953*, p. 18, and *Pertanian Tanam²an Ekspor* (*De Landbouw Export Gewassen*) 1952, p. 18.

yield record for the Philippines, though less well documented than one would wish (cf. *23*, p. 37), has been little more dramatic than Cuba's in recent years. To be sure, between the early 1920's and the early 1930's, production of centrifugal sugar rose almost eightfold, to about 1.6 million short tons, partly by diversion of cane from processing of low-grade muscovada and panocha. In the final 5 years of that expansion, sugar yields per acre reportedly rose almost 50 per cent. But ensuing crop restrictions imposed a setback, compounded by the disastrous effects of wartime occupation, from which subsequent recovery has been slow.

Of the charted regions, the interesting contrast between Java

CHART 5 B.—YIELDS OF SUGAR CANE PER HARVESTED ACRE,
SELECTED REGIONS, 1919–53*

(Short tons)

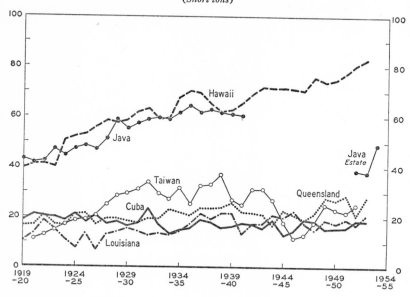

* Sources as for Chart 5 A.

and the Hawaiian Islands remains to be discussed. Throughout the 1920's, progress in agricultural performance was most striking, cane tonnages per acre rising some 50 per cent in response to varietal improvements (see chapter 6), although recovery rates actually declined for a considerable period in Hawaii. The agricultural advance is not measured from a low base comparable to that of a late starter like Taiwan, but occurred in cane economies already noted for technological leadership. In the Territory of Hawaii, progress has been sustained to date, with the result that the gap has widened between these Pacific islands and most other cane regions. In Java, despite an enormous cutback in total production, yields actually edged ahead of Hawaii's at the end of the 1930's. But World War II brought disaster, including a breakdown in the plantation organization upon which the prewar cane economy had been based. Postwar yields on the estates, though high by most standards, cannot match the prewar level when labor was cheap and operations in the field were closely synchronized with those in the factory. Resort was had also to cane

CHART 5 C.—RECOVERY OF RAW SUGAR FROM SUGAR CANE,
SELECTED REGIONS, 1919–53*

(Per cent)

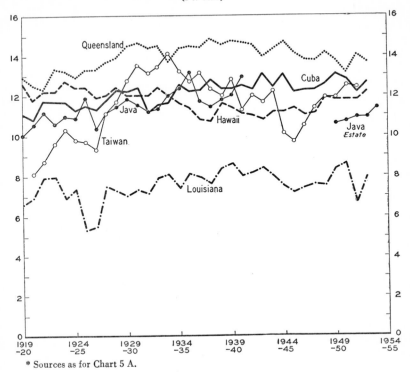

* Sources as for Chart 5 A.

purchased from independent growers employing lower standards of cultivation, for some 16 per cent of the 1954 crop (*30*, pp. 130–31).

In the world cane economy, innovations other than improved agricultural practices or higher technical efficiency in factories have recently become important. There has been a move from rail to motor haulage for transporting cane. In several regions where modernization was already well advanced, mechanization has come into the field loading and cutting of cane. Hawaii and Louisiana have led the way in the further applications of labor-saving equipment. Mechanization has meant more trashy cane, inexact cutting, rupture of the cane stalk, and consequent difficulty in the factory processes, and not all regions are prepared to pay this price. Indeed, some informed specialists maintain the view that mechanical harvesting "retains certain drawbacks and must still be considered experimental" (*11*, p. 145). About bulk ocean shipment of raw sugars, a practice that has also been spreading during this period, no such doubts are raised. In this respect as well, Hawaii has pioneered, but United Kingdom refiners have been rapidly promoting bulk methods within sugar-exporting portions of the British Commonwealth.

It is quite evident that the protected sugar suppliers, not the free-market exporters, are now the technological leaders. Labor has been sufficiently abundant in Cuba to cause organized resistance to both field mechanization and bulk shipment. As for Indonesia, its high prewar yields were based on a low wage level and a degree of labor intensity that the industry can never again expect to command. High-yield varieties, the major feature of agricultural advance between the wars, were consistent with the former structure of Java's sugar industry. But the mechanical harvester is not.

CITATIONS

1 P'ei-kang Chang, *Agriculture and Industrialization* (Cambridge, Mass., 1949).

2 E. M. Ojala, *Agriculture and Economic Progress* (London, 1952).

3 J. K. Galbraith and C. S. Solo, "Puerto Rican Lessons in Economic Development," *The Annals of the American Academy of Political and Social Science*, January 1953.

4 David Ricardo, *The Principles of Political Economy and Taxation* (Everyman's Edition).

5 P. A. Baran, "On the Political Economy of Backwardness," *The Manchester School*, January 1952.

6 B. F. Johnston, "Agricultural Productivity and Economic Development in Japan," *Journal of Political Economy*, December 1951.

7 Naum Jasny, *The Socialized Agriculture of the USSR: Plans and Performance* (Stanford, Calif., 1949).

8 David Spring, "The English Landed Estate in the Age of Coal and Iron—1830–1880," *Journal of Economic History*, Winter, 1951.

9 United Nations, Dept. Econ. Affairs, *Measures for the Economic Development of Under-Developed Countries* (New York, 1951).

10 Federal Reserve Bank of San Francisco, "The Sugar Beet Industry in the Twelfth Federal Reserve District," Supplement to *Monthly Review*, April 1951.

11 International Bank for Reconstruction and Development, *The Economic Development of British Guiana* (Baltimore, 1953).

12 Harold Wickenden, *Sugar from British Farms* (London, 1945).

13 South African Sugar Journal, *South African Sugar Year Book, 1952–53* (Durban).

14 Pieter Honig, "Developments in Cane Sugar Production since 1938," *Sugar*, September 1950.

15 Noel Deerr, *The History of Sugar,* I (London, 1949).

16 Pierre Gourou, *The Tropical World, Its Social and Economic Conditions and Its Future Status* (transl. by E. D. Laborde, London, 1953).

17 P. C. Goswami, *Sugar Industry in India and Abroad* (Calcutta, 1952).

18 U.S. Dept. Agr., Prod. and Mkt. Admin., Sugar Branch, *Beet Sugar Factories of the United States* (1950).

19 British Guiana, Dept. Agr., *Sugar Bulletin No. 20* (1952).

20 H. H. Dodds, "The World's Sugar Industries—Their Scientific and Technical Progress," *S. A. Sugar Journal* (Durban), November 1951.

21 India, Min. Food and Agr., *Annual Report of the Sugarcane Breeding Institute Coimbatore for 1950–51.*

22 India, Min. Food and Agr., *Annual Report of the Sugarcane Breeding Institute Coimbatore for 1951–52.*

23 Hawaiian Sugar Planters' Association, *Sugar Manual* (Honolulu, 1954).

24 Brazil, Instituto do Açúcar e do Álcool, *Anuário Açucareiro, Safra 1949/50.*

25 M. P. Ghandi, ed., *The Indian Sugar Industry 1949 Annual* (Bombay, 1950).

26 F. Maxwell, *Economic Aspects of Cane Sugar Production* (London, 1927).

27 International Institute of Agriculture, *Sugar Economy in the Interval Between the Two World Wars* (by E. Romolini, Rome, 1944).

28 J. D. Black, "Agriculture in the Nation's Economy," *American Economic Review*, March 1956.

29 Centro Azucarero Argentino, *Estadistica Azucarera No. 6* (Buenos Aires, 1945).

30 Bank Indonesia, *Report for the Year 1954–1955* (Djakarta).

31 W. A. Lewis, *The Theory of Economic Growth* (Homewood, Ill., 1955).

CHAPTER 4

INTERNAL STRUCTURE, ORGANIZATION, AND CONTROLS

The institutional environment of sugar production is highly peculiar. For reasons both historical and economic, there has been a degree of government intervention in peacetime that can be matched in few, if any, world commodities. For this commodity, as for others, protectionism of itself complicates the international structure of prices. But in addition, national systems of sugar control encompass a comprehensive set of internal regulations. In the agriculture sector there is frequently control of grower price, planted acreage, harvested tonnage, and even the proportion of arable land that the farmer may plant in sugar crops. The factory similarly is often restricted in its sources of supply, its volume of production, its rate of profit, and its share of various markets.

The fact that these elaborate regulations are surprisingly comparable between countries of quite different economic philosophies may be entirely fortuitous. But it invites a search for identifiable economic characteristics of this commodity that may impose certain social patterns upon the community in which it is produced. At the same time, one can certainly indicate ways in which the direction and pace of technological change have been affected by distinctive institutional arrangements to be found in different producing regions. The interaction between social organization and technological change that has accompanied the modernization of the sugar industry is a common theme of the next three chapters.

SUGAR AND THE TAX SYSTEM BEFORE 1900

Several bases for governments' attention to sugar are of long standing. For one thing, in the earlier period of low per-capita consumption and luxury status, sugar ranked with matches, salt, and tobacco as an item with a peculiarly inelastic demand, and was accordingly a fit subject for domestic excise taxation. Because practically all sugar consumed in European countries before 1800 came from overseas and passed through relatively few refineries at port centers, administration of the tax was fairly easy and evasion difficult.

Customs collections from sugar represented an important contribution to government revenues in many countries well after 1850. Besides, colonial possessions in the Caribbean area enjoyed a virtual monopoly of the European market and their interests were protected in the home country. At the same time, domestic sugar refining was doubly encouraged by tariff protection. Not only was raw sugar typically admitted at a lower rate than refined, but re-export of refined sugar enjoyed indirect tax benefits. Drawbacks on refined exports, based on yields officially accepted for fiscal purposes, tended to underrate the efficiency of manufacturing recoveries, and import taxes were refunded on a higher tonnage of raw sugar than had actually been processed for export (11). The limits of governmental intervention in the late nineteenth century were nonetheless narrow. Refining in bond, which would protect the public purse but required continuous supervision by customs personnel, was considered too radical a practice for Belgium to accept as late as 1875.

The rise of beet sugar was closely associated with tax policy. To some extent, excessive drawbacks on refined cane exports had been inadvertent, since commercial application of the polariscope to determine sucrose content scientifically did not come until the 1870's, while the Dutch Standard based on color was liable to considerable abuse. Fiscal encouragement of beet sugar, on the other hand, was deliberate from the beginning. Initially, beet sugar had automatic protection by virtue of the import duties levied against cane sugar from overseas. When beet-sugar production rose sufficiently to make the loss of customs revenue a matter of fiscal concern, an excise tax would be levied on the domestic product, though generally at a lower rate than the customs duty. Production of beet sugar for export was also encouraged, through the rebate of excise taxes, as refined cane exports had been in an earlier period. Especially in the last quarter of the century, when shipments of cereals from America were disrupting the European agricultural economy, fiscal aid to the beet-sugar industry became a means of indirectly funneling financial aid to the farm sector and at the same time provided an inducement to introduce the more rational crop rotations associated with the sugar beet (see chapter 9).

One feature of the taxation system was crucial. The excise was not levied on the weight of final sugar produced, but on some indicator of productive capacity. Germany taxed the tonnage of beets processed, Belgium assumed a standard recovery from the volume of beet juice, and Russia assumed a legal rate of sugar extraction from beets. These tax bases provided a strong incentive to improve productive efficiency. If more beet sugar was extracted, the increment escaped the excise and moreover might even recapture part of the tax by moving into export. Temporary French efforts to exact the excise on a more accurate basis seriously delayed technical improvement in the local factories (*15*, p. 259), such as the shift from hydraulic presses to the newer and more efficient diffusion process of extracting the juice. The levy on beets as such encouraged the breeding of a richer sugar beet at the same time that it promoted superior factory performance. In Germany factories were moved to develop, and able to insist that growers plant, high-sucrose varieties from an early date.

Government-subsidized disposal of surplus sugar in foreign markets, a basic feature in the expansion of beet acreage and production, overreached itself in the last two decades of the nineteenth century. After 1885, direct export subsidies were paid, for varying intervals, by Russia, France, Austria-Hungary, and Germany, the latter two even providing a differential in favor of the refined product. The degree to which the entire system depended on foreign markets is indicated by the fact that as of 1901 Germany was exporting 57 per cent, Austria 63 per cent, and France 67 per cent of their sugar production (*12*, p. 11). Ultimately the drain on the public purse became excessive. In several instances, subsidies and rebates exceeded government receipts from excise tariffs. In any case, once the United States levied a countervailing duty against subsidized-sugar imports in 1898 and Britain was finally prepared to embargo dumped sugar that was demoralizing cane production in parts of its Empire, the export basis of the beet-sugar regime was destroyed. The Brussels Sugar Convention of 1902 formalized the demise. When direct subsidies became important in the middle 1920's in Great Britain and in the middle 1930's in the United States, they no longer served as an export stimulant.

LARGE-SCALE CAPITAL IN CANE MILLING

If relations between the tax system and the European beet factory typify sugar developments between 1850 and 1900, the succeeding half-century was equally well characterized by spectacular increase in the size of the cane-processing unit. In the background was the transmission to tropical cane regions of methods already worked out for extracting sugar from beet juice, such as the evaporating equipment that assured a more uniform product. For the successful application of steam power to the actual milling of cane, however, beets clearly provided no model. Steam-powered equipment, by improving the speed and efficiency of extraction, incidentally raised the grinding capacity of the factory. Simultaneously, improved haulage by rail was expanding the region from which a single factory could draw its cane. Advance on these several fronts was well under way even before 1900.

The effect on size was spectacular. As of 1889/90, the average beet factory in Germany produced something over 3,000 tons of sugar a year (*38*, p. 31) and in Russia about 2,000 (*73*, p. 154). The Cuban mill, which was large by contemporary standards for cane, averaged some 2,000 tons of sugar (*36*, p. 130), as against only 1,000 tons in Mauritius (*36*, p. 185) and less than 200 tons in Louisiana (*39*, p. 263). But it has been the capacity of the cane mill that has expanded more rapidly ever since. Large beet units constructed in such advanced industrial nations as the United States and Germany since the end of World War II are capable of producing between 30,000 and 40,000 tons of sugar per year, whereas cane mills constructed in Pakistan, Madagascar, Mexico, and the Dominican Republic range up to 90,000–100,000 tons (*37*, pp. 34–35). What these figures mean in financial terms is suggested by the fact that the cost of constructing new mills in 1954 was estimated at two million dollars per 10,000 tons of annual sugar production. Five older Cuban mills actually produced over 150,000 metric tons of raw sugar in 1951/52 and one, Central Delicias, over 200,000 (*18*, pp. 306–12).

Under favorable circumstances, the optimum size of cane mill is clearly larger than beet factory. In both cases, ultimate limits on size are set by seasonality of operations and by the heavy cost of transporting bulky sugar crops from the surrounding region. But the beet factory must live alongside rotation agriculture, a pattern of land use reinforced by the risk of nematode damage if the crop is replanted too frequently on the same land. Accordingly, the territory in its immediate vicinity cannot serve exclusively as a source of raw-material supply, as is possible in the cane monoculture. Where beet pulp is not dried, the limited radius within which wet pulp can be hauled to farms is an additional factor operating in the same direction. If cane has the handicap of lower sucrose content in raw crop, this is more than compensated by higher sugar yields per acre. The cane harvest, moreover, continues at high efficiency in favored regions for more months than the beet campaign, and overhead costs of seasonal idleness can thus be more readily borne. Modernization of the cane industry, as a result of its later start, has been the more able to take advantage of the latest technology available to both. Finally the cane mill has been erected in certain frontier regions, where the supporting agricultural unit could be adapted to the requirements of the factory, while the beet factory had typically to adjust to the existing pattern of European farm holdings. It is of some interest that the mill reputed to be the largest in the world, Central Rio Haina in the Dominican Republic with a rated capacity of 15,000 tons of cane per day (74, pp. 42–46), and the giant Cuban centrals are located in regions where investment in large-scale manufacturing is not matched by correspondingly heavy investment in agricultural mechanization.

Tables 5 and 6 provide additional detail on the size distribution of sugar factories in selected countries. Mill production in Cuba, the Philippines, Dominican Republic, Puerto Rico, Taiwan, and Queensland (as well as Hawaii, not shown in the table) is on the high side, as contrasted with Louisiana, Jamaica, Mauritius, and India, though figures must be interpreted with some caution. In most recent years, Taiwan has operated with considerable unutilized capacity (cf. 26, p. 27), whereas Hawaii produces its sugar in mills of modest daily grinding capacity (cf. 42, pp. 4–6) that operate under advantageous conditions for a particularly long season. The relative scarcity of capital in India and the exceptionally short harvest season in Louisiana would appear to be important factors contributing to small mill size. Regions where cane production has a long history also tend to carry over a legacy of small-scale processing enterprises. Thus on the United States mainland a single mill in the state of Florida, where cane has been grown commercially for less than three decades, outproduces ten average units in the older region of Louisiana, while only two factories in Jamaica and none in Mauritius rank as large-scale by modern standards. Similarly in beet areas, a country with a short producing history has the larger units; in the United Kingdom, annual sugar production is around 35,000 tons per factory. Peacetime policies that allocate markets perpetuate small-scale units in many cane regions, but wartime destruction has the incidental effect of speeding the transition from outdated processing unit to modern plant. As a result of consolidation of older factories or construction of larger-scale units, the average sizes of factories in the cane regions of the Philippines and Taiwan and the beet regions of western Germany and the USSR are considerably larger today than they were before World War II.

Particularly in export regions, the capital requirements for modernization and expansion have frequently exceeded local resources, and heavy capital import from metropolitan areas has been necessary. Although the problem of funds has thereby been solved, the relations of the mill to the local economy have been correspondingly

TABLE 5.—SIZE DISTRIBUTION OF SUGAR FACTORIES:
ANNUAL SUGAR PRODUCTION*

(Number of factories)

Production (1,000 metric tons)	Cane								Beet
	Cuba 1951	Philippines 1952	Dominican Republic 1951	Puerto Rico 1951	Queensland 1953	South Africa 1952	Mauritius 1952	Jamaica 1951	West Germany 1952
0– 9.9	5	4	3	1	2	4	5	11	34
10– 19.9	27	4	5	8	5	5	14	8	23
20– 29.9	35	2	2	5	3	1	8	—	10
30– 39.9	30	2	1	12	6	1		—	4[a]
40– 49.9	18	4	—	1	13	2		1	
50– 59.9	16	4	1	3	1	4		1	
60– 69.9	6	2	2	1	—	1			
70– 79.9	4	—	1	2	1	1			
80– 89.9	3	1	—	—					
90– 99.9	3	—	—	—					
100–109.9	3	1	—	1					
110–119.9	3	—	—						
120–129.9	2	1	—						
130–139.9	—		—						
140–149.9	1		—						
150–159.9	2		—						
160–169.9	—		1						
170–179.9	1								
180–189.9	—								
190–199.9	1								
200–209.9	1								

* Data from Farr & Co., *Manual of Sugar Companies 1952–53* (New York, 1953), pp. 306–12, 321 [for Caribbean islands]; *Weekly Statistical Sugar Trade Journal*, Dec. 17, 1953, p. 508 [Philippines]; *ibid.*, April 15, 1954, p. 155 [Queensland]; South African Sugar Journal, *South African Sugar Year Book, 1952–53* (Durban), pp. 148, 182ff.; Mauritius, Cen. Stat. Off., *Year Book of Statistics, 1952*, p. 75; Wirtschaftliche Vereinigung Zucker, *Geschäftsbericht für das Geschäftsjahr 1952/53* (Wiesbaden, 1954), p. 46.
Sugar years beginning year indicated, except for calendar year in Mauritius.
[a] 30,000 tons and over.

complicated. In Cuba, for example, large American-owned mills tended to become mere protrusions from the United States, incidentally resident in Cuba. Located frequently near the coast, served by their own subports, operating self-contained transportation systems, importing directly most of their supplementary needs (sometimes even the labor supply), and disbursing dividends to nonresident stockholders, mills of this type were poorly integrated with the Cuban economy. Except through the tax system, these rural enterprises made little more contribution to the development of a national economy than might an isolated mining company. Several had been built

TABLE 6.—SIZE DISTRIBUTION OF SUGAR FACTORIES:
DAILY PROCESSING CAPACITY*

(Number of factories)

| Daily grinding or slicing capacity (*metric tons*) | Cane | | | | Beet |
	Cuba 1951	Louisiana 1952/53	Mauritius 1952/53	India 1949	United States 1952/53
0– 499	—	1	4	27	—
500– 999	6	7	10	77	7
1,000– 1,499	21	20	10	31	33
1,500– 1,999	37	9	3	7	17
2,000– 2,499	29	8	—	2	5
2,500– 2,999	19	4	1		11
3,000– 3,499	9	4			3
3,500– 3,999	14	1			—
4,000– 4,499	2				—
4,500– 4,999	2				—
5,000– 5,499	4				
5,500– 5,999	3				1
6,000– 6,499	1				
6,500– 6,999	3				
7,000– 7,499	4				
7,500– 7,999	—				
8,000– 8,499	—				
8,500– 8,999	4				
9,000– 9,499	—				
9,500– 9,999	1				
10,000–10,499	1				
10,500–10,999	—				
11,000–11,499	1				

* Data from *Anuario Azucarero de Cuba 1951* (Havana), pp. 99–101; Farr & Co., *Manual of Sugar Companies 1952–53* (New York, 1953), pp. 292–93, 298–99; South African Sugar Journal, *South African Sugar Year Book 1952–53* (Durban), pp. 187–92 [Mauritius]; M. P. Ghandi, ed., *The Indian Sugar Industry—1949 Annual* (Bombay), App. pp. 2–28.

primarily to supply mainland refineries. Such were Centrals Cunagua and Jaronú, sources for American Sugar Refining Company's plants on the eastern seaboard; Boston and Preston, shipping to United Fruit Company's Revere refinery in Massachusetts; and Delicias and Chaparra, associated with the Colonial refinery at Gramercy, Louisiana. Such mills could not be expected to take full advantage of a technological development like the use of activated carbon, which facilitated white-sugar manufacture at the mill and reduced the locational advantages of the bone-char refinery at metropolitan seaboard points. Moreover, in lean years, it was the American-owned mills which, by virtue of easier access to finances and more assured markets, enjoyed certain natural advantages in the struggle for survival. Foreign-owned multiple-mill companies might also adjust more readily than Cuban family proprietorships to lower levels of production. By halting operations entirely at particular factories, the former could keep their more efficient mills closer to full-capacity operation and hold down costs.

In several export regions other than Cuba, the influx of capital from metropolitan areas was associated with concentrated control of total output. Two American firms contributed three-quarters of the Dominican Republic's crop in 1952: the South Porto Rico Sugar Company (which also operated the largest mill in Puerto Rico) and the West Indies Sugar Corporation (which owned several Cuban factories) (18, p. 318). The two large mills of the West Indies Sugar Company, Ltd., produce more than a third of Jamaica's output (18, p. 321). Three companies registered in the United Kingdom, of which the largest is Booker Bros., McConnell, Ltd., grow and process practically all the sugar cane produced in British Guiana (2, p. 39). There are comparable examples in the Far East. Nearly all of Formosa's sugar factories were owned by four Japanese companies at the outbreak of World War II. The successor enterprise, the government-sponsored Taiwan Sugar Corporation, enjoys a complete local monopoly. In the Fiji Islands, all commercial sugar is produced by the Colonial Sugar Refining Company, whose activities extend to sugar mills, refineries, and related enterprises in Australia.

THE PLANTATION AND ALTERNATIVE SYSTEMS OF CANE AGRICULTURE

However valuable the productive contribution of the highly capitalistic modern mill, it has also brought difficult problems of social control, among which absentee ownership and high degree of concentration of total output are not necessarily the most important. Large industrial plants came into rural settings where economic units were characteristically small. The *central* became the economic and social core of the entire community, to the point where the manufacturing enterprise was on a par with the constituted political authorities as an instrument of local government. This was not an environment in which relationships between growers and laborers, on the one hand, and the mill, on the other, might be expected to work well on a purely market basis. In nineteenth-century United States, rural dissatisfaction over the farmer's bargaining weakness in dealing with private railroads was an important factor in the introduction of public-utility regulation. In cane regions, not only may the mill be the exclusive market for the grower's crop, but the local railroads themselves are frequently under mill ownership.

Sugar crops do not in any case readily lend themselves to pricing according to customary market practices. Final sugar can serve a national market, but perishable sugar crops must be priced locally. The independent cane farmer, who usually cannot switch readily from cane ratoons to other crops, and who harvests a rapidly wasting asset, is in a particularly exposed position, while the beet grower is an unequal match for the factory once his crop is planted. But beet factory and cane mill are also in a vulnerable position. Low unit costs can be realized only if operations are continuously at full capacity during the processing season and supplies are regularly forthcoming from year to year. In this respect also the cane economy provides the more extreme case. Because cane as much as 24 hours old is becoming a rather stale commodity, harvesting in the cane fields must be exceptionally well coordinated with the rate of factory operations and the immediate job of satisfying the mill's heavy appetite for raw material given precedence over all other community activities. In years when factory capacity is excessive either because the year's crop is short or because manufacturing facilities have been expanding more rapidly than agricultural production, even a small grower can press a considerable market advantage. If the crop is extraordinarily abundant or factory output artificially restricted, the market advantage to the processor is considerable. Even in the absence of government intervention, arrangements between factory and grower go far beyond a straight purchase transaction.

These special difficulties in coordinating factory and field operations have been important in shaping the institutional environment of sugar-crop production. In cane, one approach has been the plantation system, of which the key feature is that agricultural operations are performed by hired labor under management supplied by the factory. The large-scale manufacturing unit accordingly assures its necessary volume of supplies by engaging in a collateral large-scale agricultural enterprise. Perhaps the ideal example is provided by Hawaii, where fewer than 30 plantations grow practically all the cane, operate their individual mills, and ship the bulk of their output to a co-operatively owned refinery at Crockett on the California mainland. Fragmentary detail on the role of plantation agriculture in various cane regions is as follows:

		Plantation percentage of			
Region	Date	Acreage	Cane	Sugar	(Source)
Java	1937	...	99	...	(50, pp. 43–44)
British Guiana	c. 1951	...	98	...	(20, p. 133)
Indonesia	1953	...	...	92	(21, p. 117)
Hawaii	1954	90[a]	...	...	(19, p. 12)
Dominican Republic.	c. 1950	...	85	...	(22, p. 26)
Florida	1950/51	...	84	...	(22, p. 9)
Barbados	1954	78	...	...	(23, p. 116)
Jamaica	1939	70	...	...	(54, p. 129)
Trinidad	1951	...	67	...	(8, pp. 135–37)
Jamaica	1951	58	...	...	(54, p. 129)
Mauritius	1952	49	58	...	(24, pp. 71–73)
Trinidad	1920	...	40	...	(8, pp. 135–37)
South Africa	1951/52	...	29	...	(27, p. 157)
Réunion	c. 1952	25	...	...	(25, pp. 403–08)
Taiwan	1952/53	...	...	18	(26, p. 27)
Cuba	1950	...	9	...	(58, p. 157)
Fiji	1952	3	...	...	(46, p. 274)
Queensland	c. 1953	3	...	...	(49, p. 32; 16, p. 46)

[a] About half the cane acreage cultivated outside the plantations is leased from plantation companies.

Plantation organization has certain definite advantages for cane production. The industrial unit's easier access to capital means better yields from a crop that responds well to commercial fertilizer, and less embarrassment from waiting out the 18 to 24 months that frequently separate planting and harvesting. Manufacturing costs can be reduced by stress on a cane with superior milling qualities rather than mere cane tonnage per acre. The distance raw cane has to move is less when competition from all alternative crops is entirely excluded in the immediate vicinity of the mill, and when there is no danger of the crosshauling that is likely if independent growers are favorably situated for shipping to more than a single mill. An assured volume, similarly, permits the mill to carry the overhead costs of its local transport system. This is also a way of accomplishing a coordinated pace of milling and of harvesting, upon which depends the maximum recovery of sucrose available in raw cane. These advantages are extremely important, so much so that it would be improper to identify the plantation system for this crop as merely an attribute of quasi-colonial status. In Venezuela, for example, where modern mills have been organized by local nationals with the assistance of governmental financing, the manufacturing unit has also been assured its own cane fields (*34*, p. 58). In northeastern Brazil, the rise of the modern mill has similarly had its counterpart in expansion of factory-grown cane, and incidentally has undermined the older rural aristocracy (*59*, pp. 512–25).

That plantation agriculture is capable of extremely high cane yields cannot be disputed. Before the war the estate organization of Java and Hawaii led the world in productive performance. When the average Hawaiian figure passed 10 short tons of sugar per harvested acre in 1953, it exceeded its best prewar performance by 25 per cent (*19*, p. 16). Cane from Indonesian small holders, which had come to be the source for about 10 per cent of factory sugar by 1953, yielded only 86 quintals per hectare (equivalent to 3.8 short tons per acre), as against a national average of 122 quintals and the higher figure of 164 quintals under full-scale estate operations before the war (*45*, p. 117). There is good evidence that yields of small cane farmers in the West Indies suffer from neglect of cultivation, inadequate fertilizing, and excessive ratooning; these growers are further disadvantaged by the greater distance their cane must move to reach the mill. In Trinidad (*8*, pp. 135–40) and Jamaica, cane farmers do well to produce half as much cane per acre as the estates.

There have, however, been less satisfactory features of plantation enterprise. Large land holdings are resented in agricultural societies where land hunger prevails. Thus the Taiwan Sugar Corporation was embarrassed by the fact that it took over from the former Japanese sugar companies about 117,000 hectares of land, representing about 15 per cent of the island's arable land, and part of its holdings have been disposed of in the interests of land reform (*47*, pp. 2–5). Creation of an agricultural proletariat to work the fields raises obvious social problems, particularly in regions like the Caribbean where the demand for field labor is highly seasonal. Political difficulties in British Guiana have accordingly led to the search for means of converting to a system of independent or tenant farmers (*69*,

p. 327). A mill that grows a considerable volume of cane is at some bargaining advantage in purchasing the remaining portion of its cane from independent growers. The very contrast between advanced cultural practices in the neighborhood of the mill and the backward methods employed in raising crops less oriented to foreign markets may be a source of local irritation (*6*, p. 177), although the contrast may also perform an educational function. Overhead costs associated with the plantation system may become oppressive in periods of low prices. It was by resort to independent cane farmers that the industry was able to survive the difficult period of the 1890's, and that Fiji succeeded in maintaining output after India prohibited the export of indentured labor in 1916. More recently in Hawaii, which has enjoyed assured markets but not attractive returns on capital, the companies have sloughed off certain marginal functions and now seek, for example, to promote home ownership rather than tenancy for their year-round labor force.

THE CANE FARMER

Successful cultivation has been possible without estates as large as the 13,000 acres to be found on the east coast of Demerara (*60*, p. 172), or still larger plantations located elsewhere. The entire output in Fiji is provided by small growers, each characteristically operating a 10- to 12-acre farm, about half of it ready for harvest in any given year (*55*, pp. 8–9). The Fiji grower is so closely supervised by the factory that the arrangement has much in common with a straight plantation operation, and yet the social status of a tenant grower with secure tenure is quite different from that of a paid employee. Queensland practice is more illustrative of the scale of operation that permits efficient production by an independent grower with access to modern farm equipment. One plantation of 5,000 acres survives (*49*, p. 32), but independent cane growers farming an average of some 50 acres (*62*, p. 66) are the backbone of the industry. In Taiwan, most factory estates range between 150 and 500 hectares in size (*61*, pp. 33, 51–56), while the private farms are predominantly of 1 or 2 hectares, intensively cultivated. Yields from the estates of the Taiwan Sugar Corporation exceeded those of private farms by only 20 per cent in 1951/52.

Commercial production of sugar cane by independent cane farmers rather than by plantation enterprise has required a peculiar set of institutional devices. Some border on the coercive. Growers on rented land in Cuba have at times been forbidden to plant any acreage to other crops (*51*, p. 275). Similarly in prewar Formosa, the control of irrigation water was exercised in some regions so as to compel a fixed place for cane in competition with paddy (*50*, p. 36). But some assurance of continuity in planting is so important to the mill that comparable arrangements exist under voluntary agreements elsewhere. Thus nearly all cane farmers in South Africa commit themselves to keep a considerable portion of their area in cane for 20 to 100 years, and they have done so even in the face of short-run adversities (*52*, p. 51). It is of interest that contractual obligations in beet-sugar areas are more likely to be of the opposite sort. Thus the annual contract of the British Sugar Corporation expressly states that beets are not to be grown on land given to beets or mangolds in either of the two preceding years, as a precaution against infesting the soil with eelworm (*28*, p. 17). A few such regulations as these exist also in cane regions, but chiefly as a restraint on plantation enterprise when or where the availability of a local food supply is a matter of concern. Acreages devoted to estate cane production in Java have been closely controlled, while estates in Barbados were required to plant at least 20 per cent of their arable land to food crops during the war and the requirement was later continued at 10 per cent (*54*, p. 29).

Two other devices of a quasi-coercive sort have also been important in the successful operation of a cane-farmer system. Since the mill suffers from inactivity if the harvest starts too late or goes too slowly, from glut and deterioration of cane if the harvest moves too quickly, and from higher recovery costs if cane is cut too soon before or too long after the time of maximum sucrose, the timing of deliveries cannot be left entirely to chance. Some schedule that rations the amounts a farmer can deliver at the peak of the season and that requires a certain volume of delivery at off-peak periods is desirable. This is a device for meshing operations in any particular season. To some extent, a system of exclusive factory zones serves similar purposes: one mill need not starve for supplies while a close neighbor surfeits with an excess; there is no incentive to delay or interrupt the harvest by hard bargaining over price; and the heavy transport costs (and deterioration of cane) involved in crosshauling are avoided. A system of assigned territories for mills dates from about 1905 in Formosa, from 1915 in Australia (*4*, p. 239), from 1921 in Trinidad (*8*, p. 136), from 1938 in Jamaica (*22*, p. 20), and has also been practiced in other countries, both in expansion and in restriction stages of the sugar industry.

A system of exclusive factory zones also has a wider significance. The rural-oriented factory is an automatic center for the dissemination of advanced scientific practices and enjoys the capital required for investment in agricultural experiment. The farmers who come into continuous contact with a factory company or are tenants on factory land have access to a wide variety of services that give a strong upward thrust to agricultural yields. The factory may maintain small plots as demonstration farms, for experimentation, or for training personnel; it can provide planting stock and develop new local varieties; and its trained supervisors who go out into the farmers' fields operate as a private agricultural extension service and also ensure that cane is cut closer to the time of maximum sucrose. Direct financial encouragement is also important in raising the level of cultural practices. Thus the Taiwan Sugar Corporation provides cash subsidies for early planting, green manuring, and irrigating, and also provides loans to help meet costs of cultivating and fertilizing (*48*, p. 39). In this case, the financial arrangements not only assure that cane will be planted on better land and be better cultivated, but are vital to the farmer's decision to plant cane at all in competition with two to three crops of rice that might be grown during the 18 months while the farmer is waiting for his cane to mature.

Colonial Sugar Refining Company, which encouraged independent cane farming in Australia at an early date by purchasing large estates and carving them up for sale in units manageable as a family operation, has earned an international reputation for the system of cane agriculture developed in Fiji, especially since 1924 (*55*, pp. 8–14). Expensive disk plows, needed for heavier field operations such as removing old ratoons ("stools"), burying trash, and plowing green manure under, can be rented from the Company, which also maintains the main drainage system and supplies fertilizer at cost. While neighboring farmers have pooled their livestock to provide the draft power for deep plowing, cooperative arrangements are even more important in organizing the harvest (see chapter 6). The Company hauls the cut cane along permanent track to the mill, and provides portable line which is laid in the fields by the growers. Cooperative harvesting is a crucial factor in raising the size of farm that a Fijian farmer can successfully manage.

The pricing of purchased sugar cane is also a complicated question with an important bearing on the efficiency of agricultural operations. In India, the sugar cane is typically purchased on a tonnage basis alone. The mill must pay as much for fiber, tops, leaves, and trash as it does for sucrose, and the incentive to grow a higher-quality cane hardly exists. Australia is at the opposite extreme. Not only is the crop priced in accordance with its recoverable sugar content, but in some localities the natural seasonal pattern of sucrose content is taken specifically into account. Under the "relative percentage system" the price paid an individual grower reflects the sucrose content of his cane relative to all cane ground by the mill during that pay period (56, pp. 38–39). In this fashion, the grower has more incentive to keep the mill supplied with cane in the early part of the grinding season, when the sucrose content of all cane is on the low side. At the same time, voluntary arrangements for selecting and harvesting the ripest cane of a group of farmers permit the entire crop to be harvested closer to optimum than when each individual is responsible for delivering a fixed allotment in each period of time. Payment on the basis of sugar content is clearly more difficult to administer when there are large numbers of growers harvesting as few as 1 to 6 acres each. In Réunion, the calculation is accordingly on the basis of the average recovery of each factory (57, pp. 219–20), while Fiji once priced on the basis of the individual grower's cane but does so no longer (55, pp. 22–23).

Any international comparison of farm prices for cane, whether related to mere cane tonnage or to equivalent sucrose, is exceedingly tenuous and will not be attempted here. The return to growers depends on local practice with respect to a wide range of variables: whether the harvest is undertaken by the grower or by hired labor at the grower's expense, whether the grower or the mill bears the cost of transportation to railhead or to the factory, whether the grower does or does not share in the income from sale of by-products, whether cane is being ground for sugar or for some lower-valued use (such as invert syrup in Cuba), and whether the relations between grower and factory are merely those of seller and purchaser or also have their tenant-landlord and borrower-lender aspects. As a further complication, a number of the mills in Louisiana and Australia are cooperatively owned by the growers.

Practice also varies with respect to the amount of price risk assumed by the grower. The price may be fixed in absolute terms as in Queensland; the mill may simply grind on a custom basis in return for a share of the product, as has been the practice in the Philippines; the grower may be paid the average market price during a specified period for a portion of the milled sugar, as in Cuba; or growers may share in the net profits of the mill. Custom must adjust to changing circumstances. Under the Japanese regime, a fixed money price for Formosan cane was contracted in advance. But in the years of postwar inflation the Taiwan Sugar Corporation could meet the competition from rice only by offering growers the market value of a portion of the sugar produced, amounting to 50 per cent after 1947. In 1951/52, a minimum price per pound of sugar equivalent to the price of two pounds of rice was guaranteed as a further incentive (*48*, p. 39), a guarantee that involved considerable loss to the Provincial government when sugar prices subsequently fell and rice prices rose.

ORGANIZATION OF BEET AGRICULTURE

In contrast to the scale of operations to be found in most cane regions, average beet acreages per farm tend to be extremely small. To be sure, the earliest factories were frequently built by a large landowner to serve his own estate. Even after introduction of the diffusion process made larger units desirable, factories in eastern Germany, in Austria-Hungary, and in Russia were commonly jointly organized and financed by several large landowners who incidentally committed themselves to supplying the necessary beets. Educating the smaller farmers in the proper growing of beets took time. But as public policies choked off cane-sugar imports from overseas, commercial and refining interests in port centers became interested in supplies of raw beet sugar. With competition for beet supplies correspondingly more intense, the practice of purchasing beets from smaller growers increased. Factories were larger and purchase from independent growers more typical, the later the date at which the industry was introduced. The breakup of large estates in some countries after World War I, and the successful integration of crops and livestock on farms of relatively small acreage, only a fraction of which would be planted to beets, meant a wide disparity between the economic position of the factory enterprise and the beet grower.

Representative data on the size distribution of sugar-beet acreages are of interest. In the United Kingdom, where the scale of factory operations is as large as anywhere in Europe, more than half the growers contracted for less than 6 acres (2.4 hectares) in 1949, and the national average was about 10 acres (4.0 hectares). Some 55 per cent of the acreage was grown in contract units of 20 acres (8.1 hectares) or less; only 8 per cent was grown by the group contracting in excess of 150 acres, or about 61 hectares (*70*, p. 12). As of the same date in western Germany, more than 90 per cent of all beet growers, on farm units where the total arable land amounted to 50 hectares or less, contributed more than two-thirds of the beet acreage. For this entire group, the average size of beet acreage was under 1.5 hectares. Even for larger agricultural units, the average size of beet acreage was only 4.7 hectares in farms of 50–100 hectares, and 14.0 hectares in farms over 100 hectares in size (*71*, pp. 106–07). If one looks to the United States, where farm units are typically larger, the average beet acreage per farm according to the 1950 census was only 23.7 acres (9.6 hectares), with state averages ranging from less than 10 acres in Utah and about 13 acres (5.3 hectares) in midwestern Ohio and Michigan to 90 acres (36.4 hectares) under irrigated farming in California (*72*, pp. 505, 642).

Successful production of sugar beets by intensive methods on relatively small farm units has required a sophisticated social organization and special contractual relationships between factories and growers (*64*, pp. 38–58; *65*, pp. 230–41; *66*, pp. 532–39). To assure a raw material supply for the factory and a market for the grower's crop, specific acreages to be planted are typically contracted for annually in advance of the growing season. Generally only seed supplied by the company can be used. Such seed, if not provided free, is at least sold at cost, since the company is more interested in promoting high-quality beets than in profiting by the seed operation as such. In the interests of a good stand, the company will sometimes specify the seeding rate and a company man must in some cases approve the seed bed. The contract usually specifies or implies that practices of good husbandry must be followed, and in newer beet regions the company commonly takes a more active role in educating farmers to high cultural practices in a variety of ways, including use of a staff of field representatives. Capital advances to the grower, whether in the form of fertilizer and lime, as in Iceland, or in cash at the time of thinning, as in Denmark, are further means of assuring the crop the necessary degree of care. Since beet farmers enjoy definite crop alternatives, these arrangements operate most smoothly if relations between the factory and the growers are cordial.

Important rights nevertheless accrue to the company. As an extreme case, if the Canadian grower ignores the factory's field instructions, the company may cultivate and harvest the crop at the grower's expense. Contracts in the United States commonly specify that the grower may not sell, assign, or mortgage the crop without the consent of the company. The grower everywhere is obliged to turn over his entire crop under the contract, but the company can frequently reject beets that fail to meet minimum standards of topping, sucrose, purity, or tare. Typically some schedule of deliveries is arranged, with penalties imposed for exceeding or falling short of the agreed quotas.

The peculiar economic and technological features of sugar beets are also reflected in the price clauses of the contract. Premiums and discounts about the base price are usually specified for higher or lower sucrose content. In some countries, such as Canada, wagon or truck loads of beets delivered by the individual grower are sampled, but elsewhere the actual recovery of the individual factory is taken as the basis for implementing that provision. In view of the lower unit costs associated with processing a larger crop, Belgian factories and the Irish Sugar Company pay a higher price the higher the total sugar production, while Denmark similarly rewards a higher national acreage. Occasionally there are also premiums on early- or late-delivered beets that allow a longer campaign, though at the cost of reduced sucrose content. While growers in Central Europe had by the 1880's earned the right to link the price of beets to the price of sugar as a profit-sharing device (*14*, p. 39), in Europe today the base price about which differentials are calculated is more customarily specified in fixed monetary terms. In Canada and the United States, however, the base price is frequently related to net returns either from sugar alone or from sugar and by-products.

A few secondary provisions are also of interest. Since the factory will not dispose of its final sugar for some months after the processing season has been completed, while the grower incurs rather exceptional outlays in handling this crop, a large interim payment is generally made within a few weeks after delivery, but settlement in full is delayed some months. Growers in Europe frequently enjoy the right to take up a portion of the wet beet pulp, either free of charge or at concessional rates. As in cane contracts, transportation costs are shared between grower and factory in various proportions. In European beet regions, as in Queensland and Louisiana cane regions, a number of beet factories are cooperatively owned, and in these cases contractual arrangements are very different. Members of Dutch cooperatives, for instance, oblige themselves to deliver a minimum quantity of beets, and have a right to deliver a maximum quantity for each share in the enterprise.

Aside from cooperatively owned factories, two major exceptions to the pattern of independent beet growers may be noted. Several factory companies in the irrigated sections of western United States originally grew a portion of their beets on company farms. Though the practice has largely disappeared with the passing of time, some beets are still grown by tenants on factory-owned land. But for very large agricultural units one looks to the collective farms of the USSR, which were supplying 90 per cent of the Soviet Union's beets by the early 1930's (*67*, p. 8). A "technical" crop like sugar beets, which must find its way to market through the medium of a processing plant, facilitated the task of crop collection and was accordingly especially attractive to Soviet authorities. It is somewhat ironic that the capitalistic cane plantation, which is liable to serious social criticism (cf. *33*) and has clearly been on the wane in such regions as Cuba, Puerto Rico, and Indonesia, should find so close a parallel in the beet fields of the People's Republics, in the production of a crop that offers fewer advantages to large-scale agriculture than either the cereals or sugar cane. By contrast, Communist control of Poland and East Germany brought nationalization of factories but parceling of larger estates into units of 5 to 10 hectares. Organizing a steady supply of beets to a single factory from as many as 39,000 inexperienced growers has raised serious difficulties of a different sort (*63*, pp. 38–39).

NATIONAL SYSTEMS OF SUGAR CONTROL

The importance of group organization and governmental intervention in the internal affairs of national sugar industries is barely suggested in the preceding discussion. The centuries-old fiscal interest in sugar has tended to give way to other considerations. Internal controls have resulted from the peculiar relationship between growers and factories, from marketing problems and the urge to limit output, as a legacy of wartime controls over a basic foodstuff, and because the political influence of agrarian interests commonly exceeded their market strength in periods of agricultural depression. Interventionism has been feasible by virtue of the high degree of regional specialization characteristic of most cane regions as well as of such beet areas as Saxony and Silesia in prewar Germany; the Aisne, Pas-de-Calais, Nord, Somme, and Oise departments in France; the region west of the Dnieper in the Ukraine; Moravia and Bohemia in Czechoslovakia; and certain counties in the irrigated valleys of the American West.

Control of beet-sugar marketing on a national scale dates from the period when rapid expansion in Europe was coming to an end and factories were gradually being consolidated into more efficient units. Competitive practices in Russia, Germany, Austria, and France were undermined to a considerable degree by the action of governments (cf. *41*, p. 32): tax favors, which were responsible for creating much excess factory capacity, and prohibitive tariffs, which made domestic cartels a feasible device. The export cartel introduced by the Russian industry in 1887 had official blessing. Surpluses could be disposed of at the expense of the domestic consumer rather than of the public treasury if a domestic quota was assigned each factory, if the domestic price was held high, and if any additional sugar moved into export. Actual enforcement of the quotas was taken over by the government in 1895 (*12*, pp. 9–11). In the interest of consumers a maximum domestic price was established, and a reserve quota set to help enforce it, while the Russian government on occasion imported sugar to break a tight supply situation. The Brussels Convention, which eliminated the export subsidies and brought a reduction of import duties, meant a temporary setback to marketing controls, but the cartel movement picked up speed after World War I (*53*, pp. 161–68). Shipping difficulties resulted in a private marketing cartel for Java's sugar exports in 1918, which subsequently came under government sponsorship (*5*, pp. 282–84). As a result of marketing difficulties, similar arrangements were introduced in such exporting countries as Cuba, Poland, and Czechoslovakia after 1925. The necessity of implementing international sugar agreements in the 1930's brought control systems to several countries which had theretofore escaped them. Self-sufficient nations, like Argentina, introduced sugar-marketing arrangements of a type similar to that in nineteenth-century Russia, in order to support their domestic industry and provide for export of occasional surpluses without excessive burden on domestic consumers or the public treasury (cf. *1*).

This record of marketing controls can possibly be matched by other industries and other commodities. What is exceptional in the case of sugar is the complexity of the intervention in the internal affairs of the industry. The cartels which made their appearance in Austria-Hungary by the middle 1880's (*13*, pp. 567–68), for example, were partly an attempt at a *modus vivendi* between refineries interested in controlling supplies of raw beet sugar, on the one hand, and beet factories concerned over their position in the refined sugar market, on the other. A purpose of at least equal importance, however, was the suppression of competition in the purchase of beets. Despite grower protests, a system of exclusive factory zones was set up. This was one of the earliest attempts to organize by nonmarket means the supply of beets from independent growers. Since sales outlets were no longer attractive, the bargaining power at the time was in the hands of the factories. Farmers were able to redress the balance only after the turn of the century, when their associations began to bargain collectively with representatives of the manufacturing industry over sugar prices. Subsequently, particularly since the end of World War I, governments have come to participate directly in the group process whereby sugar-beet (and sugar) prices are set (cf. *64*, pp. 9–19). Similar trends have been operating in cane regions. Thus in Queensland, the price of sugar cane has been fixed since 1915 by the Central Sugar Cane Board, an agency with grower, miller, and official representatives (*4*, p. 239), while in South Africa the government since 1926 has concerned itself directly with grower-miller bargaining (*68*, p. 541).

THE CASE OF CUBA[1]

From the standpoint of comprehensive governmental intervention and its ramifications, Cuban experience is of particular interest. Since the Sugar Stabilization Law was enacted on November 15, 1930, the production of every mill, as well as the national total, has regularly been apportioned among four categories: local consumption, exports to the United States, exports other than to the United States, and certain special reserves. Generally the local consumer, who absorbs barely 5 per cent of the crop, has been served at a relatively low price, while prices in New York, except under unusual circumstances, are higher than for "world" sugar. Under an export quota system, American-owned firms could not pre-empt the most attractive outlet. The fact that mills well located for export shipment might incidentally be deprived of differential cost advantages seemed not too important when sugar was not in any case admitted into foreign markets on an economic basis.

Temporary withholding of the special reserves served as a moderate price support but required a reinforcing mechanism. Except for relaxation during the period of wartime and postwar sugar shortage, the total Cuban crop has been officially restricted, and individual production quotas allotted each mill. Export quotas had served to distribute among all firms the peculiar advantages and disadvantages of each category of disposition. Production quotas completely

---

[1] Material in the succeeding paragraphs originally appeared in *40*, pp. 350–55.

halted that consolidation into larger producing units that current technology was making desirable, deliberately favoring a group of smaller Cuban-owned mills that would have suffered most in a competitive struggle.

Comparable regulations were imposed in the agricultural sector, many of them formalized in the Sugar Coordination Law of 1937. The key control was the "permanent grinding factor" calculated for each grower (*colono*), entitling him to supply a fixed proportion of the cane ground by a designated mill. Cane quotas of the smaller growers were adjusted upward at the expense of crops grown by larger farmers and plantations. As additional protection in an agricultural system where many growers were tenants on mill land, each *colono* enjoyed permanent tenure so long as he paid his rent (at rates also specified in the Law) and filled his cane quota. These provisions assured the grower a market outlet and the smaller mills a supply of cane. But restrictive quotas, particularly when specified in terms of quantity of cane rather than acreage of land, gave no incentive for raising Cuban yields per acre, developing new cane varieties, or employing chemical fertilizers.

In regulations covering the financial settlement between mills and growers, there was equally little encouragement for improved agricultural practices. Mills which recovered 12 to 13 pounds of sugar per 100 pounds of cane, the standard range, were required to pay 47 per cent of the sugar to the grower's account. Where recoveries were higher, *colonos* received only 46 per cent; where lower, 48 per cent. The monetary value attached to this sugar was the official average price announced fortnightly by the Secretary of Agriculture. Since the grower's return depended on the average recovery from all cane ground by the mill, he had no incentive to increase the sucrose content of his cane. Moreover, the arrangement implied that higher sugar recoveries resulted from improved mill technique rather than from a more scientific agriculture, since the share to the *colono* was lower when recoveries were higher. Regulation was nevertheless of particular advantage to growers in the eastern portion of the island, where mills had typically been larger and more isolated, plantation production more important, competition for cane less effective, and grower returns relatively low.

After the overthrow of the Machado regime (August 1933), labor's claims for special treatment were pressed at least as vigorously as those of *colonos* and the smaller mills. As the largest single employee group in the nation, sugar workers benefited from a number of general labor laws: control of immigration, collective bargaining rights, workmen's compensation, health and maternity coverage, vacations with pay, the basic eight-hour day. Beginning with temporary legislation in 1934, minimum wage rates were specified for the sugar industry. Different patterns were applied to cutters and haulers of cane, other agricultural labor, and the various categories of industrial help in the mill itself. There were piece-rate and daily-wage arrangements; minima were specified separately for the duration of the sugar harvest and for the dead season; and some were designated in monetary terms, while others were tied to the value of a specified quantity of sugar. But all shared the characteristic of incorporating a sliding-scale principle that reflected fluctuations in sugar prices and thereby provided flexibility in costs and incomes.

This complex set of controls, initially induced by depressed foreign markets, was readily adapted to the job of apportioning within the industry the gains associated with the wartime and immediate postwar periods. *Colonos* suspected that mill owners, in crop contracts with the United States' Commodity Credit Corporation (see chapter 7), were accepting a lower return for raw sugar, which would have to be shared with the growers, but were bargaining hard on the price for blackstrap molasses, which had customarily accrued exclusively to the mills' account. By 1945, the *colonos* won the right to receive 47 per cent (later 50 per cent) of the returns from molasses in excess of 4 cents per gallon. Growers also succeeded in raising their share for raw sugar to a flat 48 per cent, valued on a 10-month basis rather than the former fortnightly average that was likely to be seasonally low. Minimum rates for field labor, 80 cents a day before the war, had been raised to $2.88 by 1947. Moreover, sliding-scale provisions were not allowed to apply downward after sugar prices passed their postwar peak, and accordingly an important element of cost flexibility in an export-dependent industry was lost. Labor has also successfully pressed its claim to a share of the returns from improved mill efficiency. Indeed, public policy now tends to be settled by open contest between organized groups—associations of mill owners, of *colonos*, and of sugar workers—each financed by mandatory levies.

CONCLUDING OBSERVATIONS

The broad purpose of government controls has been to divide up the returns from sugar within the industry on some nonmarket basis, and particularly so as to favor the more vulnerable groups. This has implied intervention against the very large mill and on the side of the planter, whose social position modern technology was tending to undermine. Similarly, smaller mills have been protected against the competition of the large-scale units, older producing regions (like the northeast provinces of Brazil) against newer and modernized regions (like São Paulo). Governments have found this industry more amenable to regulation than many others because ownership was concentrated at the refining end, growers necessarily lacked the protection of a national commodity market, and sugar crops could not be disposed of commercially except through the medium of a limited number of factories.

What is perhaps remarkable, however, is the degree to which the pattern of control applies also to the United States, despite its free-enterprise tradition. Since May 9, 1934, except in the years of war-time dislocation, total marketings have been restricted to a figure determined by the Secretary of Agriculture, while the shares accruing to various producing regions (foreign and domestic) must conform to statutory criteria. The market is shared not only among the different regions but implicitly also as between cane refiners and producers of beet sugar, and the quantity of direct-consumption imports is specifically limited. Quotas are implemented, when necessary, by marketing allotments to individual processors and acreage allotments to growers. There are separate statutory provisions which empower the Secretary of Agriculture to set minimum prices grower-processors must pay for purchased cane or beets, and minimum wages for field labor, including adjustment for differences in regional conditions. The legislation is less in the spirit of agricultural price supports than of the industrial codes of the National Recovery Administration.

Analysis of the internal organization of the sugar industry might therefore seem to support the broad generalization that "private economic power is held in check by the countervailing power of those who are subject to it" (*35*, p. 118). But patently our study denies Professor Galbraith's inference that "countervailing power" is a distinctive attribute of American capitalism. It is even an open question whether this process of horizontal sharing of markets and vertical price-fixing by group bargaining is properly considered an attribute of capitalism at all. Certainly the United States sugar control is best described as a government-sponsored cartel, in which the division of the market has legislative sanction and returns to various interests in the industry are largely determined by executive decree under statutory mandate, though ownership of producing units remains in private hands. The prototype of these arrangements is less capitalism and private enterprise than the corporate state. "Countervailing power" is as good a rationalization of sugar organization in Nazi Germany, Fascist Italy, and Communist Poland, as in the United States.

CITATIONS AND SELECTED REFERENCES

1 O. W. Willcox, *Can Industry Govern Itself?* (New York, 1936).

2 K. N. Stahl, *The Metropolitan Organization of British Colonial Trade* (London, 1951).

3 W. L. Holland, ed., *Commodity Control in the Pacific Area* (Stanford, Calif., 1935).

4 J. B. Brigden, "Control in the Australian Sugar Industry," in *3*.

5 C. G. H. Rothe, "Commodity Control in Netherlands India," in *3*.

6 Erich H. Jacoby, *Agrarian Unrest in Southeast Asia* (New York, 1949).

7 R. Ramos Grau, "News of the Cuban Sugar Industry," *Sugar*, November 1948.

8 T. B. Wilson, "The Economics of Peasant Cane Farming in Trinidad," *World Crops*, April 1954.

9 E. C. Freeland, "The Sugar Industry of Peru," *Sugar Journal* (New Orleans), April 1953.

10 C. S. Griffin, "The Sugar Industry and Legislation in Europe," *Quarterly Journal of Economics*, November 1902.

11 Gr. Brit., *Papers on the Proceedings of the Conference Held at Paris, on the Question of Drawbacks on Sugar, Parliamentary Accounts and Papers, 1863*, Vol. 67, No. 470.

12 Gr. Brit., "Correspondence with the Russian Government . . . ," *Parliamentary Accounts and Papers, 1903*, Vol. 75, C.1401.

13 F. Walker, "The Sugar Situation in Austria," *Political Science Quarterly*, December 1903.

14 Josef Siegel, *Die tschechoslowakische Zuckerindustrie* (Berlin, 1928).

15 G. Martineau, "The Statistical Aspect of the Sugar Question," *Journal of the Royal Statistical Society* (London), June 1899.

16 Australia, Commonwealth Bur. Census and Stat., *Primary Industries, Part I—Rural Industries 1952–53*.

17 E. A. Weber, "The Sugar Industry," in L. J. Hughlett, *Industrialization of Latin America* (New York, 1946), Chap. XVI.

18 Farr & Co., *Manual of Sugar Companies 1952–53* (New York, 1953).

19 Hawaiian Sugar Planters' Association, *Sugar Manual* (Honolulu, 1954).

20 International Bank for Reconstruction and Development, *The Economic Development of British Guiana* (Baltimore, 1953).

21 Java Bank, *Report for the Financial Year 1952–53* (Djakarta, 1953).

22 A. C. Barnes, *Agriculture of the Sugar-Cane* (London, 1953).

23 Caribbean Commission, *Monthly Information Bulletin* (Port-of-Spain), December 1953.

24 Mauritius, Cen. Stat. Off., *Year Book of Statistics, 1952.*

25 E. Hugot, "Le Sucre à la Réunion," *Etudes d'Outre-Mer* (Marseilles), October 1953.

26 *Weekly Statistical Sugar Trade Journal,* Jan. 21, 1954.

27 South African Sugar Journal, *South African Sugar Year Book 1951–52* (Durban).

28 Gr. Brit., Min. Agr. and Fisheries, *Sugar Beet Cultivation* (Bull. No. 153, 1953).

29 *Weekly Statistical Sugar Trade Journal,* July 15, 1954.

30 *South African Sugar Year Book 1952–53.*

31 *Weekly Statistical Sugar Trade Journal,* April 15, 1954.

32 *Weekly Statistical Sugar Trade Journal,* Jan. 31, 1952.

33 R. Guerra y Sánchez, *Azúcar y Población en las Antillas* (3d ed., Havana, 1944).

34 R. Ramos Grau, "The Venezuelan Sugar Industry," *Sugar,* February 1952.

35 J. K. Galbraith, *American Capitalism, The Concept of Countervailing Power* (Boston, 1952).

36 Noel Deerr, *The History of Sugar,* I (London, 1949).

37 Pieter Honig, "Technical Progress in the Sugar Industry," *Sugar,* October 1954.

38 F. O. Licht, *World Sugar Statistics, 1937* (Magdeburg, 1937).

39 J. C. Sitterson, *Sugar Country, The Cane Sugar Industry in the South, 1753–1950* (Lexington, 1953).

40 B. C. Swerling, "Domestic Control of an Export Industry: Cuban Sugar," *Journal of Farm Economics,* August 1951.

41 R. Liefmann, *Kartelle und Trusts* (4th ed., Stuttgart, 1920).

42 Hawaiian Sugar Planters' Assn., Exp. Sta., *1953 Factory Report* (Special Release 96, Honolulu, 1954).

43 U.S. Dept. Agr., Prod. and Mkt. Admin., Sugar Branch, "Production, Consumption, Foreign Trade . . . in 59 Countries" (1953, mimeo.).

44 M. Lynsky, *Sugar Economics, Statistics and Documents* (United States Cane Sugar Refiners' Assn., New York, 1938).

45 Bank Indonesia, *Report for the Year 1953–1954* (Djakarta, 1954).

46 "The Sugar Cane Industry in Fiji," *Tropical Agriculture* (London), October 1954.

47 Taiwan Sugar Corp., *News Letter* (Taipei), Sept. 20, 1952.

48 C. S. Loh, "Taiwan's Sugar Development Program," *Sugar,* February 1953.

49 T. A. G. Hungerford, "Fairymead—Australia's Thriving Sugar Plantation," *Sugar,* July 1954.

50 Saburi Ebi, "Sugar Industry of Java and Formosa—A Comparative Study" (Econ. Coop. Admin., Mission to China, Tokyo, 1947, mimeo.).

51 Foreign Policy Assn., *Problems of the New Cuba* (New York, 1935).

52 *South African Sugar Year Book 1950–51* (Durban).

53 P. de Rousier, *Les Syndicats industriels en France et à l'étranger* (Paris, 1919).

54 Gr. Brit., Col. Off., *British Dependencies in the Caribbean and North Atlantic 1939–1952* (Cmd. 8575, 1952).

55 Gr. Brit., Col. Off., *The Sugar Industry of Fiji* (by C. Y. Shephard, Col. No. 188, 1945).

56 "The Relative Percentage System of Cane Payment," *Sugar*, June 1950.

57 François Ripert, "L'Industrie Sucrière a l'Ille de la Réunion," *Revue Internationale des Produits Coloniaux* (Paris), December 1953.

58 Anuario Azucarero de Cuba 1951 (Havana).

59 T. Lynn Smith, *Brazil: People and Institutions* (Baton Rouge, La., 1946).

60 Gr. Brit., Col. Off., *Report of a Commission of Inquiry into the Sugar Industry of British Guiana* (Col. No. 249, 1949).

61 U.S. Mutual Sec. Admin., Mission to China, *A Survey of the Taiwan Sugar Industry*, Part V (Consultant Rep. No. 1-A, Taipei, 1952).

62 Queensland Cane Growers' Association *et al.*, *Australian Sugar Year Book 1951* (Brisbane).

63 J. Frejlich, "Poland's Sugar Industry—Behind the Iron Curtain," *Sugar*, August 1949.

64 Food and Agriculture Organization of the United Nations, Econ. Div., "The Economic Relations Between Growers of Sugar Beets and Factories in Certain Selected Countries" (Rome, 1952, mimeo.).

65 R. H. Cottrell, ed., *Beet-Sugar Economics* (Caldwell, Ida., 1952).

66 R. A. McGinnis, ed., *Beet-Sugar Technology* (New York, 1951).

67 V. P. Timoshenko, *The Soviet Sugar Industry and Its Postwar Restoration* (Food Research Institute War-Peace Pamphlet 13, Stanford, Calif., August 1951).

68 J. M. Tinley and B. M. Mirkowich, "Control in the Sugar-Cane Industry of South Africa," *Journal of Farm Economics*, August 1941.

69 International Sugar Journal (London), December 1953.

70 British Sugar Corporation Ltd., *Report on Mechanisation of the Sugar Beet Crop in Great Britain During the Year 1949* (Peterborough, 1950).

71 Germany, Statistisches Bundesamt, *Statistisches Jahrbuch für die Bundes-republik Deutschland 1952* (Stuttgart-Cologne).

72 U.S. Dept. Comm., Bur. Census, *United States Census of Agriculture: 1950, Vol. II, General Report, Statistics by Subjects* (1952).

73 K. T. Voblyi, *Essay on the History of the Beet-Sugar Industry of the USSR*, II (Moscow, 1928) [in Russian].

74 Sugar, March 1955.

CHAPTER 5

MECHANIZATION IN THE SUGAR-BEET FIELDS

Spectacular changes have occurred in sugar-beet agriculture since the outbreak of World War II. Some were almost entirely the result of wartime conditions; others were merely accelerated and a few even inhibited by the war atmosphere. Together they provide a systematic case history of dynamic agricultural technology, reaching from the growing of sugar-beet seed through to the factory itself. Initially the process was one of extending mechanized agriculture in the United States, though there was rather heavy borrowing from scientific work overseas. Transmission of the new techniques, throughout the United States as well as internationally, proceeded at an uneven pace which itself is a subject fit for study. The specific kinds of practices that were introduced or abandoned, and the very direction of scientific research, offer interesting interregional comparisons.

DOMESTIC SEED PRODUCTION IN THE UNITED STATES

Commercial production of sugar-beet seed began in north central Europe, where the sugar beet was first successfully developed. Central Germany retained a leading position for many decades, though secondary supply areas in France, Poland, and Czechoslovakia were of some importance. As recently as the early 1930's a large part of the seed entering world commerce came from a single firm operating in Saxony (1, p. 18). A century and a half of careful selection had produced reliable varieties, high in sugar content and in beet tonnage per acre. Importers, however, clearly ran certain risks. Dependence on imports for seed during the First World War had been almost as serious for the United States as dependence on Continental beet sugar had been for Great Britain. Considerations of security, attractive sugar prices, and rapid expansion of domestic beet acreage resulted in a considerable seed production in the United States, especially between 1916 and 1920 (2, p. 21). In the year when imports were least available, domestic seed was sufficient to have met all planting needs in 1914, though beet acreage had increased some 80 per cent in the interim.

Beet seed was not at that time a satisfactory crop under American agricultural conditions. Small fields had to be broadcast thickly with beet seed in the spring. The resulting dense stands of small beets ("stecklings") were harvested by hand in the fall; roots lifted out of the ground and tops removed; and roots kept in earth-covered mounds over the winter. Losses occurred as the result of freezing or decay in storage. In the spring, the healthy roots had to be sorted out and planted, and the biennial beet in its second year could be expected to "bolt" into seed instead of storing sugar. Once European agriculture recovered from its postwar disorganization, American farmers were happy to abstain from these labor-intensive activities. In the decade beginning with 1922, the dependence on imported seed was again absolute.

Not security considerations but outbreak of disease eventually made that arrangement unsatisfactory. Two beet diseases are endemic in the United States, curly top in the West and Cercospora leaf spot or "blight" east of the Rockies. European seed producers had been urged to produce resistant varieties ever since 1900 but, since curly top does not occur in Europe and leaf spot is rare in the breeding areas, suppliers could not be persuaded to take an interest in the problem (*3*, pp. 636–38). A serious outbreak of curly top hit the United States in the late 1920's. Beets were completely eliminated from the crop pattern in infested regions, while acreages planted elsewhere in the western states declined markedly. As a matter of survival, the United States Department of Agriculture had, through selection of the more resistant types, developed a curly-top resistant variety, "U.S. 1," by 1929. Soon it was displaced by the superior varieties U.S. 33 and U.S. 34, as well as by others developed by the western beet-sugar companies. Work on leaf-spot resistance was begun in earnest in 1925. In this case, selection alone was not enough, but inbreeding succeeded in producing such varieties as U.S. 217, U.S. 200 x 216, and U.S. 216 x 222 (*4*, pp. 1333 ff.).

The gap which separates successful experimental results from commercial practice can be extremely wide for agriculture with its multitude of small producing units. Especially for the sugar beet, the gap is a matter of time and biology as well of as of cultural resistance. Even when the years of slow, plodding experimentation are over, mere reproduction of an adequate stock of commercial seed is a considerable task. To sow the million or so acres of beets each year in the United States during 1938–40 at the prewar rates of seeding required more than 15 million pounds of seed. With seed yields of about 2,000 pounds per acre, some 7,500 acres had to be actually planted to seed for a single year's sowing of beets. Since the plant is a biennial, planting for seed production must take place two years ahead of planting commercial beets; continuous operations required 15,000 acres in cultivation for seed. The biological lag is shorter than for tree crops, but twice as long as for annuals. U.S. 1, though originally produced in 1929, did not come into extensive use until 1934 (*3*, p. 637), partly owing to problems of multiplying the elite seed stock. In general, five to seven years had to pass before commercial seed on any large scale was available from the mother beet (*5*, p. 22).

Successful production of seed in the United States required a method of avoiding the high labor costs of harvesting, storing, and replanting stecklings. The major innovation was a change in cultural practice, specifically the time of seeding. Mild winter conditions of the southwestern United States were found adaptable to fall sowing and overwintering of the plants in the field. Commercial production by this method proved successful in 1927/28, even before improved varieties had become available (*2*, pp. 22–23). Not only was the interval between planting and harvesting cut almost in half, but the process now lent itself easily to mechanical operations. The seed beets could be cut by standard mowers and reapers, and the stalks, once dried, could be threshed after only minor adjustments had been made in standard threshing equipment (*6*, pp. 334–37). Application of further agricultural improvements resulted in pronounced increases in seed yields per acre, which were by 1950 almost triple those of 1932 (*7*, p. 65). Without these developments, loss of the German source of supply during World War II would have been a serious blow to the beet-sugar industry of the United States. As it was, small quantities could even be spared for export to the United Kingdom and Canada after 1940. Moreover, exports could be stepped up as temporarily required by postwar Europe, and fully 60 per cent of seed production went abroad in 1948 (*7*, p. 65).

As is true of the introduction of any new crop, farmers in the Southwest were inititally reluctant to turn to the production of beet seed. That resistance was overcome in part by the beet-sugar processors, individually and collectively. It was they who, by purchasing the entire seed production of growers under contract, assured the crop a market and eliminated much of the commercial risk. Since it was customary for the companies to supply all seed used by commercial sugar-beet farmers, the new varieties could be introduced in the sugar-beet fields as soon as seed became available. After the mid-thirties, all seed sown in the western states was of domestic varieties, and this became true also in the eastern regions early in the war. Since domestic seed proved more viable than the imported product, the shift meant an incidental lowering in seeding rates by about one-quarter (*8*, p. 29).

Conquest of disease and success in commercial seed production were only the first two steps in the transformation of American sugar-beet agriculture. Interest in breeding led to fairly rapid proliferation of new varieties intended for particular purposes, including "U.S. 15." In frost-free climates, this nonbolting type could be planted in the fall and harvested in the spring, thereby "reversing the seasons." It permitted the spread of irrigated sugar-beet culture to the subtropical Imperial Valley of southern California (*4*, p. 1335). Further advances had the effect not merely of permitting sugar beets to survive, but of revolutionizing the customary practices in the beet fields.

TRANSFORMING BEET AGRICULTURE: SPRING WORK

Standard whole beet seed is a small ball containing three to five individual seed kernels, tightly interlocked. From each seed ball sown, several plants can normally be expected to emerge (*9*, p. 427). That prolificity is a decided complication for the grower. Not only must he cultivate intensively, as he would other row crops, but he must also perform two special tasks. At an early stage in growth, seedlings must be (1) blocked, i.e., spaced to a proper succession of clumps by elimination of a part of the sown row, and (2) thinned, to the point where a single vigorous seedling remains from each of the former clumps. These operations, customarily performed by hand methods and requiring "stoop" labor equipped only with a short-handled hoe, are exceedingly demanding of man-hours. Growers planting anything more than a few acres found it necessary to hire farm hands for the job, frequently migratory workers from foreign lands.

The tight labor market associated with the onset of World War II precluded use of labor on the extravagant scale required by the orthodox methods. Fortunately, the stage had already been set for some systematic changes in the old ways of doing things. Tractors had been coming into the fields in quantity after the late 1920's. The potential source of mobile power they represented had thus far been applied mainly in plowing and general preparation of the soil for planting. Intensive research into more comprehensive mechanization of field operations had also commenced. The most fruitful work was begun in 1931 at the California Agricultural Experiment Station at Davis, in cooperation with the Agricultural Engineering Division, United States Department of Agriculture. After 1938, substantial financial assistance came from the national trade association of the beet-sugar processors. Technical contributions were soon also to be made elsewhere: by agricultural machinery and chemical firms, by individual inventors, and by machine-minded growers. In 1945, for an even broader attack on the problems of beet agriculture, the sugar processors of the United States and Canada organized the Beet Sugar Development Foundation.

Attention was first devoted to mechanization of the blocking operations. Field trials soon demonstrated that a uniform distribution of seedlings was a prerequisite to progress in that direction (9, p. 425). The focus of research accordingly shifted from postemergence equipment to the planting drill itself. By 1938, a planter had been designed capable of dropping a single seed ball at a time. This became the basis for "precision planting" which, by giving control over the space between successive clusters of seedlings, offered the theoretical possibility of completely eliminating the blocking operation. Not only was the distribution of clusters more uniform, but single seedlings occasionally emerged, with consequent savings in thinning. By 1940, several types of precision planters were already in the fields, though performance was still imperfect. The new planters might still bunch the seed, resulting in too heavy a plant population; there might be skips or the seed might be mangled by the planting apparatus, with reduction in plant population and loss in yields (9, p. 427).

While some of the early disadvantages were overcome by gradual improvement in the working efficiency of the planting drills, the problem also suggested the possibility of adapting the beet seed to the needs of the planter. Both irregularity of size and shape and its multiple-seed-germ characteristic gave grounds for dissatisfaction. By mechanical processing of the seed, it was found possible to break up the seed ball into components that were predominantly single-germ cells. "Segmented seed," successfully developed by 1941, was first introduced commercially on a large scale in 1942. It quickly became common in field use, as beet factories installed the necessary seed-processing equipment and performed their usual function of supplying seed to all contracting growers.

The introduction of segmented seed was crucial to the planting of even such low acreages of beets as those of 1943–45 in the United States. Where the whole seed averaged about 1.9 germs per unit, the segmented product had only 1.1 (*9*, p. 427). As early as 1942, under optimum conditions 90 per cent germination could be expected, of which 90 per cent would be singles (*10*, p. 32). Field experiments demonstrated that even occasional double or triple seedlings were no cause for concern, and could be left to harvest with no significant loss in yield (*11*, p. 258). There was also a saving in planting rate. Instead of the previous sowing of about 15 pounds per acre, 4 to 5 pounds could now do the job and even 2 pounds was not impossible (*12*, pp. 36–37). The great reduction in plant population now permitted fields to be thinned by long-handled hoe. Even if performance was less skillful, there was, aside from the saving in labor, a simultaneous improvement in speed. The farmer could consequently almost entirely avoid the risk of greatly reduced yields which resulted, under former conditions, whenever thinning could not be done at precisely the right time (*13*, p. 4).

Many of the advantages of segmented seed were, however, illusory. Since growers were eager to take full advantage of the labor-saving features of their new seed, they took the potentialities of low seeding rates seriously. At such light rates, the operating efficiency of the existing drills was inadequate, while bunching or skipping became a more serious matter. Grower dissatisfaction with poor stands due to these factors alone gave an added impetus to improvement in planting equipment, and results in 1943 seem to have been more satisfactory than in 1942 (*10*, p. 32). Even efficient planting did not preclude poor final stands of beets. To get as much as 90 per cent germination, moisture conditions had to be satisfactory and the seed bed well prepared, while crusting of the soil would prevent seedlings from actually emerging at anything like the expected rate. Even light infestation of seedling disease would seriously cut back the beet population below the size necessary for good yields (*10*, pp. 32–33; *12*, pp. 36–37). Machined seed also gave a certain proportion of abnormal plants. Indeed, had the war labor shortage not been so intense, growers' patience would surely have been exhausted by the poor initial performance of seed and drills alike, and the new techniques might have been sufficiently discredited that resistance to further innovation would have been stronger.

Beet processors had their own additional reasons for being dissatisfied with segmented seed. There was in fact no saving over the use of whole seed between 1942 and 1945, merely because the recovery of segmented seed from the mechanical processing was so low (*14*, pp. 625–39). To obtain something of the order of 85 per cent single-germ seeds from whole seed, one incurred a loss in processing of about one-half. Indeed, from the standpoint of the mechanics of segmenting as well as the field results, those who were most closely associated with segmented seed were of the opinion that such processing was a sheer expediency, adopted commercially before it was ready, under pressure from a demand for labor-saving devices that outran the pace of technological accomplishment (*15*, pp. 120–24).

More elaborate processing provided one means of overcoming the difficulties. Poor seedling stands in the Middle West in 1943 brought the introduction of "pelleted" seed. Segmented seed was coated with a water-soluble layer of beneficial and inert material. The improved shape of the seed permitted it to pass much more smoothly through the drill, assuring superior planting. The coating itself permitted experimentation with the addition of fungicide, fertilizer, insect repellant, or even hormones, in order to speed germination, improve emergence, and get a head start in the race against weeds (*10*, pp. 32–33). Under the auspices of the regional Farmers' and Manufacturers' Beet Sugar Association, large-scale field tests were made in 1944 (*16*, p. 21), but even in Michigan only between 1 and 2 per cent of the beet acreage was planted with pelleted seed during the following season (*17*, p. 14).

Pelleting never became general practice, partly because the inert material delayed germination but mainly because of a retreat from the segmentation process. By 1946, convinced that the savings in field labor did not justify the previous high wastage in seed processing, the companies had begun to lower their sights and to satisfy themselves with as many as 40 per cent doubles (*15*, pp. 120–24). Within two years, the idea of segmentation was abandoned in favor of the different process of decortication. Instead of cracking the seed ball and setting free the naked germ, seeds were sized, the outer corky layer was worn down by a milling process, and a smooth ball of uniform size obtained (*18*, p. 52). The graded size and spherical shape permit improved drilling and uniform spacing. Besides, decorticated seed germinates as well as whole seed and emerges somewhat better (*15*, pp. 120–24). The more severe the field conditions, the better is its relative performance (*19*, pp. 114–16). Not only do more of the potential plants emerge, but they do so more quickly, getting a head start on the weeds and promising a longer period of root growth as a result of earlier thinning dates (*18*, p. 52). Recoveries from whole seed are much better than for segmented—about 60 to 70 per cent (*14*, pp. 625–39; *15*, pp. 120–24). Only in one respect is there a deterioration in performance. Decorticated seed gives more seedlings per viable unit than segmented (*14*, pp. 625–39), so that slightly more thinning is required.

This sacrifice of gains previously anticipated is a good reflection of the mixed progress made in mechanization of blocking and thinning generally. From 1929 on, the practice of drawing a cultivating implement at right angles across the rows, in order to form the desired blocks, had become common in the Red River Valley of Minnesota (*20*). In spite of clear savings in labor use, such cross-blocking and further cross-cultivation of the beet fields did not become popular in many regions. There was risk of reducing a light stand so far as to seriously jeopardize yields, or of burying small plants. In irrigated districts, damage to raised beds or irrigation furrows made the practice unattractive (*8*, p. 36). Growers resorted to cross-cultivation mainly when weather or labor conditions delayed thinning too long beyond the optimal date, or in regions seriously short of labor for spring work.

To the extent that processed seed and precision planting assured a more uniform distribution of seedlings, the way of mechanical operations appeared easier. As early as 1942, the revolving cutter head of the Dixie cotton chopper had been adapted for down-the-row blocking in the Imperial Valley of California (*21*, p. 18), permitting mechanical procedures where beets are grown in raised beds. This innovation spread gradually even to regions where cross-cultivation had earlier been held in some favor (*8*, p. 37). Indeed, with regular spacing of single or double seedlings, the distinction between blocking and thinning began to lose most of its meaning. There seemed good reason to believe that "the year 1945 has witnessed a definite solution of the thinning problem . . . The operation of the Dixie thinner and its variants is practically perfect" (*22*, pp. 33–34).

Spread of the new techniques was nevertheless slow, and there was much backsliding. Seedling stands continued to be so uneven that many growers were reluctant to thin mechanically. In some regions, as in northern Colorado, even with processed seed, a seeding rate that would ensure satisfactory stands and adequate yields was high enough to warrant blocking by hand. Some growers in that state turned to cross-cultivation in 1947 when a wet spring delayed hand thinning until beets had become overgrown and fields weedy, only to revert to hand operations when light stands occurred in 1948 (*23*, pp. 17, 20). Even in the Red River Valley, where cross-cultivation was practically universal in 1947, poor stands required half the acreage to be blocked and thinned entirely by hand the following year (*24*, p. 20).

For thinning and for further weeding and hoeing, distinctively new techniques may be almost within reach. There are reports of an electronic thinner which operates in a strictly selective manner, producing the desired spacing without widening gaps and capable of operating around the clock (*25*, p. 50). The beet leaves do not as yet lend themselves to selective types of chemical weed killers, but pre-emergent sprays allow seedlings to sprout in weed-free soil and speed their growth (*26*, p. 51). Dusting by airplane has been used by some growers (*23*, p. 23). Tractor-drawn weeders, equipped with springy, pliable tines, are capable of unearthing the fibrous-rooted weeds while leaving the deep-rooted beets intact (*8*, p. 37). The mechanical transformation of beet agriculture is, in these respects, still incomplete and very much in progress.

MECHANIZING THE BEET HARVEST

Mechanization of harvest operations is more a matter of results already achieved than of advances still in progress. The task of harvesting includes lifting the beets out of the ground, topping each beet to remove crown and leaves, removal of excess sod or dirt, and windrowing beets or loading them into a vehicle. Given the power unit represented by the tractor, beets could readily be loosened with the aid of a simple subterranean blade, and such lifters had been in common use for some years before the war began. At the end of the 1930's, mechanical loading of hand-windrowed beets was also coming into practice, except in California where cloddy fields complicated the job of the field loader (*27*, p. 61). In these operations as in others, the war promoted improvisation with various types of equipment. In the Salinas Valley of California, for example, the Spreckels Sugar Company offered for rent a 20-row cross-conveyor type, modeled after equipment used in the harvest of lettuce, which moved just in advance of the harvesting crews (*28*, p. 36).

The difficulties involved in mechanization were serious ones. Mechanical beet diggers and lifters might bruise the root, tear it too abruptly from its tail, or require the actual spiking ("sticking") of the beet. There was the problem of separating lifted beets from adhering clods, dirt, or stones, and the more so under wet conditions. Failure to deliver clean beats adds heavily to the transportation load for a commodity which yields at best only some 15 per cent by weight in main product. Additional cleaning apparatus must be installed by the factory, if the recovery of sugar is to be sustained and damage to equipment (especially the slicers) avoided. Dirty beets deteriorate much more quickly in storage, with consequent risk to the sugar content of cleanly harvested beets in the same pile. The risk is greater if the beets have been bruised or punctured, and in these cases there are also definitely higher processing costs as well as lower recovery of sugar.

At that, the main bottleneck blocking mechanization was the imperfection of machine topping. Hand work resulted in a topped beet that was of high quality as well as clean. Beets are of uneven size and shape, and the penalties for careless topping are heavy. If the knife is overzealous, beet tonnage per acre will suffer; if the cut is too high, or merely at the wrong angle, sugar extraction will be seriously affected. Not only are the crowns themselves low in sugar content and therefore more expensive to process, but the impurities which they contain impede recovery of sucrose from the richer portions of the beet. While the tops were detrimental in the factory, they were generally much in demand by farmers for feeding livestock, and equipment that ruined them cost the growers a valuable by-product.

The entire institutional framework of the American sugar-beet industry was set to favor higher quality in beets supplied to factories. Beet-purchase contracts typically specified a sliding-scale price that included premiums for higher sugar content. Penalties were provided for excessive tare, while factories generally retained the prerogative of entirely rejecting beets considered so low in quality as seriously to jeopardize average recovery. Processors' profits depended to a high extent on their skill in extracting the maximum possible sugar from their raw material. Furthermore, the price of sugar crops tended to be relatively favorable in the decade before the war, growers were eager to negotiate acreage contracts with factories, and at times growers or processors operated under governmental restrictions on planting or marketing. Factories were in a strong bargaining position to demand a high-quality, carefully handled beet, the more so because the financial interests of growers worked in the same direction.

Factory attitudes on acceptable quality of delivered beet, quite as much as grower approach to field operations, had to be decidedly modified before mechanical harvesting stood any chance of finding general acceptance. Beginning with a crisis in the beet fields in the fall of 1942, the hard economic facts of the war period brought about the required psychological shift. Urged on by government prodding and by a favorable relative price for sugar beets at time of planting, growers in the spring of 1942 seeded record acreages. Serious as were the difficulties of getting labor for spring operations, it was at harvest time that the lack of local or migratory labor made itself most severely felt. Merely getting the crop out of the ground was a burden that embittered growers against planting the crop again in the near future. As a final blow, the price prospects for sugar beets also became unattractive in the years immediately following. It was the factories and not the growers who then found themselves in the weaker bargaining position. Although agricultural resources are popularly considered to be extremely inflexible, most growers found it possible to switch fairly readily to other, higher-paying and less troublesome crops, whereas the processors were saddled with excess factory capacity. Of the 84 factories required to handle the large 1942/43 crop, 24 stood idle the following season (*29*, p. 155). Any beets, even those that were dirty, bruised, or punctured, looked better to the factories than none at all.

A few machine harvesters were sufficiently developed to allow field trials in the autumn of 1942, but successful commercialization really began the following year (*9*, p. 429). As might be expected, there was a wide diversity in the early types of equipment used (cf. *17*, pp. 6–12), and, indeed, regional conditions vary sufficiently that there continues to be a place for several quite different models. Beets may be pulled by their tops, or raised by a spiked wheel or other lifting device. Topping may take place either before or after lifting, and in northern Idaho and eastern Oregon the tops may merely be flailed off by a beater-topper which precludes their further use except as green manure (*8*, p. 45). In some cases beets are loaded directly into a following truck; in others, they go first to a linked trailer for reloading into a truck at the end of the row; in still others, they are windrowed for subsequent mechanical loading.

The early harvesters suffered from many operational shortcomings. Heavy costs for servicing and maintenance were a rather temporary matter, but the problems of conserving the tops, reducing the volume of trash hauled, and separating beets from clods had not been solved to complete satisfaction as late as 1950 (*13*, p. 7). Top windrowers were being added to several models by 1949, while one implement company introduced a top harvester that, conforming to European-made units on the market by that date, removed leaves and crowns and loaded them before roots were lifted from the ground.

Though screens and sorting tables were tried for cutting down on extraneous material, beets continued to be delivered less clean than formerly, and the limits of acceptable topping became a contentious issue between growers and processors (*30*, p. 217). The average tare of beets received by the Spreckels Sugar Company, which had been only 2.15 per cent in the two years 1939 and 1940, had risen to 5.13 per cent in 1947 (*31*, p. 50). Under the circumstances the lead taken by the processors in developing improved models and in promoting their wider use was quite crucial. Aside from publicizing new methods (cf. *32*) and experimenting with them in the field, these companies for a time made the new units available to growers on a rental basis, and at charges that did not cover the full costs of servicing and maintenance (*17*, p. 6), let alone allow for the rapid obsolescence rate to which the earlier harvesters were subject.

Against some of the purely agricultural risks, the new harvesters gave mixed results. Early rains were a particular hazard in California, where wet fields seriously hampered mechanized harvest of the 1950 and 1951 crops. Indeed, a portion of the beets in the latter year had to be left in the ground unharvested until the following spring, with delay in planting the fields to succeeding crops. By contrast, mechanical operations need not be interrupted even by a several-inch fall of snow (*8*, p. 6), while the possibility of operating around the clock under emergency conditions brought an entirely new element of flexibility, especially against the frost hazard.

Even when confined to daylight, harvesting tended now to be a considerably speedier operation. Roots could therefore be lifted later than formerly, closer to the date of optimum yield (*8*, p. 43). This advantage showed up more in beet tonnages than in sugar content, since beets were being delivered in fresher condition and less dehydrated than when left a day or so in the fields (*33*, p. 23). To the extent that dirtier beets obstructed factory processes, field gains were not directly reflected in extraction rates.

THE PACE OF MECHANIZATION IN THE UNITED STATES

Although the first sugar-beet harvesters came into commercial use in 1943, mechanization of fall work won wide acceptance only after the end of the war. Less than one-tenth of the national acreage was machine-harvested in 1945, though in California the figure had already climbed above one-quarter, as indicated in the following tabulation (*13*, p. 2; *33*, p. 3; *35*, pp. 752–63):

Region	Percentage of acreage harvested by machine				
	1945	1946	1947	1948	1949
United States	7	17	29	...	52
California ("Area 1")	28	52	60	71	78
Utah, Idaho, Oregon, Washington, W. Colorado, W. Wyoming, W. Montana ("Area 2")	...	4	28	...	63
W. Nebraska, E. Colorado, E. Wyoming, E. Montana ("Area 3")	...	10	23	...	49
Red River Valley and Iowa ("Area 4") ...	...	1	7	...	39
Wisconsin, Michigan, Indiana, Ohio ("Area 5")	...	4	15	...	36

Not until 1949 did the national figure reach one-half. By that date, a good part of the acreage still harvested by hand was being mechanically loaded.

Several features of this transformation are noteworthy. The speed with which mechanization proceeded in California from the stage of field trials in 1942 to that of comprehensive use by 1948, notwithstanding the difficulties experienced by the innovator and the high costs of obsolescence, is surely impressive. The beets that continued to be harvested by hand beyond that date were generally grown on a small scale by farmers with means too limited to support a combine, or where the local labor situation happened to be nontypical (*34*, p. 994). States progressively east tended to be under less pressure to switch to machines, and the contrast between California and other regions was particularly striking in 1946. Improved models were sufficiently attractive, however, to persuade all regions to make the switch. While the turn to mechanization started in the different regions at different dates, the process moved with considerable dispatch once the initial decisions had been made. By 1952, less than one-fifth of the crop in the Rocky Mountain area (Montana, Wyoming, Nebraska, and Colorado) was being lifted and topped by hand (*36*, p. 51), while even in Michigan the portion mechanized had passed 60 per cent by 1950 (*37*, p. 39).

The pace of domestic transmission of the innovations associated with spring work is also of interest. Data on the percentage of total acreages handled by the newer methods in 1947 are as follows (*13*, p. 2; *35*, pp. 752–63):

Technique	United States	"Area 1"	"Area 2"	"Area 3"	"Area 4"	"Area 5"
Planting:						
Segmented seed ..	89	74	99	96	94	69
Decorticated seed..	2	11	...	2	...	...
Precision planter .	45	15	60	46	26	62
Mechanical blocking:						
Down the row.....	2	...	...	2	17	...
Cross-blocking	3	...	...	...	39	...
Total	5	...	...	2	57	...
Thinning:						
Short-handled hoe .	83	97	91	85	42	63
Long-handled hoe..	10	2	7	11	...	37
All spring work (1946)	8[b]	3	3	2	62	...
All spring work (1949)	30[b]	1	20	32	58	...

Mechanization of spring operations 1947[a]

[a] About 10 per cent of national acreage not reported.
[b] Excluding "Area 5."

One experimental acre planted with segmented seed in 1941 was succeeded by 10,000 acres of plantings in 1942 and 300,000 acres in 1943 (9, p. 427). By 1947, processed seed was being used in over 90 per cent of the United States' plantings. The sheer speed with which the results of laboratory experimentation can, under American conditions, become universal field practice is here again impressive, and a sharp challenge to those agricultural economies where institutional deficiencies make the process of internal transmission painful and tardy. Improvement of seed, of course, involves quite modest investment of capital, and commercial acceptance meets far lighter resistance than is true of mechanization. In the switch from segmentation to superior processing of beet seed, the highly commercial agriculture of California again took the lead.

Changes in such spring operations as blocking, thinning, and weeding have tended to be adopted much more slowly and with wider interregional contrasts. Persistence of the older practices was promoted by the considerable labor economies from the lighter seeding rates generally associated with processed seed, and by the conversion of a stoop-labor job into one that could be performed in upright position with a long-handled hoe. As late as 1949, hand work predominated by a considerable margin except in the Red River Valley. Although mechanical harvesting becomes increasingly common as one proceeds west, the reverse is true of mechanical blocking and thinning. The explanation is in large part the seasonal availability of labor. The Middle West could generally count on a fall supply of farm workers, who could find work on other crops, but the situation was tight during the spring. In California, the sugar-beet harvest had to compete with various truck crops for labor, but the thinning and blocking occurs at a seasonal trough in farm-labor requirements (*34*, p. 994). In that state, precision planting was liable to result in light seedling emergence because of the problem of soil crusting. Advantage was taken of processed seed, but seeding rates continued to be relatively heavy, and the higher yields associated with hand thinning more than covered the wage bill (*33*, p. 13).

Once the war labor shortage had passed, California growers totally abandoned mechanical thinning (*33*, p. 4) and, by resort to migrant Mexican labor, could even afford the luxury of the short-handled hoe. Growers elsewhere also became far more deliberate in evaluating the advantages of further savings in labor costs against risk of lower yields. To be sure, field experimentation continues, with active support from the beet-sugar companies. In 1950, more than 30,000 acres were thinned by machines which the Great Western Sugar Company loaned to its growers (*39*, pp. 29–30). By contrast, the spectacular airlift that flew 5,500 Puerto Ricans into Michigan to avert a labor crisis in the beet fields took place in the spring of that same year (*40*).

Several studies of the effects of field mechanization on grower costs have been published (cf. *23; 24; 33*). The difficulty of making farm cost estimates is considerable, and fortunately our particular interest can be served quite readily by collateral evidence. It can certainly be said that cost reductions were not sufficient at first to increase greatly the profitability of beets relative to other crops, as compared with the prewar position. At prices prevailing in the period 1946–51, the output of beet sugar averaged less than 10 per cent above that in the middle and late 1930's, while acreage planted to beets had actually fallen. Although the postwar tonnage of beets harvested was relatively high, the decline in sugar recovery was itself one feature of a mechanized harvest. Moreover, in the 1930's beet production was pressing hard against statutory ceilings and in the 1940's holding considerably below them.

The effects of the new processes on labor requirements do, however, reflect progress made in the beet fields in raising man-hour productivity within a single decade. In Yolo County, California, as recently as 1947, about 75 man-hours per acre were required in the beet fields if operations were by hand, as indicated in the following estimates of man-hour requirements per acre of sugar beets (*33*, p. 17):

Operation	Spring[a] and harvest unmechanized	Harvest mechanized	Spring and harvest mechanized
Plow, level, prepare	3.5	3.5	3.5
Plant	.5	.5	.5
Cultivate and furrow	2.7	2.7	2.7
Hand block and thin	16.9	16.9	9.0
Mechanical block and thin	—	—	.6
Hand hoe and weed	11.2	11.2	6.0[b]
Irrigate	10.5	10.5	10.5
Total preharvest labor	45.3	45.3	32.8
Lift	1.4	—	—
Hand top and load	28.0	—	—
Machine harvest	—	5.4	5.4
Total harvest labor	29.4	5.4	5.4
Grand total	74.7	50.7	38.2

[a] Spring work includes blocking, thinning, hoeing, and hand weeding.
[b] Assuming pre-emergence spraying where feasible, and general improvement in cultivation and weed-control practices.

Some 60 man-hours were provided by contract labor, divided about equally between spring and harvest work. This pattern differed little from the situation as it had been in 1922 (*41*, p. 30). At that earlier date, the figure was over 100 hours in Colorado, which did not in the main come to enjoy the advantages of tractor power in preparation of the soil, planting, and lifting until after 1935. Acceleration in productivity gains has come only within the last decade. Processed seed, even if followed entirely by hand blocking and thinning, appears to save about 20 per cent of the spring work (*13*, p. 3), or some 15 per cent of the total. An equivalent saving is possible by mechanical blocking and thinning. Mechanical harvesters bring even more striking reductions, saving as much as three-quarters of the labor used in the fall (*23*, p. 30) and representing in the California case a cut of almost one-third in over-all labor requirements. Full mechanization by equipment available in 1947 would have placed man-hour requirements per acre in California below 40. Labor productivity per ton of beets has been climbing even more rapidly, with yields per acre on the increase. Nationally, yields in the early 1920's were below 10 tons of beets per acre, and now are above 14 (*7*, p. 84; *42*, p. 139).

REPERCUSSIONS OF FIELD MECHANIZATION

Within a decade, a new technology has converted sugar beets from a labor-intensive crop based on migratory labor into one requiring skilled machine operators and considerable capital investment in such specialized machinery as harvester combines, as well as in general-purpose seed drills, cultivators, loaders, and other equipment. It is still too early to judge the full impact of these changes, or their implications both for agriculture and beyond. For that very reason, it is challenging for students of economic dynamics to try to distinguish patterns of reactions as yet dimly reflected, and to anticipate the direction in which further change might be likely to proceed.

Chiefly because unexpected problems frequently resulted from success at a particular point, there have been a stirring and ferment that continue to bring a re-evaluation of former field practices. Just as precision planting forced improvement in preparation of the seed bed if germination and emergence were to be satisfactory, mechanical harvesting creates a need for longer rows and superior cultivation procedures if combines are to work with top efficiency. Increased use of mobile equipment in the fields promotes wider spacing between the rows themselves, resulting in a smaller plant population and lower yields per acre (*43*). At the same time, a new danger arises, in the form of damage to the soil structure under the pressure of tractor and harvester wheels or tracks.

Breeding research has been reoriented. To facilitate machine harvesting, geneticists are progressing toward a shorter, more globe-shaped, and more standardized variety by hybrid crossing, without sacrificing the high sugar content for which existing varieties are notable (*44*, pp. 187–91). Reduction of spring labor requirements is also aimed at by varietal experimentation. Monogerm beet varieties, which would make machine processing of seed unnecessary, would completely eliminate the need for thinning and reduce blocking to routine proportions. Such varieties are now actually available, and important progress is being made in raising their resistance to endemic disease (*45*, p. 36).

The character of the sugar-beet farm enterprise has been substantially altered. With spring and fall labor peaks leveled off, there is a new prospect for steady year-round operations, especially in those regions where the trend to irrigated pasture opens up improved opportunities for converting feed into animal products (*23*, pp. 45–47). To the extent that the labor force now becomes a stable group of regular employees, the social and community problems associated with migratory help—much of it frequently of an alien tongue—disappear, while at the same time the possibilities of unionizing agricultural workers improve. The investment sunk into specialized equipment might be expected to promote greater stability also of acreages planted to beets from year to year, though this does not appear to have in fact resulted, and in any case it could be counteracted by the changing relative profitability of alternative crops.

Related crops have in turn been affected in a variety of ways. With processed seed, growers in the Salinas Valley have less trouble pulling spring labor away from other lines of work, and for the first time they can pay as low a rate for thinning beets as for thinning lettuce (*46*, p. 19). While mechanized harvesting equipment in California made possible the lifting of a crop for which labor was simply not available, the march of mechanization in the Middle West, cutting down on the job opportunities open to fall migrants, can create new problems for growers who depend on that group to harvest their tree fruits. In Colorado, machines for beets put potatoes, a closely competing crop, at a disadvantage, presumably to result in lower potato acreages or the advance of mechanization for that crop as well (*23*, p. 47).

The size of investment required for mechanical equipment might be expected, further, to raise the average size of farms or at least the average acreage planted to beets. The trend has definitely been in this direction. The 1,014,000 acres planted to beets in 1950 were farmed by some one-third fewer growers than the 1,048,000 acres in 1942 (*47*, pp. 97, 773). Within individual states, however, competent analysts find little evidence of rising acreages (*33*, p. 11; *23*, p. 46).

What has come about, rather, is a marked shift in the competitive position of different beet-growing states, to the advantage of those with very high average acreages (especially California) at the expense of those with lower unit operations. The number of recipients of government subsidy payments was in 1950 less than one-half the corresponding 1942 figure in Ohio, about two-thirds in Utah, and about three-fifths in Michigan. At the earlier date, all three of these states had averaged less than 10 acres per payee. In California, by contrast, more growers were in beets in 1950 than in 1942 and the state's total beet production rose correspondingly. Instead of being barely above that of Colorado, California now exceeded the next two states combined. Prices received by growers also bear out this shift in comparative advantage. The expansion in California has come despite the fact that growers there received the lowest price per ton of any major producing state in 1950. Competition from other crops available to this highly flexible agricultural economy had required a price per ton of beets near the top of the list in 1942 (*48*, p. 90; *47*, p. 773).

At that, the strong tendency toward both year-to-year stability in beet acreage and an even further increase in average size would have been substantially greater but for three additional factors. First, lifters and loaders were a halfway house open to small growers, who could if necessary also share the costs of purchase (*24*, p. 34). Second, the ingenuity of those responsible for the development of mechanical equipment was directed toward invention of machinery appropriate to the farm organization in particular regions. While large California operators felt the wartime labor shortage to an extreme degree and had accordingly an urgent necessity to mechanize, growers elsewhere could hold back a little until harvesters appeared that were practical on as low as 20 to 25 irrigated acres in Colorado, or 20 to 30 in the Red River Valley (*4*, p. 1348). Third, ownership of a mechanical harvester represented not merely a sizable investment of capital, but considerable overcapitalization and excess capacity. The short harvest season in any particular region put a severe cramp on economies of large-scale operation. One way of spreading this overhead was by a new division of labor. As early as 1947, the number of harvesters owned by growers or rented from processing companies in California had declined, in favor of ownership by commercial operators who offered themselves and their machine for hire (*49*). In place of field workers who followed the harvest season as it moved north, there was now the migratory harvester combine.

Largely because of the speed with which the harvest now proceeded, the old pattern by which factory and field operations were coordinated now proved unsatisfactory. Several lines of policy were forced on the processors by the faster delivery rates and the change in quality of hauled beets. The less favorable bargaining position of factories, the upward pressure on their costs, and the resistances to a higher volume of sugar production provided strong incentives for improving manufacturing processes. At the same time, a new set of choices opened up. Should new factories be built to the higher peak load, or would it be better to find means of speeding the operations in existing plants? As an alternative to increasing peak capacity, might storage procedures be improved to allow a buffer between field and factory? And if storage practice came under review, were stockpiles of whole beets, of sliced cossettes, or of the mother liquor likely to offer the greater technological opportunities?

Improvement of the system of handling beets at receiving points was a matter of necessity, given the new speed of mechanical harvest (*27*, pp. 123–33). The old practice of windrowing beets from one to ten days after topping had gone out with the advent of field loaders and harvester combines. With a less dehydrated beet, the volume hauled per ton of sugar produced was correspondingly higher. Long lines of trucks waiting at weighing stations meant general irritation and a heavy drain on trucking facilities at best. Where truck and combine worked closely together in the field, delay in unloading might bring the harvest to a temporary halt. Hoists and conveyors have become standard equipment at railroad sidings to speed operations at these points, while pilers operated from swinging booms have come into use in the factory yards. Since beets also arrive in dirtier condition than formerly, various types of screens, shakers, and trash catchers have become necessary. More recently, interest has been revived in the use of hydraulic handling at the factory, a method long in general use in Europe (*50*, pp. 58–59) but not attractive to American factories at the low tare rates of prewar days (*51*, p. 32).

But if the beets now come in less clean, they also arrive fresher, cooler, and therefore in better condition for storage (*1*, pp. 103–21). Natural conditions for storing whole beets outdoors are reasonably favorable in the Intermountain States (*30*, p. 22); and indeed a brief harvest and a sizable stock of beets were the norm in Idaho even before mechanization of the harvest (*52*, p. 4). In warm California, the new practice of forced ventilation of storage piles by cool night air has about doubled the period during which beets can safely be kept in outdoor piles. The general importance of storage has made this stage in factory operations a matter of research interest to the United States Department of Agriculture since October 1947 (*4*, pp. 1346–47), and further innovations, such as preharvest foliar spray to inhibit respiration and new growth of stored beets, seem on the way (*53*, p. 48).

Some improvements in operations within the factory are in the traditional lines of saving on labor and fuel, and increasing sugar recovery. There has been a resort to automatic process controls, with incidental benefit to the quality of factory performance (*54*, pp. 24–27). Commercial application of the ion-exchange method for more complete extraction of sugar from the juice has fallen somewhat short of its earlier promise (*55*, p. 28). But the most spectacular development inside the factory—continuous diffusion—is an answer to the higher peak load to which the factory must now adjust. So overburdened were the factories when mechanical harvest first became fairly general practice that California factories in 1948 found it necessary to enforce quotas on daily deliveries by growers, at the sacrifice of much of the cost advantage of field mechanization (*33*, p. 21). Building additional factories was not an attractive alternative, except in a very few districts, since the national output of beet sugar at first showed no tendency to rise. The technique of continuous diffusion fortunately had advantages of both an impressive saving in man-hours and increased daily capacity of the plant. This process has accordingly been rapidly introduced.

More intensive daily operations tended to mean a shorter processing season and a longer period of inactivity for the factory. The combined effect of greater factory capacity through continuous diffusion, larger area made available to a single factory by truck haulage, and stabilized national sugar production (at least in good times) may be expected to lead to some reduction in the number of factories operated in the United States. Much of this physical plant appears to have been retained by the industry thus far on a stand-by basis, presumably in expectation that growers will switch to beets—under the existing government program—once farm prices generally declined.[1] Of the 83 American beet factories in 1948/49, only 72 were actually operated (*29*, pp. 155, 270).

Innovations in beet fields and factories have been made possible, it must be recognized, by an extremely favorable public policy. However impressive their technological virtues may be, economically they represent to a considerable degree what Taussig called "that disadvantageous application of labor and capital which is the ordinary consequence" of protective measures (*56*, p. 64). The pace of technical change was quite slow in the late thirties when legislative quotas were placing a distinct limitation on markets available to domestic beet sugar. After 1942, with marketing quotas inoperative during the war and with statutory figures temporarily more of a target than a ceiling, all segments of the industry could give full rein to their inventive genius.

At that, the innovations were initially almost purely of a defensive nature. At price levels prevailing in the 1940's, they were required merely to prevent further declines in beet acreage, and did not then give any basis for a real expansion of the domestic industry. This was mainly because, in the dynamic agriculture of the United States, tech-

[1] This actually occurred. Under statutory authority, the United States Department of Agriculture found it necessary to impose marketing quotas on factories in 1954, and acreage restrictions on growers the following year.

nology of crop production advances all along the line. The competing demands for use of irrigated bottom lands in the western states are many, and the pull by crops alternative to beets is strong. The danger of mechanized beet agriculture to the markets for offshore cane sugar is clearly exaggerated when one recognizes that a decline in the absolute cost of growing beets does not preclude a deterioration in its competitive position, at least during years when prospects for American agriculture generally are good. Whether American beet production would even in a future wartime period fare better than during World War II is entirely questionable. With demands for farm produce extremely heavy at such times, and expansion of sugar output in Cuba so easy, any rational system of allocating scarce resources might establish low priorities for the petroleum, rubber, implement steel, and skilled labor, not to mention the fertilizer, now required in the sugar-beet fields.

ADVANCES IN SOVIET SUGAR TECHNOLOGY[2]

There has been considerable similarity between the pace of beet mechanization in the United States and that in the USSR. But the contrasts between the two countries are highly revealing. That the overriding task of the Soviet sugar industry after 1941 was to overcome the disaster of invasion (see chapter 10) is by no means the crucial factor in these contrasts. Of special interest are the differences in the problems that mechanization sought to solve and the solutions that were considered satisfactory, and the strikingly different environment in which the Soviet innovator was required to operate.

The USSR inherited from Czarist days not merely a major sugar industry but also an associated complement of technical personnel, research staff, scientific laboratories, and agricultural experiment stations. Before the First World War, Russia imported only 5 to 6 per cent of its beet seed, but about half the seed was obtained from German selection stations installed in Russia. Although beet-seed selection had been developed to the point of complete self-sufficiency in the period 1914–17, the Revolution disrupted the work of the selection stations. The Soviet government had consequently to depend somewhat on imported seed, and supplies were particularly tight during the expansion of beet acreages in 1930–34. By the middle 1930's, however, the state farms were using half their beet acreage to fulfill the entire national seed requirement. While World War II, therefore, cut off no foreign sources, invasion of the Ukraine was equally effec-

---

[2] Much of the material in this section was previously published in *38*.

tive in temporarily disrupting the domestic supply. By evacuating personnel, laboratory equipment, and some breeding materials to temporary locations in Central Asia, seed production was effectively put beyond the reach of the invading armies (*57*, pp. 107–08; *58*, pp. 30–35).

The network of selection stations operated by private industry before the Revolution was gradually reactivated and extended to the new sugar-beet regions. There were 12 selection stations before World War II (*80; 81*, pp. 12, 22–30). By 1941, there were also 52 separate seed-testing stations, one located in practically every "microregion" throughout the beet areas. Seed selection was originally directed toward higher over-all yields. But expansion of plantings in the East called attention also to the need for developing varieties adapted to local conditions. Such collateral characteristics as early ripening, or resistance to cold, drought, and heat became worth developing. Although Soviet sources claim that locally developed varieties perform better than imported types under experimental conditions, there is no evidence of spectacular or even modest improvement in yields as compared with more general-purpose "cosmopolitan" varieties. From experimental farms and selection stations, elite (selected) seed goes to the state farms, both for trial and for later commercial production. The state farms indeed are the field testing ground for new varieties in old regions or old varieties in new regions, as they are for most advances in agricultural practices, including mechanization and rotation systems.

According to Soviet agronomists, the idea of using segmented seed was originally proposed by Professor Tishchenko as early as 1933; the first machine for splitting the seed ball was constructed at the Central Research Institute of the Sugar Industry in 1934; and experiments with planting such seed were performed during 1934–37 (*59*, p. 510; *60*, pp. 29–37). So far as priority of scientific discovery is concerned, these are valid claims. Similar experiments were not begun in Germany until 1937 (*61*) and the successful demonstration of the technique in California came only in 1941.

Of special interest, however, are the purposes of the Soviet innovators and the gap between invention and application of the new technique. The original experiments set out to economize on the use of beet seed (*62*, p. 49), not on the labor required for blocking and thinning. When the immediate bottleneck in beet-seed production had been eliminated by 1937, experiments with segmented seed were abandoned. Only in 1945, after the practical application of segmented or processed seed had been fully demonstrated in the United States, was the work renewed. As late as 1948, planting of segmented seed in collective farms is reported in terms of only *tens* of thousands of hectares (*63*, p. 41; *58*, p. 14), a small fraction of the total area, though the United States figure had reached 90 per cent in 1947/48. Not until 1951 is there evidence of a swing toward precision planting of processed seed. Progress was delayed in large part because successful use of segmented seed required improvement in the general level of agronomic practices in Soviet agriculture (*60*, pp. 29–37).

Soviet and American practices in beet agriculture were most similar in the use of tractors in heavy spring work. According to Soviet claims, the 70,000 tractors available in the beet areas (*64*, pp. 85 ff.) were taking care of practically all deep plowing, 93 per cent of the planting, and 73 per cent of cultivation between the rows by 1938. As a result, thinning the beets and cultivation within the rows continued to be the main function of manual labor, as it was also in the United States. The claim that machine blocking of sugar beets was widely applied in the USSR even before the war, and that American growers merely borrowed this technique (*59*, p. 519) is not so easy to judge. It seems more likely that prewar blocking was merely across-the-row, while down-the-row blocking was barely begun by 1948 (*63*).

Improvement in the efficiency of operations proceeded much more slowly than the above figures imply (*65*, pp. 11–12). Workers and managers on collective farms were not favorably disposed to plant sugar beets, which required extra attention but received no commensurate price. The reaction to large compulsory acreages and a fixed price was a decline in yields, in large part due to poor timing of field operations. There was much copying of American and German machine attachments. Machines would frequently be put into mass production before they had been properly adapted to Soviet conditions. A foreign disk plow was introduced on a large scale, only to be discarded after overwhelming evidence that it plowed too shallow and gave a poor soil texture. After five years of intensive experiment, the type of plow selected for general use was one common in Russia before 1917 but which had been discarded in the rush to surpass American technology. Since incentive payments to tractor operators were related to economy of fuel consumption rather than quality of work, full advantage was not taken of the new opportunities for deep plowing.

After the German armies had been driven out of the Ukraine,

there remained a tremendous shortage of tractors, as well as of draft animals, all agricultural implements, and fertilizer. As late as 1946, the Ukraine had less than 60 per cent of the prewar number of tractors (*66*, p. 58) and recovery in numbers of horses and machinery was slower still. Yields suffered from a further deterioration in the quality of work, with respect to both timing of operations and depth of plowing. Not until 1949 did the supply of tractors and equipment permit operations on a full prewar basis.

If the tractor was common to Soviet and American fields, it also was a key symbol of differences in social organization and in capital supply. In the United States the tractor was the instrument, in sugar beets as in other crops, for expanding the amount of land subject to the control of a single farm operator; in the USSR, the Machine Tractor Station was a means for state control of the people and the output of collective farms. Under American conditions, furthermore, there was considerable underutilization of tractor capacity, despite an increase in the acreage farmed by one man and the introduction of less powerful machines for farmers with smaller acreages. Soviet tractors tended to be used intensively, with labor applied in shifts and sometimes even around the clock.

Adaptation of the tractor to harvest operations was also proceeding in the 1930's. By about 1938, practically the entire crop was being removed with the aid of some 17,500 tractor-drawn lifters maintained by the MTS (*67*). Retrogression as a consequence of war must have been even greater than in spring work. Plans for 1947 called for the use of lifters, mainly horse-drawn, on only 30–40 per cent of the crop, but performance fell short of this target (*68*). The following year, no more than 20 per cent of the beets could have been harvested by tractor-drawn lifters (*69*, p. 24), and the prewar level had not been regained as late as 1951.

More interesting, because less affected by the war as such, were the difficulties involved in developing the combine harvester. The pressure of labor shortage had never operated as intensely in the Ukraine as in the United States, though the incentive to mechanize was somewhat greater in the newer beet areas in Siberia where the land-labor ratio was high, the grain harvest still under way at the time the beets had to be removed from the fields, and the danger from early frost great. While mechanization of the wheat harvest was consistent with large-scale organization, certain contrary tendencies operated in the case of beets. Yields per acre had been raised on collective farms in the thirties by the device of assigning small plots of beets to individuals or small groups of workers for an entire season. Intensive cultivation was obtained by awarding premiums for yields exceeding stated tonnages. Some experts prefer to wait for improved methods of mechanical harvesting before abandoning this system of personal responsibility for beet agricultural practices (*69*, pp. 24–30; *70*, pp. 25–26). It is worth noting that Soviet systems of organization encounter the most serious obstacles in such matters as intensive cultivation, rotation systems, and animal husbandry, which lend themselves less readily to mass mechanization but which are the essence of beet agriculture in the West.

Work on a combine was proceeding nevertheless in the 1930's (*65*, pp. 14–17). Against the advice of technical experts, the central authorities ordered 300 units of an experimental model developed in 1932. When the model proved subject to clogging, and its use resulted in improperly topped and excessively dirty beets, the machine designers who had counseled against its introduction in the fields were held responsible and given prison terms. Despite this setback to the mechanization program, by 1935 a fresh group of designers was working on new types that pulled the beets by their leaves before topping. Political risks continued to be an inhibiting factor. Not only might a technician be accused of deliberately delaying the program by obtaining pessimistic results, or be held responsible for the ineptness of a politician's decision, but the government had to be persuaded to give up a favorite engineering target. Because flat topping meant the loss of as much as 150,000 tons of sugar a year, and the government was not seriously interested in correspondingly heavy tops to be fed for the most part to privately owned animals, the government persisted in demanding a device capable of conical topping. It was a brave technician who convinced the authorities that flat topping was immediately practicable, and conical still a long way off. But the war intervened before field advantage could be taken of the decision to compromise.

In 1946, a John Deere harvester, as well as one of domestic construction, was tried on a few state farms, but with unsatisfactory results (*71*, pp. 42–48). A domestic one-row model had been improved sufficiently by 1948 to permit its mass production in the factories, while mass field trials took place the following year on state and collective farms in practically all important sugar-beet regions. The number of combines involved may have reached into the *thousands*, as suggested in some sources (*70*, pp. 23–31), though other reports indicate a more modest figure. One-row harvesters evidently saved about 50 per cent of the labor required by three-row lifters, and about 80 per cent of the roots were satisfactory for delivery to the factory without additional hand topping, cleaning, or sorting. With further improvement, these types were expected to harvest about three acres of beets per 12-hour day. In 1951, machine factories delivered 1,000 three-row combine harvesters, and a larger number was planned for 1952 (*72*, pp. 51–64). But the proportion of the beet acreage harvested by this means in the latter year could hardly have exceeded 6 per cent (*73*).

The customary bottleneck in the Soviet beet-sugar enterprise was not, however, in the purely agricultural operations but in hauling beets from the fields to factory dumps. The strain on truck and rail facilities in the Soviet Union is characteristically heavy, and the more so for a crop which has to be shipped in great volume first as a raw material and then as final product. One incentive for eastern expansion of acreage was the desire to avoid the long haul of refined sugar from the Ukraine. It was nevertheless possible in 1938 to transport about half of the beets from the fields in approximately 35,000 trucks (*74*; *64*, pp. 88 ff).

Though the volume of beets actually grown in the 1920's was well below the processing capacity of factories inherited from Czarist days, factory expansion became necessary when beet production increased in the 1930's. In the old regions, it was not deemed necessary to add to capacity until the processing season for existing factories exceeded 160–170 days (*75*, pp. 5–10). Soviet factories in the 1937 campaign operated for an average of 157 days (*76*, pp. 3–4); in Central Asia, where a good many of the newer factories were built, 200–230 days was the rule. The campaign, in short, might last until April in the USSR, though Western Europe usually finished processing by December.

So long a season might on the face of it appear as an entirely rational utilization of factory capacity otherwise left idle. The adjustment was not, however, an entirely happy one. Shortage of capital in the form of transport facilities, especially while the grain harvest was being moved, and of factory capacity had to be compensated for by storing beets in the fields far into the winter months. Loss of sugar content in storage is a serious matter in the Soviet Union, and nowhere more than in the new production regions of Central Asia where seasonal and diurnal variations of temperature are wide. Longstanding concern with problems of storage provided the incentive for discovering the technique of ventilating storage piles with cool night air. While there is no evidence that this has become wide-scale practice in the Soviet Union, the United States readily borrowed the technique once the speed of mechanized harvest created a new problem of beet congestion.

The Soviet transport bottleneck also meant that harvester combines were generally of the windrowing type, with subsequent piling and loading by hand. Even when bunkers were supplied for side loading, beets were generally dumped in the field for piling, cleaning, and loading onto vehicles. Consequently, there was less attraction in a combine that, in a single operation, lifts, tops, and loads into a truck. Here is a further explanation of the slow rate at which combines came into use. Hand labor evidently has continued to perform the job that might have been relegated to field loaders, although "tractor shovels" speed the loading of railroad cars at dump stations.

These problems of coordinating harvest and factory operations persisted throughout the war and postwar period. Although restoration of damaged processing facilities proceeded more rapidly than beet output between 1943 and 1946, transport services lagged until 1949, and losses in storage were at least as serious as before the war. After 1947, better beet crops again strained transport facilities and factories alike.

Reconstruction of the 55 sugar factories (about one-quarter of the national total) completely destroyed by war (*62*, p. 25) involved a heavy capital expenditure that countries outside the theaters of military operations were spared. But it also created certain opportunities. In 1944, plans were announced for radically transforming the processing industry (*77*, pp. 19–24). Beets were to be sliced into cossettes, dehydrated, and pressed into briquettes at a large number of small factories near the beet fields. A smaller number of central factories were then to be relied on for the succeeding stages of processing. Since beets would thereby be reduced about 80 per cent in weight and more than 90 per cent in volume (*78*, p. 27), a major economy was expected in transportation, as well as in factory capital requirements, with incidental benefits through reduced losses in storage. Actually, only a few experimental briquetting plants were ever built (*58*, pp. 12–14), and reconstruction proceeded instead on the basis of the orthodox pattern and generally in the old geographical locations.

The opportunity to modernize the industry in the course of its restoration was not wasted. Early in 1946, a conference of sugar technicians was organized for the specific purpose of selecting the most modern and efficient equipment available for installation. The work of this conference was based on special reports of Soviet technicians who had already completed a study of methods used in Germany, Czechoslovakia, and the United States (*79*, pp. 3–21). In re-equipping its plants, the USSR could call upon not only its own machine-building industry but also those in Germany and Czechoslovakia, the two countries on the Continent most specialized in the construction of beet-sugar factories. Particular attention was paid to the mechanization of operations requiring heavy manual labor, such as unloading beets and hauling beets from dumps to the factories. Automatic control of various manufacturing processes was also introduced. There is no question that by 1950 the factories had been fully restored to their pre-war capacity, at a much improved level of scientific technique.

WESTERN EUROPEAN VARIATIONS AND RATIONALE

In many important respects, the environment of technology is strikingly different in Western Europe from that in either North America or the USSR. With land rather than labor the limiting factor, any practice that jeopardizes high yields per acre will be strongly resisted. High seeding rates provide better assurance of large plant populations, even if the burden of thinning is increased. If full adaptation to mechanical equipment requires wider spacing between rows, the consequent loss in beet tonnage can outweigh any possible advantages. More than 95 per cent of the plantings continued to employ ordinary seed in most European countries as recently as 1951 (*82*); only in Sweden and Germany were the risks of poor germination from processed seeds accepted for as much as one-fifth of the plantings, as compared with a complete shift to the new methods in the United States and Canada. Growers seem prepared to endure their spring labor requirements until research on the biological front can turn out entirely reliable monogerm varieties (*85*, p. 7).

Problems of improving sugar-beet agriculture are closely tied up with the general organization of European farming. The size of farm unit is such that many growers in a country like Belgium will contract for as few as two or three acres of beets (*86*, pp. 25–27). The problem of buying specialized equipment that will go under-utilized except for short periods of peak load is accordingly serious. While a machine may occasionally be purchased for its "psychological" effect in pacifying labor, even though it remains unused (*84*, p. 12), expenditure on tractors or machinery represents a heavy commitment for most individual operators, and caution becomes the rule. At the same time, fuel is relatively expensive, so that horse-drawn or even hand equipment is more common. Improvisations include use of hand hoes or simple "topping sledges" for removing crowns and leaves, lifting by one-row plows, and removing the dirt by harrowing (*87*, pp. 24–26).

Slow adoption of a beet harvester rests on additional grounds of a somewhat special sort. In a feed-deficit economy, any mechanical damage to beet tops becomes a serious defect. Natural conditions at harvest time are far less favorable than under irrigated culture in the American West (*17*, pp. 15–17). Wet fields at harvest time mean dirty beets, while soggy soil tends to keep trucks close to the roads. The mechanical harvester will accordingly give less satisfactory results than in drier regions, and direct loading into motor vehicles is less common. Moreover, the European beet-sugar companies as a group were not in the critical position of their American counterparts, who had promoted the introduction of imperfect harvesters in order to maintain a flow of beets and stay in operation.

In most respects, Great Britain is somewhat better placed for mechanization than the rest of Western Europe. The average beet acreage is higher—about 10 acres—while some 10 per cent of the growers with acreages in excess of 20 produce half the beets (*88*, p. 12). Damage to tops is a less important matter, since 40 per cent are plowed in rather than fed (*90*, p. 13). A relatively tight labor supply even before World War II, and the establishment of minimum field wages, provided a considerable incentive for raising man-hour productivity. Great Britain has accordingly attained a higher density of tractors per acre of cultivated land than any other country, while a certain ferment was affecting beet practices before 1939 (*89*, pp. 207–11). Two- to six-row lifters hauled by tractors were available, though not in common use. Lifters were officially reported on some two-thirds of England's beet farms in 1946 (*91*, pp. 180–81); a quarter of the crop was being sowed by self-propelled or tractor-driven drills in 1947 (*84*, p. 3); a third of the crop was being mechanically hoed in 1948.

Several important advances in beet technology abroad proved to be little suited to British needs. Professor Roy Bainer, associated with the transformation of beet agriculture in California, visited the United

Kingdom early in the summer of 1945 and a Sugar Beet Mission reciprocated the visit in the fall (*17*). Few lasting results appear to have followed from exposure to American practice in the two processes of most immediate interest: seed segmentation had only a temporary flurry of favor in Great Britain, and American harvester-combines proved to be unsatisfactory. Even the domestic Catchpole harvester, which comprised two-thirds of all those available on farms in 1948 (*84*, p. 11), was too large-scale a unit for representative British conditions.

Technological borrowing from the European Continent turned out to be more extensive. At national demonstrations of mechanical equipment in October 1948, British farmers in large numbers had their first look at small, light models invented in Denmark, where some 1,400 units were reportedly already in operation (*86*, p. 27). These models, which represented a direct adaptation of mechanical equipment to the small-scale farm economy of Western Europe, operated on a distinctive two-step system. In recognition of the importance of tops as feed, a topping unit harvested the leaves and crowns as a separate crop, to be cleared from the field while the roots still remained in the ground. On a second tour of the fields, the beets themselves were lifted and shaken clean, to be windrowed in the field or elevated into a following trailer. Whereas standard combines were selling for £400 to £500, the two-stage units could be had for as low as £150 (*83*, p. 97). The small grower could reduce his costs further by contracting his services on neighboring farms as well as by purchasing the lifting unit alone. Within a single year, the Roerslev harvesters had become the most popular make in use, while the Mern, also of Danish construction, ranked third (*88*, pp. 6–8). Moreover, the great majority of these units served farmers growing not more than 15 acres of beets. As a result, in 1949 the degree of mechanization was highest in four factory districts where the average grower contracted for fewer than 8 acres. But the relative position of these smaller units has since declined, while the percentage of the total British crop harvested by mechanical means has risen steadily, as follows (*90*, p. 3):

1946	1947	1948	1949	1950	1951	1952	1953
0.9	1.4	3.4	10.8	16.2	21.7	25.0	30.4

Mechanical loaders ("elevators") have also been introduced in large numbers to reduce the manhandling of beets from roadside clamps into motor vehicles.

CITATIONS

1 C. J. Robertson, *World Sugar Production and Consumption* (London, 1934).

2 G. H. Coons, "U.S. Sugar Beet Seed Meets War Crisis, Part I," *Sugar*, January 1943.

3 G. H. Coons "Improvement of the Sugar Beet" in U.S. Dept. Agr., *Yearbook of Agriculture, 1936.*

4 U.S. Cong., House, Com. Agr., *Research and Related Services of the United States Department of Agriculture*, Vol. II, 81st Cong., 2d Sess., Dec. 21, 1950.

5 G. H. Coons "U.S. Sugar Beet Seed Meets War Crisis, Part II," *Sugar*, February 1943.

6 U.S. Dept. Agr., *Yearbook of Agriculture, 1934.*

7 U.S. Dept. Agr., *Agricultural Statistics, 1951.*

8 S. B. Nuckols, *Sugar Beet Culture in the Northern Great Plains Area* (U.S. Dept. Agr., Farmers' Bull. 2029, 1951).

9 H. B. Walker, "A Resumé of Sixteen Years of Research in Sugar-Beet Mechanization," *Agricultural Engineering*, October 1948.

10 Glenn R. Larke, "The Status of Pelleted Beet Seed," *Sugar*, April 1945.

11 G. H. Coons, "Space Relationships as Affecting Yield and Quality of Sugar Beets," *Proceedings of the American Society of Sugar Beet Technologists, 1948* [hereafter cited as *Proceedings . . .*].

12 Roy Bainer, "Use of Sheared Beet Seed on the Increase," *Spreckels Sugar Beet Bulletin* (1942), Vol. 6, as abstracted in *Sugar*, February 1943.

13 P. B. Smith, *A Survey of Sugar Beet Mechanization* (Beet Sugar Development Corporation, Fort Collins, Colo., 1950, mimeo.).

14 Roy Bainer *et al.*, "Processing Sugar Beet Seed by Decortication," *Proceedings . . . , 1946.*

15 Roy Bainer, "Present Status of Sugar Beet Seed Processing," *Proceedings . . . , 1948.*

16 " 'Pelleting' Segmented Beet Seed," *Sugar*, February 1944.

17 Gr. Brit., Min. Agr. and Fisheries, *Mechanisation of the Sugar Beet Harvest in North America*, Agriculture Overseas Rept. No. 4, 1947).

18 "Beet Seed Decortication," *Proceedings . . . , 1949*, as abstracted in *Sugar*, February 1950.

19 Robert C. Brown, "Experience with Decorticated Seed on 30,000 Acres," *Proceedings . . . , 1948.*

20 J. B. Bingham, "Methods for Reduction of Hand Labor in Spring and Summer Work," *Proceedings . . . , 1946.*

21 *Sugar*, December 1942.

22 O. W. Willcox, "The Mechanization of Harvesting," *Sugar*, January 1946.

23 H. G. Sitler and R. T. Burdick, *The Economics of Sugar-Beet Mechanization* (Colorado A. and M. Coll., Agr. Exp. Sta., Bull. 411-A, Fort Collins, 1950).

24 R. M. Gilcreast, *Sugar Beet Production in the Red River Valley* (N. Dakota Agr. Coll., Agr. Exp. Sta., Bull. 363, 1950).

25 Sugar, September 1950.

26 W. W. Robbins and Roy Bainer, "Weed Control in Sugar Beets," *New Agriculture* (1947), Vol. 30, No. 2, as abstracted in *Sugar*, July 1948.

27 R. A. McGinnis, ed., *Beet-Sugar Technology* (New York, 1951).

28 J. E. Coke, "Utilization of Beet Harvesting Machinery," *Spreckels Sugar Beet Bulletin* (1943), Vol. 7, No. 8, as abstracted in *Sugar*, January 1944.

29 Farr & Co., *Manual of Sugar Companies, 1949/50* (New York, 1950).

30 R. H. Cottrell, ed., *Beet Sugar Economics* (Caldwell, Ida., 1952).

31 William Conner, "The Importance to the Grower of Delivering Clean Beets," *Spreckels Sugar Beet Bulletin* (1948), Vol. 12, No. 4, as abstracted in *Sugar*, January 1949.

32 Utah-Idaho Sugar Company, *Machine Harvesting Marches On!* (Salt Lake City, 1948).

33 W. R. Bailey, *Economics of Sugar Beet Mechanization in California* (U.S. Dept. Agr., Circ. 907, 1952).

34 U.S. Senate, Com. on Labor and Public Welfare, Subcom. on Labor and Labor-Management Relations, *Migratory Labor*, Hearings, Part 2, 82d Cong., 2d Sess., 1952.

35 P. B. Smith, "Statistical Data Collected for 1947 [on Sugar-Beet Mechanization]," *Proceedings . . . , 1948.*

36 Sugar, November 1952.

37 Sugar, March 1951.

38 V. P. Timoshenko, *The Soviet Sugar Industry and Its Postwar Restoration* (Food Research Institute War-Peace Pamphlet 13, Stanford, Calif., August 1951).

39 Sugar, April 1951.

40 Leonard Jackson, in *The Bay City Times* (Mich.), Aug. 6–8, 1950.

41 L. K. Macy *et al.*, "Changes in Technology and Labor Requirements in Crop Production—Sugar Beets," *National Research Project*, Rept. No. A-1, August 1937, as quoted in *23*, p. 12 and *33*, p. 2.

42 U.S. Dept. Agr., *Agricultural Statistics, 1940.*

43 Bion Tolman *et al.*, "Row Width and Sugar Beet Production," *Proceedings . . . , 1948.*

44 G. W. Deming, "Breeding Sugar Beets with Root Conformation Adapted to Machine Harvest," *Proceedings . . . , 1948.*

45 U.S. Dept. Agr., Agr. Res. Admin., *Report of the Chief of the Bureau of Plant Industry, Soils, and Agricultural Engineering, 1952.*

46 Sugar, April 1943.

47 U.S. Dept. Agr., *Agricultural Statistics, 1952.*

48 U.S. Dept. Agr., *Agricultural Statistics, 1944.*

49 Austin Armer, "A Review of the Mechanical Harvest in California," *Proceedings . . . , 1948.*

50 G. E. Hrudka, "Elfa Beet Fluming System," *Sugar*, September 1952.

51 Sugar, March 1952.

52 Hobart Beresford, *March of Mechanization of Sugar Beet Production in Idaho* (Univ. Idaho, Agr. Exp. Sta., Circ. 111, Moscow, Ida., 1946).

53 Sugar, August 1951.

54 R. H. Cottrell, "Developments in Processing of Sugar Beet," *Sugar,* May 1948.

55 Sugar, May 1950.

56 F. W. Taussig, *Some Aspects of the Tariff Question* (3d ed., Cambridge, Mass., 1931).

57 N. I. Orlovskii, "Development of the Selection of Sugar Beet Seed in the USSR," *Agrobiology* (Moscow), November-December 1947, No. 6.

58 I. P. Lepeshkin, "Advanced Technique in Action," *Sugar Industry,* October 1948.

59 D. N. Pryanishnikov and I. V. Yakushkin, *Field Crops* (Moscow, 1936).

60 P. Petersen et al., in *Socialist Agricultural Economy* (Moscow), June 1946.

61 F. Fleher and W. H. Fuchs, University of Halle (Germany), unpublished reports (1945).

62 V. P. Zotov, *Development of Food Industry in the New Five-Year Plan* (Moscow, 1947).

63 P. Petersen, in *Socialist Agricultural Economy,* March 1948.

64 N. Nazartsev, in *Socialist Agricultural Economy,* February 1939.

65 M. Menjeha, unpublished manuscript.

66 N. Klimenko, "Agriculture of the Soviet Ukraine on the Way to Prosperity," *Socialist Agricultural Economy,* November 1947.

67 P. Kolpakov, in *Socialist Reconstruction of Agriculture* (Moscow), September 1938.

68 Socialist Agriculture (Moscow), Sept. 20, 1947 and Sept. 24, 1947.

69 V. Mikhnovsky, in *Socialist Agricultural Economy,* March 1949.

70 V. Mikhnovsky, "Organization of Harvesting of Sugar Beets by Combines," *Socialist Agricultural Economy,* May 1950.

71 P. Petersen, in *Socialist Agricultural Economy,* July 1947.

72 M. Gazdarov, "Organization of Sugar Beet Harvesting by 3-Row Combine," *Socialist Agricultural Economy,* April 1952.

73 I. Sipiagin, "Sugar Beet Cultivation in the Fifth Five-Year Plan," *Socialist Agriculture,* Nov. 1, 1952.

74 Polovenko, in *Socialist Agricultural Economy,* January 1939.

75 I. V. Zilberman, "Rational Duration of the Sugar Campaign," *Sugar Industry* (Moscow), July 1950.

76 B. A. Rubin, *Storage of Sugar Beet* (Moscow, 1946).

77 Socialist Agricultural Economy, August-September 1944.

78 R. A. Davies, "Russia Tackles Reconstruction Job," *Sugar,* July 1945.

79 Sugar Industry, July-August 1946.

80 E. Iu. Zalensky, *Sugar Beet in Western Russia and Poland* (Petrograd, 1919).

81 A. Arkhimovich, *Sugar Beets in the USSR: Selection and Seed-Raising* [in Russian] (Institute of the Study of the History and Culture of the USSR, Munich, 1954). Abridged translation in English, as *Selection of Sugar Beets in the USSR* (Fort Collins, 1955, mimeo.).

82 J. Baratte, "Spring Mechanization of Sugarbeet Production," *Fif-*

teenth General Meeting of the International Institute of Sugarbeet Research (Paris, 1952).

83 Claude Culpin, "British Farm Mechanisation Programs," in Gr. Brit., Min. Agr., *Agriculture*, May 1951.

84 British Sugar Corporation Ltd., *Report on the Mechanisation of the Sugar Beet Crop in Great Britain During the Year 1948* (Peterborough, 1949).

85 "Home-Produced Sugar," *Ministry of Food Bulletin* (Gr. Brit.), Apr. 29, 1950.

86 "Beet Harvester Shows in Europe," *Sugar*, March 1949.

87 Gr. Brit., Min. Agr. and Fisheries, *Green Crop Conservation in Germany* (Agriculture Overseas Rept. No. 9, 1949).

88 British Sugar Corporation Ltd., *Report on the Mechanisation of the Sugar Beet Crop in Great Britain During the Year 1949* (Peterborough, 1950).

89 Claude Culpin, *Farm Machinery* (London, 1938).

90 British Sugar Corporation Ltd., *Report on Mechanical Harvesting of Sugar Beet in Great Britain During the Seasons 1952 and 1953* (London, 1954).

91 Gr. Brit., Min. Agr. and Fisheries, *Agricultural Statistics 1945–49, England and Wales*, Part I (1952).

CHAPTER 6

MODERNIZATION OF CANE AGRICULTURE AND TRANSPORT

In the cane regions of the world, the decades since the end of the First World War are notable for major developments in the agricultural sector. To be sure, modernization of milling was effected in the Philippines in the 1920's and was going forward in Brazil, Mexico, and the British West Indies in the 1930's and 1940's. But these are examples of known technology penetrating belatedly to less industrialized locations. Cane agriculture was in ferment everywhere. Newly bred varieties of cane sharply raised the level of per-acre yields. The gradual adoption of the tractor (and the truck) brought important changes in preparation of the soil, in planting, in cultivation, and in the organization of the harvest. In those regions where labor shortage was a factor to be contended with, increased attention came to be paid to mechanical handling of cane and of sugar. A fully mechanized harvest has been accomplished in Louisiana and on some Hawaiian plantations, and is regarded with favor in Queensland. This process of modernization, by which cane has incidentally been a major carrier of agricultural science to tropical regions, warrants serious attention.

THE VARIETAL REVOLUTION

While the breeding and commercial introduction of improved seed are common to modern cultivated plants, the development and transmission of new cane varieties have displayed highly dramatic features. As might be expected from the example provided by other crops, improved varieties meant higher sugar yields per acre, greater resistance to disease, better adaptability to natural hazards like drought or frost, and more response to fertilizer applications. Indeed, among the countries that stand in the scientific vanguard, better varieties have been the most important single factor responsible for higher sugar yields during the past four decades. But their function has been strikingly diversified. To the extent that breeding succeeds in lowering fiber content and reducing impurities, the efficiency of factory operations is improved. Combinations of early and late maturing varieties perpetuate high sucrose results over a longer grinding season, and spread overhead factory costs at the same time that they increase yields. Good ratooning qualities cut the heavy expense of replanting; self-trashing canes reduce the costs of the harvest; erect canes lend themselves more readily to mechanical cutting than do those that lodge heavily. A modern cane economy marshals its front-line and reserve varieties much as a general calls up his troops.

The period of quiescence.—For cane, the primitive stage of plant selection is not long past. From the sixth century to the eighteenth, all cane grown in the Western world was of a single variety (*15*, pp. 18–25), the Puri (or "Creole") cane that migrated westward from its original home in the South Pacific by way of India. It was not until 1768 that a second variety, known as Bourbon or Otaheite, was introduced under French auspices into Mauritius, the first cane migration to the West directly from the South Pacific. "Selected in the first instance by some untutored Polynesian," Bourbon dominated cane fields for almost a century, as Creole had for the twelve centuries previous. Under the name Lahaina, it was practically the only irrigated plant in Hawaii from 1854 until 1915.

The successor to Bourbon was borrowed from Southeast Asia, rather than from the South Pacific. So-called "Batavian" cane reached Mauritius and Réunion from Java in 1782, and slowly came into favor over a wide area under a number of names. Displacing Otaheite after 1840, Crystalina (Light Praenger) gained almost exclusive control of the Cuban fields until the 1920's and in 1948 it still ranked second locally, with over 13 per cent of the planted area (*21*, p. 155). Canes of this group were standard in Java until the 1880's, and in Louisiana (especially the Purple variant) for a decade longer. Variously designated, as Rose Bamboo in Hawaii, White Transparent in the British West Indies, Crystalina in Cuba and Santo Domingo, and Rayada (Striped) elsewhere, it was considered responsible for more of the world's sugar in the middle 1920's than any other variety (*3*, p. 63).

Several other transplanted canes have come to dominate the fields of their adopted habitat for lengthy periods. Tanna reached Hawaii from the New Hebrides via Mauritius in 1881. As Yellow Caledonia, it had become responsible for almost half the Hawaiian acreage by 1919, while White Tanna represented two-thirds of the Mauritius acreage in the middle 1920's (*13*, pp. 33, 37). Badila, a high-sucrose cane brought to Australia from New Guinea in 1895, was not outranked by any other variety harvested in Queensland until 1949/50

(*23*, p. 75). Uba cane had a similar history in South Africa. Though its origin is obscure, it somehow reached Brazil, was brought back to India and Mauritius, and was introduced into South Africa in 1882. Not until 1932, in which year 99.96 per cent of the commercial cane harvested in Natal was still Uba, did this variety begin to be displaced (*22*, p. 192).

Of these long-dominant canes, several proved highly susceptible to disease. The failure of Otaheite in Mauritius in 1848 explains the interest of that island in varietal selection in the nineteenth century and its role as transmitter of varieties to other regions. Shortly before 1900, the sereh disease destroyed Batavian canes in Java, and Bourbon succumbed to red rot in the British West Indies and British Guiana. From this twin disaster dates the modern period of scientific selection and breeding as well as recent achievements in cane pathology and genetics. Disease continued to be a major factor in the international transmission of new varieties at least until the 1920's. Of the aboriginal varieties, only Badila and Uba could continue their supremacy for a longer period.

Seedlings and the accomplishments of POJ.—In 1887–88 it was demonstrated almost simultaneously in Barbados and Java that sugar cane yields viable seeds. Entirely new horizons were now opened up to the plant breeder. Formerly all cane had been reproduced vegetatively, and for commercial planting, "seed cane" today still refers to the "setts," or pieces of cane stalk, that are placed in the soil. In vegetative reproduction, all plants are genetically the same individual, and hence the new crop duplicates the properties of the planted cane. But cane does not breed true from seed. The hundreds of minute blossoms in each flowering arrow, when successfully pollinated, yield as many new varieties as there are seeds. After careful selection, the one prized seedling in 10,000 that meets the tests of commercial feasibility can fully compensate for the ten-year gap separating initial breeding from full-scale planting in the fields. What had been the most stable sector of the sugar economy was suddenly in ferment, as the search for improved varieties went forward on the private lands of estate owners and plant breeders or at experiment stations sponsored by governments or by planter associations.

The succeeding record of cane-breeding programs reads like the genealogies of the Old Testament and is almost equally unamenable to summarization (*3*, pp. 42–108; *7*, pp. 582–91). Three lines of early development, however, could claim notable successes. First, several seedlings of the older standard canes had outstanding properties. Thus D 74[1] and D 95, seedlings of White Transparent originally bred at Barbados, came to occupy an important place in Louisiana until struck by the virus mosaic and other diseases in the middle 1920's. Similarly H 109, a seedling of Lahaina, took over from its parent after 1918. Yielding at its best more than 18 tons of sugar per acre on 120 acres of Hawaiian land in 1924 (*3*, pp. 79–80), it nevertheless could not earn a place outside the Islands. Second, a succession of high-yielding canes was obtained in Java in the period 1886–1911 by crossing improved (or "noble") varieties already in cultivation (*Saccharum officinarum*). Notable in this group were POJ 100, 247 B, DI 52, EK 2, and EK 28. By the early 1920's, the results of this systematic genetic work could already be seen in the succession of new varieties moving across the cane fields, each wave raising sugar yields but soon to be succeeded by another. In Java, 247 B and POJ 100 contributed 84 per cent of 1915 plantings, but 8 years later 66 per cent of the total was of the varieties EK 28 and DI 52 (*13*, p. 31).

A third line of breeding proved most promising of all. This was the attempt to combine the disease-resistant features and less exacting soil and climatic requirements of vigorous wild species, such as *Saccharum spontaneum* and *Saccharum robustum*, with the attractive sucrose quality of the "noble" cultivated canes. From around 1910 in Java, there began the process of crossing noble canes with a semi-wild upland cane known as Kassoer, itself evidently a natural cross between the noble Dark Cheribon and a variety of *Saccharum spontaneum*, and of "nobilizing" or upgrading the cane by backcrossing with further noble parents. These efforts were rewarded when POJ 2878 began to be planted commercially in 1926. Though responsible for only three-fourths of 1 per cent of Java's acreage that year, its share had risen to 66½ per cent within two years, and for a decade not more than 6 per cent was planted to other varieties (*24*, p. 77). The result was a 50 per cent increase in sugar yields per acre, in a country that was already outstanding in this respect.

Ironically, the great superiority of the Java canes resulted in their coming to dominate the cane fields of various countries during the very years when Java's sugar was being displaced from international

---

[1] Standard abbreviations in varietal nomenclature include the following: D— Demerara, H—Hawaii, POJ—Proefstation Oost Java, B (suffix)—Bouricius, DI— Dimak Idjoe, EK—Edward Karthaus. The terms represent location of breeding station or name of individual plant breeder.

markets. How large a role canes of the POJ designation came to play in other lands is illustrated by the following tabulation:

Place	Date	POJ varieties as percentage of:		(Source)
		Planted acreage	Harvested acreage	
Formosa	1930	99	...	(*26*, p. 41)
Peru	1952	...	75+	(*40*, p. 16)
Tucuman	1943	70+	...	(*14*, p. 30)
Cuba	1948	68+	...	(*21*, p. 155)
British Guiana	1943	...	68	(*17*, p. 55)
Puerto Rico	1951	65+	...	(*29*, p. 40)
Jamaica	1939/40	...	31	(*36*, p. 47)

In a desperate effort to retain the competitive advantages of local scientific achievements, the Dutch East Indian government in 1930 prohibited the export of all plant materials, but the world-wide process had already been set in motion.

Varietal development elsewhere.—Canes from Java were important not only for direct use in cane fields abroad, but also as parents of successful varieties bred in various parts of the world. Many advances that resulted from more systematic breeding programs undertaken after 1928 at Canal Point, Florida (*19*, p. 354), at the Experiment Station of the Hawaiian Sugar Planters' Association (*35*, pp. 48–49), and in Barbados, British West Indies (*36*, p. 240), developed out of POJ ancestry. The first successful variety bred in Formosa, F 108, was of POJ parentage (*26*, p. 41). In Mauritius, the Sugar Cane Research Station obtained its own M 134/32 by crossing POJ 2878 with D 109. M 134/32 came to be almost universally planted on the island and was largely responsible for doubling sugar production within a 25-year period (*25*, pp. 219–20). PSA 96, a cane of some importance in postwar Philippines, was a self-seedling of POJ 2878 originally produced at Canlubang in 1930/31 (*33*, p. 119).

Paralleling the global dispersal of nobilized POJ "blood," however, were several developments of a rather contrary character. The late POJ canes were adapted to a wide range of tropical conditions and had admirable qualities of disease resistance. But their spectacular performance placed a new emphasis on the successive replacement of older varieties, not to salvage cane fields from rampant disease, but simply because the new varieties provided higher yields (*12*, p. 97).

Cane breeders concerned with raising yields soon found it desirable to develop varieties uniquely adapted to soil, climate, and other distinguishing local characteristics. There was implied a more decentralized approach, and indeed the orientation of cane-development programs soon became almost as nationalistic as sugar policies. Hawaii came to depend heavily on seedlings propagated there, particularly during 1931–37. In Queensland, over 75 per cent of the 1952 crop was from locally produced varieties as against about one-quarter in 1943 (*2*, pp. 147–48). The overwhelming portion of the acreages cultivated in the British West Indies, amounting to as much as 97 per cent in Trinidad in 1951, came to be planted with varieties developed at the B.W.I. Central Sugar-Cane Breeding Station, originally established on Barbados in 1932 (*17, pp.* 39–40).

Though some of the very early Java crosses proved useful in such subtropical regions as Louisiana, Argentina, and northern India (*7*, pp. 587–88), India itself proved a collateral center of cane research that made important contributions toward the local adaptation of new varieties. The Coimbatore Station in Madras, which began operating in 1912, developed an impressive list of its own varieties, designated as "Co." Most were trihybrids obtained by crossing *S. officinarum,* the common domesticated varieties, with *S. spontaneum,* wild canes, and *S. barberi,* varieties cultivated locally. The results were well adapted to rigorous conditions: the risk of frost in Louisiana, hot winds and dry spells in Natal, even attack from jackals and other wild animals in India itself (*20*, p. 33). Indeed, the sudden rise in South Africa's sugar yields in the late 1930's resulted from the direct borrowing of satisfactory Co. varieties, since successful germination of Natal-grown seed was not accomplished before 1944 (*12*, p. 107). Improved varieties bred for Louisiana and Australia have relied about as much on Co. parentage as upon POJ (*19*, p. 354; *16*, p. 66). So great is the differentiation of microclimates down the long stretch of Queensland that each of 12 varieties was responsible for at least 2 per cent of the 1949 crop, and none for more than 18 per cent (*23*, p. 181). Hawaii has customarily planted separate varieties on irrigated and on nonirrigated land, but now rates varieties also according to their performance under sunny and under cloudy conditions, at high or low elevations, and on better or poorer soils (*10*, pp. 44–45). The forces causing decentralization of research are such that the British West Indies region can no longer rely on the seedlings germinated and selected exclusively at the Barbados station. In British Guiana, for example, these varieties have suffered severely from leaf scald (*31*, p. 141).

Some implications and repercussions.—The importance of varietal improvement is not only in overcoming disease, in adaptation to peculiar local conditions of soil, water, and climate, or in the greater tonnages per acre that are obtainable, but also in the wide range of subsidiary purposes that special breeding can serve. Barbados found it possible to take an average of three ratoons in 1954, as contrasted with only one in 1938 (*30*, p. 116), with consequent saving in the costs of replanting. Varieties that germinate quickly, and thereby offer better natural protection against weeds, can cut the labor costs involved in cultivation (*38*, p. 108). Experiments with 35–1933 in Hawaii raise the possibility of converting the sugar economy from a 24-month to a 12-month growing-period basis (*34*, p. 148).

Milling qualities are also important. If high sugar yields per acre result from better tonnages of cane at the expense of lower sugar recovery per ton, factory costs are correspondingly raised. But in South Africa, Uba was a particularly difficult cane to mill, high in fiber and with recalcitrant impurities, so that efficiency in the factory could not be advanced until superior varieties had been introduced in the cane fields.

By 1950, as much as 85 per cent of all cane harvested throughout the world was grown from improved plant stock developed by experiment stations (*39*, p. 23). In view of the wide territorial distribution of the world's cane fields, the primitive conditions under which much cane is grown, and the varietal quiescence that had characterized a highly commercial, highly capitalistic plantation industry for three centuries, this degree of success within four decades is a remarkable achievement in pure science and applied research. Even in India, where the inertia and poverty of the peasant is considered a strong barrier against better agricultural practices, nearly 90 per cent of the cane acreage was reportedly planted with improved varieties by 1950/51 (*11*, pp. 18–20).

What factors have contributed to this widespread accomplishment? Catastrophe, in the form of disease outbreaks, certainly has played a part. But so has colonialism. The long-sustained interest of metropolitan countries in the sugar production of the West and East Indies is one major explanation of the fact that highly talented scientists concerned themselves with the problems of this particular plant as early as the 1880's. Local endowment of land and labor has also been important. In Java, an industry that was limited in the acreage it could plant and the period for which it could occupy the land had every incentive to raise yields per acre, and the same was true in those sugar islands of the British West Indies that had long ago reached the limits of extensive expansion. Hawaii also was short of land but even less well supplied, relatively, with manpower. Higher-yielding varieties represented an economy in land use at the same time that they increased the return from the man-hours invested in land preparation and planting, though without sparing labor costs at harvest. Finally, the conspicuous dependence of the mill on the cane lands in its immediate vicinity has meant a more rational approach to agronomic problems, a readier availability of capital for the agricultural sciences, and a prompter spread of new varieties (or new practices) that experimenters develop, than can be expected when the connection between industry and agriculture is more remote. It is therefore not surprising that many varietal improvements have been developed at cane-breeding stations supported by producer associations or by individual sugar-mill enterprises.

MECHANIZATION OF CANE AGRICULTURE

Although the hoe, the wooden plow, the machete, and the oxcart are still to be found in the cane fields of the world, no other tropical crop has witnessed a more rapid introduction of the tractor and mechanical farm equipment. As in mechanized agriculture elsewhere, the field operations have benefited in many ways. The mobile power unit represented by the tractor, by permitting deeper plowing and subsoiling of heavy clay soils, assures a better preparation of the seed bed. The plowing, harrowing, and other field work can, moreover, be accomplished more speedily, so that the seed bed is more likely to be ready for planting when weather conditions are most favorable. On both accounts, higher yields can be expected. Field power also creates a basis for adapting improved equipment to the requirements of cane agriculture, and for performing various operations simultaneously instead of in separate stages. Thus plowing, disking and furrowing, or planting, fertilizing, and covering can be done in single runs by multiple-purpose units. Weeds flourish as well as cane under tropical conditions, so that improved weed control through mechanical cultivation (or chemical herbicides) is especially important during the period before the cane has "closed in" over the field.

Certain natural characteristics of sugar cane complicate the process of mechanization. "Seed cane" or setts are segments of the cane stalk, usually containing three or four eyes from which the new shoots are expected to germinate. Vegetative planting material requires careful handling if it is not to be damaged and the subsequent stand reduced. The chute of a planter may injure the eyes. Poorly cut seed pieces are more liable to suffer from fungus infection or pests, and the very position of the eyes in the furrow means the difference between uniform germination and a stand that emerges unevenly over a period of as much as two weeks (*18*, p. 78). Similarly at harvest time, this crop introduces special field problems. The stalk must be cut near ground level, stripped of its leaves, and the immature tops removed. If cut too high above the ground, a rich portion of the cane plant is lost to the mill; if topped too high, impurities present in immature cane complicate treatment of the juices and inhibit recovery of sucrose actually present. If trash or dirt are not removed in the field, transportation costs on a necessarily bulky material become higher still, while grinding unclean cane means slower milling rates and greater loss of sucrose in the bagasse, as well as excess wear and tear on factory machines. Unquestionably the quality of manual work is higher than that of machine operations both in planting and in harvesting cane.

A third complicating characteristic of cane agriculture in most regions is the practice of taking several ratoon crops. While ratooning reduces the annual cost of plowing, harrowing, and cultivating, it also means that the advantages of good and the disadvantages of poor seed-bed preparation will affect yields at successive harvests. Yields from ratoon crops will also suffer if, at harvest time, the plant stubble is uprooted inadvertently, the soil is excessively compressed, or the "stool" is damaged by mobile field equipment. Tractor power is useful, however, in plowing out old ratoons that have completed their crop cycle, in incorporating the ratoons and field trash into the soil by disk plow or harrow, and in ensuring that a cover crop can be planted and plowed under as green manure preparatory to replanting.

Hawaii: mechanical harvesting.—There are interesting contrasts in the extent to which various producing regions have exploited the advantages, endured the disadvantages, and improved the efficiency, of cane mechanization. In applying heavy capital equipment to field operations, Hawaii has led the way. Commercial production there barely predates the modern sugar factory, with local introduction of centrifugals coming as early as 1852, steam-driven mills in 1859, and the vacuum pan in 1863 (*47*, p. 7). Much of the cane land became available for planting only after heavy investment had been made in irrigation facilities. The role of capital in mill and in land placed a premium on cultivation in large agricultural units, while proximity to the United States mainland (and access to the United States sugar market) facilitated the job of raising the necessary funds. Labor was as short as capital was abundant. The indigenous population succumbed to the principal diseases of civilization, and manpower had to be recruited at considerable expense from China, from Madeira, from Japan, and from the Philippines (*53*, pp. 56–64). Lacking the flexibility provided elsewhere by a seasonal labor supply, the Hawaiian planter had every incentive to adopt machines for field use. Hawaiian conditions permitted full-time use of plantation labor by allowing planting, cultivating, and harvesting to continue practically throughout the year. This organization of agriculture also made it easier to bear the overhead cost of expensive equipment, as compared with regions that had to anticipate more months of inactivity.

Heavy machinery adapted to conditions in the agriculture of western United States quickly found its way to Hawaii. In 1881, the steam plow was introduced; in 1910, the gasoline tractor. The excellent seed bed resulting from deep plowing and improved implements was particularly important in a region where the cane would be growing almost two years before the first harvest. Tractors, including heavy-duty crawler types, were commonplace on the Islands by the middle 1930's, for the example of their effective use in West Coast timberland was close at hand. Only harvesting and planting remained predominantly manual operations on the eve of World War II.

Several local conditions were special deterrents to mechanical harvesting. A single machine could not be expected to perform equally well on irrigated cane, typically grown in furrows, and on unirrigated, typically grown on the flat. On unirrigated land, moreover, the harvest had frequently to proceed on wet fields, with consequent difficulties for mechanical equipment. Even though trash might be removed by burning on irrigated land, the extremely heavy tonnages of cane per acre put a severe strain on any but the most rugged machines. Moreover, a two-year growing period caused Hawaiian cane to topple over of its own weight. Mechanical harvest of a recumbent crop involved formidable engineering difficulties which had to be overcome with little assistance from agricultural experience elsewhere.

Even before World War I, pioneer attempts were made to develop a machine for harvesting cane, both in Cuba (*3*, pp. 12–14) and in Queensland (*18*, p. 189). The latter model, which became the Falkiner harvester in the early 1920's, was given field trials not only in Australia but also in Cuba and Florida. After the Labor Savings Devices Committee of the Hawaiian Sugar Planters' Association became seriously interested in machine cutting in 1925 (*57*, pp. 33–35), the Falkiner was brought in from Florida for trial. Though the high expense of this unit was less of a deterrent to plantation operations either in Florida or Hawaii than it had been to the Queensland cane farmer, this early combine was not a success. While it cleaned and loaded as well as cut, the principle of operation, which involved chopping stalks into short lengths and separating tops and trash by a strong air blast, had definite disadvantages. The blast involved loss of cane as well as trash, and cutting to short lengths meant loss of juice and much more rapid deterioration of harvested cane. Moreover, the machine had not been developed to handle the heavy tonnages of the Hawaiian fields.

Nonmanual harvesting in Hawaii developed out of improved methods of loading rather than from a perfected mechanical combine. By the early 1930's, some plantations had begun to load cane from the field into railroad cars by means of grabs suspended from mobile derricks. In addition, hand-cut cane might be pushed or pulled into position for the grab by heavy-duty rakes. In 1937, the grab was adapted for the first time to the job of biting up large mouthfuls of uncut cane for direct loading (which might now be into trucks). By such a grab and crane device, the Ewa plantation became in 1938 the first plantation in history to harvest practically its entire crop by mechanical means. This device was shortly followed by the Waialua rake which, when shoved into standing cane, broke the stalks off at the roots and left them in piles for the loader.

Both grab and rake harvesting gave such unsatisfactory results that harvesting remained predominantly a hand operation on the eve of World War II. The cane was not topped, and the load transported to the mill was as much as one-quarter trash, rocks, and mud (59, p. 3). But the customary processes of scientific experimentation and gradual technical improvement had to be foregone with the outbreak of war, and mechanical harvesting was prematurely hastened (54, p. 82). The needs of the Armed Forces pushed wages up and stripped the plantations of their personnel, to the point where the regular labor force in 1945 was little more than half that in 1939 (47, p. 15). With sugar short on the mainland, and return cargo space available in vessels that had carried military supplies to the Pacific Theater, every improvisation had to be relied on for gathering in the current harvests. About two-thirds of the 1942 crop was harvested by grab or rake, and the proportion soon went higher. Under the circumstances, the yield advantages from improved varieties were lost, the quality of cane and the efficiency of mill deteriorated, valuable topsoil was removed from the fields, and even the tending of crops for future years was neglected. Only in wet areas and on rocky fields did hand cutting hang on.

Interesting ramifications of mechanization fanned out in various directions. Cane-cleaning equipment had now to be installed at the mill. As late as 1950, the Ewa plantation lost 400 hours of grinding time as a result of rock damage to shredding knives, while sand, coral, and loose metal continued to cause excessive wear on crushers and rollers. Much of this difficulty was eliminated by the introduction of a sink-float process for rock separation (*58*, pp. 120–21). The problem of disposing of trash at the mill called attention to improved methods of composting (*10*, p. 24). Soil conservation practices became important as a result of removal of topsoil from field to mill, as well as the puddling and water erosion that result from movement of heavy traffic across wet fields. Soil compaction from mobile equipment, which can distort root growth, similarly drew scientific attention to root structure and characteristics. The diverse problems of cane mechanization have accordingly called into play the research talents not merely of agricultural engineers but of a wide range of scientists employed at the Experiment Station, H.S.P.A. Even the customary system of cropping has come under review. Since a recumbent 20–22 month crop is difficult to harvest without excessive trash and damage to ratoons, experiments have been made with cropping the vigorous variety 37–1933 after only 12 months' growth, with encouraging results (*60*, p. 40).

Since damage to ratoon plants continues to be serious, replanting has become a much more important part of the field operation. Although by 1951 total acreage in cane was more than 10 per cent below that of three decades earlier, the year marked the third successive one of record-breaking plantings (*61*, p. 58; *62*, p. 57). About 20 per cent of the cane fields were being replanted each year. At this rate, a full crop cycle lasts only 5 years, as compared to the 6–10 years customary under premechanization practices that called for two to four ratoon crops of roughly two years' growth. While one reason for accelerating plantings has been to take earlier advantage of the higher-yielding varieties available for commercial use, another major factor is the damage to ratoon stools resulting from present harvesting methods.

Heavier planting programs have in turn forced revision of planting techniques. Mechanically assisted operations were spreading before Pearl Harbor, whether by planting machines that opened the furrow, dropped the seed piece, and covered with soil, or at least by furrowing and covering in separate tractor operations (*63*, p. 12). A good deal of manual labor continued to be associated with planting even after the war, not only in feeding the mechanical planters, but also in placing the cane in the ground, particularly in wet areas or when planting the new high-yielding varieties, the bulging eyes of which are prone to machine damage (*64*, pp. 50–52). Poor germination resulting from uneven covering of the seed has also in some quarters brought reversion to manual covering.

The processes which have longest resisted mechanization are the harvesting and cutting of seed cane, which remained predominantly a hand operation as late as 1949 (*54*, p. 49). The need for planting larger acreages has tended to alter these practices. While seed cane continues to be harvested by hand, machines have been developed to cut it to length and the system of supplying the mechanical planter has been improved. The fear of damage from mechanical cutting continues to be a deterrent to the use of the combine cutter-planter (*65*, pp. 50–51).

Despite the considerable progress achieved, the problems of adapting machines to cutting (and to planting) cane in Hawaii have not as yet been fully overcome. The elimination of the grab and the rake, which have tended to offset the gains in yields that might have been expected from scientific advances at other points, is a long-run objective of the local scientists (*10*, p. 31). But the notion of a combination harvester-loader, which had been considered the desideratum for 15 years, has had to be given up (*55*, p. 56). The latest advances have come from more limited-purpose machines, which are primarily ground-level cutters that may permit cutting and windrowing in irrigated fields or cutting and infield transport under wet conditions. The Duncan harvester, designed to pick up a continuous blanket of recumbent cane, cutting it loose from the ground by a rotary blade and from adjacent cane by a vertical knife, is reportedly the first machine to surpass hand work in the quality of its performance (*66*, p. 204). It is along lines such as these, as well as in the development of a cane cleaner suitable for operating in the fields, that future improvement may be expected.

Local variations and rationale.—While the Hawaiian Islands by no means exemplify cane agricultural technology in general, the shift to tractors for cane land preparations has been proceeding rapidly in almost every cane region. The tractor is not only a labor-saving device but a mobile power unit that, by permitting speedier work and improved cultural practices, ranks along with new cane varieties as promoter of recent increases in cane yields. In Florida, Peru, South Africa, and Australia, the tractor is entrenched. So rapidly has the process been advancing in the West Indies that the value of tractor imports into Jamaica was more than double that of all sugar-factory machinery during 1951–53 (*73*, p. 62), without taking account of an increase in supplementary agricultural equipment. The job of land preparation is particularly heavy in a region like Formosa, where only 5 per cent of the cane is ratooned. On farms of the Taiwan Sugar Corporation, Heath steam plows (adapted from Hawaii) which had been purchased before the war were still in use in 1952, but were being replaced by wheel-type and crawler tractors (*74*, pp. 35–37). On that island, private farmers continued to use animal draft power almost exclusively, but in Fiji (*75*, p. 275) and in Mauritius (*76*, p. 70) small farmers might exercise the option of having plowing done on a contract basis. In Fiji, indeed, the Colonial Sugar Refining Company was in the habit of renting out disk plows even when draft power was supplied entirely by animals (*77*, p. 10). The device of the government-operated tractor station is also found in a few cane regions, as in the Brazilian state of São Paulo (*79*, pp. 52–54). Finally, where extensive development of cane lands is still going on, as in Queensland, bulldozer methods of land clearing speed up the process of obtaining the first crop, though at the risk of losing valuable topsoil (*18*, pp. 59–62).

Especially in the older, highly specialized cane regions like the West Indies, the shift from animal to tractor power represents a major transformation of the local economy. Draft cattle (which provided barnyard manure, could feed on produce from the land, and could be sold off for meat after their working days were over) provided the closest approach to crop-and-livestock farming that sugar cane seemed to allow. To be sure, the land required to support the full stock might be equal to as much as 50 per cent of the acreage in cane (*12*, pp. 131–34). But pasturage could only partially be diverted to cane, whereas reliance on imports for fuel, tractors, and spare parts increases the trade dependency of economic units already highly specialized on the basis of external commerce. Inasmuch as cattle can feed on the cane trash and tops provided at harvest time, which to some extent coincides with the time of planting, animal draft power can be expected to survive for hauling cane even after the tractor takes over the work of preparing the fields.

Although the tractor lends itself readily to mechanized planting and cultivation, the trend to mechanical planters has been extremely slow. This is not only the case under primitive agricultural conditions as in Venezuela, where most cane is planted in holes dug one meter apart without other preparation of the land (*78*, p. 36), or on small farms in Taiwan, where seed pieces are germinated in nurseries for transplanting as a means of intensifying the use of land; cane continues to be placed in furrows by hand even under the highly mechanized conditions to be found in Louisiana and Florida. By comparison with the man hours saved by using the tractor for opening the furrow, adding fertilizer, closing and packing the furrow, and for subsequent mechanical cultivation, the small gain from placing the cane sett by machine appears not generally to compensate for the risk of damage to the awkward seed piece and the consequent danger of a poor stand. If the crop cycle is a long one, a weak plant endangers future ratoons as well as the initial harvest; if the cycle is short and replanting frequent, the savings from employing the tractor for preparing the seed bed are all the more important.

Only in Queensland does the move to machine planting appear to have been entirely deliberate. Horse-drawn drop planters, which fed seed pieces into open furrows and covered the setts, were standard equipment in the early 1930's (*18*, pp. 72–78). Advent of the tractor permitted the use of combine planters, which drilled, fertilized, dropped the cane, and covered in a single run. After 1932, cutter-planters which cut the whole stalk to proper size came into commercial use. The necessity of dipping seed cane into a mercurial solution as a precaution against pineapple disease somewhat reduced the attractiveness of the cutter-planter, but supplementary devices promised to permit this additional step. Other than on limited acreages in wet regions, hand planting in Queensland is mainly confined to scrub areas recently opened up for cultivation.

So far as harvesting is concerned, mechanization has proceeded at least as rapidly in Louisiana as in Hawaii, though on a very different basis. In the former region, climatic conditions unfavorable to the cane crop actually contributed to the progress of the mechanical harvester. The job can be done there by lighter equipment than that required by high Hawaiian yields, the danger of frost puts an extra premium on speed (though the short season means the harvester can be utilized for only a few months), and the practice of stubble shaving to remove frost-damaged shoots has familiarized the growers with nonmanual devices for cutting the cane at ground level. As a matter of engineering development, the modern harvester was developed after 1938 from a machine originally designed for windrowing cane (*69*, p. 39), a practice formerly employed for reducing the risk of frost damage. Louisiana is especially fortunate in that its crop grows in uniform stands of erect cane (this in itself a result of light yields), on flat land, and on the ridge. Devices that cut simultaneously at the ground and below the unusable tops by an average setting for an entire row can therefore be used. Under the emergency circumstances created by the wartime labor shortage, mechanization was pushed forward rapidly. In contrast to merely experimental operations in the fields in 1940, mechanical harvesters were cutting about 50 per cent of the crop in 1945 (*70*, p. 35), and the percentage quickly rose above 90 in the early postwar years (*71*, pp. 295–96).

Previous discussion has already suggested the kinds of inconveniences associated with the new harvesters. The old schedules for coordinating field and factory operations were unsettled; milling speeds, extraction rates, and transport costs suffered from inclusion of trash and dirt with cane; the incentive to top the crop somewhat too high resulted in greater impurity in the derived juice, with consequent difficulties in the process of clarification. Loss in working efficiency was a severe blow to a milling industry already saddled with a short processing season, low sugar yields per ton of cane, and the perpetual risk of losing crystallizable sugar in frost-damaged cane. One result was a decline in the number of mills from 75 to 54 within a 10-year period. Moreover, while mechanization in Hawaii had meant adaptation to the existing unit of plantation production, the changeover in Louisiana was accompanied (even made necessary) by a considerable shift in the internal organization of agriculture. Mainly by reduction in the lowest-size groups, the number of farms growing cane declined from 10,918 in 1939 to 4,010 in 1953, despite an increase in total acreage. The average cane acreage accordingly rose from 24.5 to 75.8 acres (*72*, p. 84).

The substantial disadvantages of mechanical cutting, like mechanical planting, limit its adaptability in every region where the labor supply is at all ample. Special local conditions add further complications. In Formosa, where ratoons are rarely taken and high per-acre yields crucial, the individual cane plants are actually dug out by spade (*74*, p. 46). South Africa is handicapped by its rolling fields, Mauritius by stony soils and by the fact that the stools lie in the bottom of trenches rather than on ridges (*5*, p. 61). Flood and storm damage in Florida frequently results in recumbent and tangled cane, but resort to Hawaiian expedients is avoided by the availability of a migrant labor force from the British West Indies. Peru, another region of advanced agricultural technology, has also been slow to abandon manual harvesting (*40*, p. 22).

The contrast between Louisiana and Queensland is particularly striking. Both are keen on labor-saving devices, but for different operations. In Louisiana, although the harvest is almost completely mechanized, cane continues to be placed in furrows by hand. But in Queensland, which has pioneered in the development of mechanical planters, machine cutting has made little progress and not more than 2 per cent of the crop appears even to have been mechanically loaded as late as 1953 (*4*, p. 278). Louisiana's relative backwardness in planting technology seems to be explainable by several special factors (cf. *69*, pp. 18–21). The harvest season is so short that most cane is planted before cutting (and the heavy work-load associated with it) has begun. A significant saving results from the fact that seed cane is mechanically cut. Moreover, the disadvantageous growing conditions with which Louisiana is handicapped call for every insurance against a poor stand. Seeding rates have to be heavy; seed cane is planted in whole stalks, which are best handled manually; and there is even a disinclination to "whack" stalks into pieces after placement in the furrow, because this would make them more vulnerable to adverse moisture conditions and disease organisms. Any further damage to setts from a mechanical appliance is evidently intolerable.

The explanation of the slow pace at which Queensland has moved toward mechanical harvesting is more complicated. Certainly serious interest in it is of long standing, with workable models engineered as early as 1910. Queensland was the originator of the Falkiner harvester that subsequently received field trials in Cuba and Florida, while a reasonably satisfactory local harvester, the Fairymead, was developed there in the 1930's (*18*, pp. 189–90). But Queensland has lacked the emergency stimulus that compelled Louisiana to resort to machines during World War II despite all their technical deficiencies. While Louisiana throughout the war years had an assured market for every ounce of sugar produced, and was in a favorable position to receive allocations of scarce tractors and equipment, Queensland was cut off by sea from its normal export outlets and sources of supply (including fertilizers). In the postwar years, rapid increase in cane acreage has exceeded the rate at which milling capacity has been expanded, with consequent prolongation of the grinding season. The speedier cutting associated with mechanical harvesting would only have resulted in glut at the mill end. To match the mill's preference for a longer processing season rather than the more rapid enlargement of milling capacity that mechanization would have required, the growers had their own special reasons for favoring clean hand-cutting. Close attention to cane quality has been a mainstay of a cane economy employing white labor in tropical Queensland. Clean and selectively cut cane could not be had by the newer machine processes.

There was also the important matter of size of farming unit. The speed at which cane can be cut by hand has been a limiting factor on the amount of land that may be farmed by the independent farmer. At the rate of 2 to 3 tons per day, which is rather typical of Caribbean labor productivity, a 35-acre field yielding 20 to 25 tons per acre would require about nine months for cutting alone. Such low productivity and so long a season were not suitable for the 50-or-so acre units upon which the Queensland industry had been based. The solution has been found in the device of engaging gangs of migratory cutters on contract. Small groups of farmers, under mill auspices, are served by a cutting team that harvests successive portions of each farmer's cane so as to reap it all as close as possible to optimum ripeness and to supply an assigned mill quota (*9*, pp. 56–57). By efficient division of duties and with the incentive of piece rates, output per man-day is thereby raised as high as 13 tons and averages around 7 tons (*12*, p. 37). This high productivity has been an important factor in the survival of the independent Queensland cane farmer. However, the Kinnear harvester, which is by no means the largest model available, is capable of cutting some 25 tons per *hour* (*23*, p. 1). At that rate, a farmer could harvest his 1,000 tons of cane in about five days, but clearly could not himself afford an expensive machine that operated so few days per year. There is a real question whether the size of cane farm made suitable under a regime of manual cutting has now become a barrier to the introduction of a technology which, for all its limitations, is certainly here to stay in Hawaii and Louisiana. Perhaps Queensland should be looking toward contract harvesters rather than contract cutting gangs. If social inventiveness has been less impressive at this stage than earlier in the century when the industry was learning to do without Kanaka laborers, the answer may lie in the fact that the times may now call for the awkward process of consolidating cane farms into somewhat larger units, with consequent elimination of many individual farmers.

At this point, reference is also appropriate to the well-publicized system of harvesting in Fiji. Here, where farms are 10 to 12 acres in size and individual cane outputs 120 to 180 tons per year, steady supplies of quality cane are obtained by gangs of neighbors harvesting each other's cane, under the general supervision of a field officer supplied by the mill. Each grower is expected to supply man-hours equivalent to the time necessary for harvesting his own cane (75, pp. 278–79). It is not surprising that one firm, the Colonial Sugar Refining Company, should have been identified with the development of these harvesting arrangements both in Australia and in Fiji, but in both cases the farm organization presumes the older technology of cutting by hand.

The reluctance to introduce mechanical harvesting in the cane regions of the world is reinforced by the possibility of investing in efficient transportation facilities for cane instead of in harvesting equipment. Local conditions may require special arrangements, as in British Guiana where the mills are supplied by trains of punts borne on the canals that supply irrigation water. But generally speaking, modern means of transport are easily adaptable to various countries, and enable the mill to speed the harvest without necessarily becoming involved in agricultural operations as such. Rail transport facilities, frequently the property of the mill, laid the basis for the modern cane factory by expanding the area from which a single mill could speedily obtain its sugar cane. Though track maintenance was somewhat costly, rail shipment permitted the use of collateral devices that considerably reduced the handling of cane. Portable track might be laid and light rail cars brought into the fields, as in Australia or Peru; field-loaded oxcarts could transfer their cane at railhead by means of stationary derricks, as in the Caribbean; and side-dumping features could be built into the rail car for tilt unloading at the mill. As recently as 1952/53, Australia continued to move 95 per cent of its cane by rail (68, p. 154).

Introduction of the motor truck for cane transport creates new opportunities for reducing costs, especially on short hauls, and for eliminating manhandling of cane. Direct shipment from field to factory may eliminate the necessity for loading and unloading an intermediate vehicle. Cuba has experimented with the "Portacana" system employing portable cane baskets which may be removed from the truck bed, spotted around the field for filling, reeled back onto the individual truck, and delivered to the mill. Rubber-tired trailers may similarly be deployed, and later be reassembled into truck or tractor trains, as in Florida (37, p. 37). In South Africa, developments along such lines as these have been given a high priority in postwar planning (67, p. 5). In addition a wide variety of self-propelled equipment for loading cane provides a mechanical means of sparing available labor for the trickier job of cane cutting. Louisiana, which employed mule-powered loaders before 1900 and the gasoline engine for this purpose shortly after 1900, finds a place in its fields for several models of mobile grab loaders (*69*, pp. 42–45). The grab has penetrated parts of the Cauca Valley of Colombia (*1*, p. 38); some cane is piled by push rake in Peru (*40*, p. 22); Florida has developed continuous cane loaders of a slanting-platform type (*37*, p. 36), presumably on the model of some Hawaiian harvesters. So simple a device as a wire rope sling facilitates unloading of trucks or rail cars by stationary cranes. It is to techniques such as these that Puerto Rico has recently been turning in order to overcome a serious drain of labor to the cities and to the mainland (*44*, p. 38). Progress in these directions in various regions has been swift and varied.

BULK TRANSPORT OF RAW CANE SUGAR

With so large a portion of the world's raw cane sugar moving overseas to be refined and consumed, recent innovations in methods of ocean shipment warrant discussion. Until quite recently, the standard container for shipping raw sugar over the high seas was the burlap bag. Though cheap and reusable at least once, burlap had several important disadvantages (cf. *45*, pp. 18–19). Bags of 100 pounds to 350 pounds in size required repeated manhandling, at the raw sugar mill, in loading aboard ship, in unloading, and at the refinery. Since raw sugar is prone to pack in storage, the woven fabric was likely to wear through, and some loss by seepage was also to be expected. Bags had to be washed and brushed to avoid further wastage. Even the advantage of cheapness began to decline after 1940 since the jute from which burlap is made is grown only in the Indian subcontinent, and the long ocean shipments were made in the face of climbing freight rates. Moreover, there was the threat that supplies might be cut off by the spread of hostilities.

In the course of the 1930's, the several advantages to be had from shipping staple commodities in bulk were becoming widely recognized. Not only was the expense of the container eliminated, but labor costs were considerably lower, more efficient use was made of cargo space, and less freighter time was lost in loading and unloading, with the result that more trips by a given vessel became possible each year. To be sure, raw sugar was slightly moist and somewhat sticky, so that it could not flow as readily as such dry materials as grains, coal, ore, and certain chemicals. The relative cheapness of labor in most cane-sugar exporting regions was a further deterrent. In some respects, the ocean tanker for liquid cargoes like petroleum was an even more appropriate model than the dry bulk carrier. Molasses, the major by-product of raw-sugar manufacturing, had to be shipped in liquid form. Furthermore, the rising portion of refined sugar being absorbed by industrial users was creating a new market for liquid sugar by the late 1920's. For several years this product was being shipped to the United States mainland from Cuba and the Dominican Republic in considerable volume (*42*, pp. 3–4).

The pioneer in applying bulk methods to raw cane sugar was evidently the United Fruit Company's Revere refinery at Charlestown, Massachusetts (*41*, pp. 41–42). Several circumstances were of importance in this initial case and in the subsequent spread of the practice. Sugar was sent to Revere not in odd lots from small shippers but chiefly from two large mills of its own, both of which shipped out of Port Tarafa on the north coast of Cuba. Since United Fruit was itself in the shipping business, several vessels could be set aside and adapted for bulk sugar cargoes exclusively, without the risk that the specialized capacity might stand idle. Moreover, continuity of relations between shipper and refinery meant that appropriate equipment for handling the cargo and sufficient bulk space for storing it could be counted on at the point of arrival.

By the Revere system, portable bucket elevators and belt conveyors at the dock, and winch-operated drags to sweep the hold, permitted the 6,000-ton cargo to be unloaded in about 20 hours. The same cargo, bagged, would have required four days. Similar equipment practically eliminated manhandling of all sugar fed into the refinery, whether directly from the vessel or after storage. In the warehouse itself, clamshell buckets suspended from traveling cranes had the same effect. At the mill end, however, raw sugar continued to be packed in bags, which were slit open to permit sugar to be dumped directly into the hold. Because of this, handling costs could not be cut as much as would otherwise have been possible, though it was later estimated for Puerto Rico that the time required for stowing alone was cut from .75 man-hour to .05 man-hour per ton (*46*, p. 46). Bags could now be used as many as five times for the trip between mill and docks, whereas they were reusable only once if shipped overseas.

Cuba and the eastern American seaboard, though associated from the early 1930's with ocean shipment of both bulk raw sugar and liquid sugar, did not maintain their leadership in the two subsequent decades. At a time when East Coast refiners were saddled with excess productive capacity, the kind of investment in fixed installation and equipment that would be required to take full advantage of bulk shipment did not appear particularly attractive, especially when excellent facilities for dealing mechanically with bagged sugar existed within the plant. Moreover, the Cuban sugar economy was operating under severe restraints throughout the 1930's, and trade unions were highly resistant to introduction of labor-saving methods. Bulk storage of refined crystals and even bulk delivery in railroad tank cars were nevertheless becoming fairly common on the East Coast and in the American beet regions even before World War II (*41*, p. 45).

The next major advance in ocean shipment of raws came only after 1939, and was centered on the Hawaiian Islands. Hawaiian industry had broad advantages for bulk installations comparable with those enjoyed by the Revere plant earlier. There were again proprietary links between plantations and mainland refinery, and between sugar and shipping interests. The heavy initial investment in facilities could accordingly proceed at a coordinated rate at loading and unloading points, and with some assurance of continuous use. But several special incentives had also come into play. Defense preparations put increasing strain on shipping facilities in the Pacific area throughout 1940, raised freight rates, and pushed up the price of containers. The cost of burlap bags, of which Hawaii used close to 20 million each year, was itself a matter of considerable concern, and the very supply of jute was in serious jeopardy. Moreover, labor had been less abundant in Hawaii than in other sugar-exporting regions, and defense activity imposed a further burden on manpower.

The war increased the incentive to do without burlap bags, but temporarily impeded the shift to bulk handling methods. The first bulk unit on the islands was located at Kahului on Maui, a port through which more than 150,000 tons of sugar was shipped by four plantation companies. These facilities came into use shortly after Pearl Harbor, but further construction was out of the question during the wartime shortage of manpower and equipment. The second unit did not come into operation until 1949. This one was at Hilo (Hawaii) where a disastrous tidal wave in 1946 had largely destroyed bag warehousing and loading facilities. A third, at Nawiliwili (Kaui), shipped for the first time the following year (*47*, p. 7A). Bulk storage facilities at Crockett were for the most part erected between 1947 and 1949. At this stage, about 70 per cent of the Islands' crop could be handled in bulk, roughly the portion that was refined in California.

Not until 1951 had East Coast refineries moved ahead with receiving facilities to the point where the first bulk shipment could be made to Philadelphia.[2] Plant scheduled to be completed on Oahu by March 1955 would leave only one remote plantation not equipped to ship in bulk (*48*, p. 147).

Modernized ocean shipment, Hawaiian-style, involved moving sugar in bulk at all stages from mill to refinery. The economies achieved in the loading operation alone are impressive. Man-hour productivity was reportedly raised to 20 tons, as compared with only 1.43 tons when bags were taken on board with auxiliary use of hoists and conveyors (*49*, p. 36). With equipment capable of handling 600 tons per hour, Hawaii could expect to load 9,000 tons in 15 hours. Cuba had to allow ten days for the equivalent volume of bagged sugar (*43*, p. 89). Vessels on the Hawaiian run could also discharge freight or load pineapple, at the same time that raw sugar was being fed into the lower hold. Aside from the handling at the export point, however, methods were not basically different from those developed earlier at the Revere refinery.

After 1949, the English firm of Tate and Lyle also promoted a rapid shift to bulk shipment of raw sugar destined for its London and Liverpool refineries. While only 64,000 long tons were involved in the year ending March 1950, three years later the volume had reached 848,000 tons (*43*, p. 89). Bulk sugar was shipped successfully from the British West Indies, Australia, Natal, Mauritius, Fiji, and the Dominican Republic. If the large number of originating points complicated the process somewhat, the firm also had a renewed stimulus from the jute price inflation and high Indian export tax on burlap associated with the Korean war. Tate and Lyle, matching the integrated activities of the Revere refinery and the Hawaiian industry, found it desirable to cooperate with the United Molasses Company, an associated firm with tanker experience and an obvious interest in sugar by-products, and the West Indies Sugar Company (its own subsidiary), in developing a fleet of specially designed ships to take full advantage of the economies available through bulk transport (*50*, p. 290).

The degree to which Cuba failed to keep pace with these developments is notable. Instead of saving labor costs by eliminating bags altogether, union pressure there has promoted use of a 200-pound

---

[2] Evidently the shipping companies were ready to proceed with bulk-loading facilities in Puerto Rico before the East Coast refiners were prepared to provide receiving equipment (*46*, p. 46).

bag in place of the older 325-pound bag in order to increase the amount of handling. As one step in reducing the annual expenditure of $20 million on burlap bagging, considerable attention is being given to the possibility of growing kenaf as an alternative fiber (*43*, p. 89). The concern for local industrialization in primary exporting countries is so strong that a Mission organized by the International Bank for Reconstruction and Development was recommending establishment of a bag mill for Cuba as late as 1951 (*51*, pp. 1031–36), although the feasibility of bulk shipment had been clearly demonstrated elsewhere. Lacking the assured market and the sustained volume of sales enjoyed by nationalistic sugar exporters, Cuba has little incentive to invest in bulk-handling facilities, and new economies in use of labor have no particular advantages under Cuban conditions. Nor did Cuban producers find encouragement in their experience as technological leaders twenty-five years earlier, when the trend toward ocean shipment of liquid sugar had been halted by import restrictions under American sugar legislation.

CONCLUDING REMARKS ON AGRICULTURAL PROGRESS

The achievement of sugar-cane agriculture is indeed impressive. Between 1840 and 1940, the yield of sugar in Java rose from 1.8 tons to 18 tons per hectare. Part of the increase, to be sure, came from improved milling facilities; yet a distinct jump of 30 per cent during the late 1920's came exclusively as the result of introducing POJ 2878 (*6*, pp. 34–35). Perhaps cane sugar should be considered in some sense the tropical counterpart of animal production in temperate agriculture, for it represents a degree of intensification of the rural economy not generally associated with field-crop agriculture alone. For all the achievements of cane agriculture, the gap between research results and general practice is still wide enough that we find here, rather than in improved milling technology, the future potential for reducing production costs (cf. *27*, p. 26).

The analogy with animal products is all the more apt because in both cases the division between agricultural and manufacturing enterprise tends to be blurred. Better milling performance depends on a rich cane from the fields. While factory control over planting material speeds up the rate at which new varieties are introduced, the advance of mechanical harvesting has complicated the factory's job of extracting sugar. Mill investment in transport facilities becomes a direct alternative to investment in agricultural harvesting equipment; capital in the form of irrigation in Peru and Hawaii, by permitting almost a year-round harvesting schedule, becomes a direct alternative for increasing the daily grinding capacity of sugar mills. The tractor and other agricultural equipment bring into the farming sector generally the problems of interest charges, overhead costs, and even seasonally idle capacity that have typically been associated with the milling sector.

In agriculture even more than in milling (which had more generalized engineering experience to fall back upon), advance has come as the result of enlightened and organized research. Concerted effort toward mechanization or varietal improvement requires a staff trained in a variety of scientific specialties and able to direct their attention to basic or applied research as circumstances require. If the Experiment Station of the Hawaiian Sugar Planters' Association can count on an annual budget in excess of one million dollars a year, potential accomplishments also loom large. It is estimated that the extra yield from POJ 2878 in a single year repaid all the funds invested in the Java Sugar Experiment Station over a period of 40 years (*6*, p. 35). Research organizations sponsored by members of the sugar industry may run the risk of putting too high a premium on short-term results, but the liaison they provide between research workers and growers speeds the field testing and application of laboratory results. Governmental experiment stations run contrary risks: they may become mere depositories of breeding material (cf. *10*, p. 41), or they may be attracted by problems of scientific interest that are beyond the prospect of local application (as has to some extent occurred in India). Both governmental and private arrangements can nevertheless claim important successes when continuity of activity was assured. What has been accomplished for sugar cane may suggest the results to be expected from the application of science to the study of tropical plants hitherto neglected because they served no external market.

Innovations differed strikingly in the ease with which they could be transmitted to the various cane regions. On the milling side, the itinerant sugar engineer has for many years personally helped to keep practices on a par over a wide territory. Milling machinery, the steam engine, and subsequently the tractor were the products of manufacturing enterprises in advanced industrial countries which found ready export outlets in the various cane areas of the world. With varietal advance, the story is more complicated. Here tropical regions led the way, though the European scientist played a crucial role. Java's leadership for some years was the basis of a world-wide upgrading of cane agriculture; Java-developed varieties were borrowed outright, and the island also supplied parent stock for development of cane to meet particular local conditions. The Dutch advances in sugar-cane breeding in Java have subsequently come to be more important to the sugar islands of the Caribbean, and Indian contributions more useful to growers in Natal, than to the cane producers of the originating country. Some local canes give disappointing results when tried in foreign fields; others flounder at home but flourish abroad. Many regions preferred to select from the wide range of varieties developed elsewhere, rather than bear the initial costs of experimentation. There has been outright social resistance to some of the more recent improvements. Like the old-style planter who fought against the modern mill, the Caribbean labor union resists the advance of labor-saving devices, while the enlightened Queensland farmer clings to a farming unit that available agricultural machinery may be making obsolete.

"Research" creates new economic problems, even as it solves others. Some advances may be more spurious than real. The mechanical advances in agriculture of the last 15 years do not appear to have lowered costs to the degree that the introduction of new cane varieties did in the 1920's. For all Hawaii's technological leadership, the $3.00 per ton of sugar the industry has recently been devoting to research may not have been well spent: "Our comparatively low earnings on investment have made the industry less attractive to capital than mainland industry and Hawaii has been experiencing an outflow of investment" (*28*, pp. 2–3). New risks arise. Mechanization of cane agriculture and of raw sugar transport in Hawaii has been carried so far that the Islands' economy is extremely sensitive to any interruption of activities. While certain similarities between the agricultural and the manufacturing sectors have been stressed, there is no counterpart in industry to the losses suffered by Hawaii's sugar cane from a labor strike. In a 22- to 24-month crop cycle, failure to maintain the schedule of new plantings, damage to growing cane from lack of irrigation water, and deterioration of ripe cane left unharvested mean that three annual harvests will be reduced.

There are also dangers inherent in the organization of sugar research. Industry-sponsored research agencies and private committees established to advise on the direction of research activity (cf. *32*) may make the scientific work more realistic and relevant. But industry competition for government research budgets by no means assures the best use of public research expenditures. Moreover, the extent to which the "nonprofit" research agency can become an agency of business combination, able to evade many of the restraints on monopolistic practice, seems not to be widely enough appreciated.

It is frequently argued that periods of intense competition for world markets have offered the necessary incentives for technological improvement (7, p. 577), and that "shelter from foreign competition . . . [may] perpetuate avoidable inefficiency and technological backwardness" (52, p. 88). For sugar, the truth seems to have been otherwise. The major developments in beet-sugar production during the nineteenth century came in European countries able to exclude foreign imports and with the financial strength to subsidize large-scale exports. Successes in cane breeding in Java during the 1920's capitalized on strong export markets, and the incentives disappeared once competition for markets became severe. Local varietal development in Barbados, Hawaii, Canal Point, Queensland, and South Africa; mechanization in Hawaii, Louisiana, Queensland, and the beet fields of the United States and the USSR; modernization of cane milling in India, Brazil, Mexico, Formosa, and the British West Indies; the shift to bulk transport—each of these recent trends has proceeded behind the shield of protected markets. Heavy fixed investment, the long view that research requires, and attention to "those time-consuming routine chores which are the back-bone of science" (56, pp. 4–5) seem best fostered by assured markets, and that is precisely what competitors for the world's free export trade cannot rely on.

This is not to say that sugar cane entirely substantiates the Schumpeterian view, that some relief from competition is an important contributor to economic advance because it nurtures research (8, pp. 88–90). For there is an important sense in which sugar research accomplishments have been uneconomic, the advanced industrial countries employing their scientific endowment to the disadvantage of more promising regions in the tropics. To be sure, beet factory technology developed in Europe proved adaptable to processes of sugar recovery in the cane regions, and contributed to modernization of cane milling even before 1900. But varietal research has become increasingly nationalistic; research personnel apply their talents to solving particular national problems, whether the storage of sugar beets or the mechanization of cane harvesting. Cuba, for all its natural advantages for sugar-cane production, lacks any real incentive to raise yields, while the market misfortune that befell Java served as a highly perverse reward for a major scientific contribution. Commercial policies of metropolitan countries have certainly stimulated better production methods in favored regions, but the indirect burden placed on exposed tropical peoples has been heavy. It is to a discussion of these national policies that we now turn.

CITATIONS

1 B. E. Long, "Cauca Valley: Sugar Frontier," *Sugar*, December 1952.

2 L. T. Sardone, "Sugar Cane Breeding in Australia," *World Crops* (London), April 1954.

3 F. S. Earle, *Sugar Cane and Its Culture* (New York, 1946).

4 N. King, "The Australian Sugar Industry," *World Crops*, July 1954.

5 *International Sugar Journal* (London), March 1954.

6 C. van Dillewijn, "Some Basic Principles of Cane Growing," *Sugar*, August 1953.

7 E. W. Brandes and G. B. Sartoris, "Sugarcane: Its Origin and Improvement," in U.S. Dept. Agr., *Yearbook of Agriculture, 1936*.

8 J. A. Schumpeter, *Capitalism, Socialism, and Democracy* (New York, 1950).

9 T. A. G. Hungerford, "The Australian Sugar Industry," *World Crops*, February 1954.

10 Hawaiian Sugar Planters' Association, Exp. Sta. Com., *1953 Report* (Honolulu).

11 India, Min. Food and Agr., *Annual Report of the Sugarcane Breeding Institute Coimbatore for 1950–51*.

12 A. C. Barnes, *Agriculture of the Sugar-Cane* (London, 1953).

13 F. Maxwell, *Economic Aspects of Cane Sugar Production* (London, 1927).

14 W. E. Cross, "The Sugar Cane Smut Crisis in Tucuman," *Sugar*, October 1947.

15 Noel Deerr, *The History of Sugar*, I (London, 1949).

16 Queensland, Bur. Sugar Exp. Sta., *Fifty Years of Scientific Progress* (Brisbane, 1950).

17 P. E. Turner, *Report of Research Work on Sugar-Cane Agriculture in the British West Indies 1950–51* (Trinidad, 1952).

18 N. J. King, R. W. Mungomery, C. G. Hughes, *Manual of Cane-Growing* (Sydney, 1953).

19 G. B. Sartoris, "New Kinds of Sugarcane," in U.S. Dept. Agr., *Science in Farming, The Yearbook of Agriculture 1943–1947* (1947).

20 N. L. Dutt, "The Present Cane Varietal Position in India," in Uttar Pradesh, Dept. Agr., Bur. Agr. Inf., *Agriculture and Animal Husbandry*, December-January 1950–51.

21 *Anuario Azucarero de Cuba 1951* (Havana).

22 South African Sugar Journal, *South African Sugar Year Book 1951–52* (Durban).

23 Queensland Cane Growers' Association *et al.*, *Australian Sugar Year Book 1951* (Brisbane).

24 H. C. Prinsen Geerligs and R. J. Prinsen Geerligs, *Cane Sugar Production 1912–1937* (London, 1938).

25 *World Crops*, June 1953.

26 C. S. Loh, "Sugar Cane Breeding in Taiwan," *Sugar*, July 1951.

27 C. van Dillewijn, "Research and Cost Price," *Sugar*, January 1950.

28 Hawaiian Sugar Planters' Association, *The President Reports, 1953* (Honolulu).

29 A. Riollano, "A Larger Sugar Quota for Puerto Rico?" *Sugar*, August 1951.

30 Caribbean Commission, *Monthly Information Bulletin* (Port-of-Spain), December 1953.

31 International Bank for Reconstruction and Development, *The Economic Development of British Guiana* (Baltimore, 1953).

32 U.S. Dept. Agr., Agr. Res. Admin., "Report and Recommendations of the Sugar Research and Marketing Advisory Committee Meeting of February 23–25, 1955," April 29, 1955 (mimeo.).

33 F. V. Deomano, "PSA 69 Cane Variety," *International Sugar Journal*, May 1953.

34 *International Sugar Journal*, May 1951.

35 "Sugar Cane Varieties in Hawaii, Old and New," *Sugar*, July 1945.

36 Caribbean Commission, *The Sugar Industry of the Caribbean* (Crop Inquiry Series 6, Washington, D.C., 1947).

37 "Sugar Empire in the Everglades," *Sugar*, September 1953.

38 *South African Sugar Year Book 1950–51*.

39 Pieter Honig, "Developments in Cane Sugar Production Since 1938," *Sugar*, September 1950.

40 E. C. Freeland, "The Sugar Industry of Peru," *Sugar Journal* (New Orleans), April 1953.

41 "Bulk Handling to Date," *Sugar*, September 1952.

42 U.S. Dept. Agr., Prod. and Mkt. Admin., *Marketing Liquid Sugar* (by F. J. Poats, Marketing Research Rept. 52, 1953).

43 *International Sugar Journal*, April 1954, quoting *Mendoza's Sugar Review* (Havana), 1954, No. 825.

44 A. Riollano, "Puerto Rico Falls Short of Its Quota," *Sugar*, August 1953.

45 *Sugar*, March 1943.

46 *Sugar*, February 1949.

47 Hawaiian Sugar Planters' Association, *Sugar Manual* (Honolulu, 1954).

48 *International Sugar Journal*, May 1954.

49 *Sugar*, June 1950.

50 *Weekly Statistical Sugar Trade Journal*, Aug. 2, 1951.

51 International Bank for Reconstruction and Development, *Report on Cuba* (Washington, D.C., 1951).

52 Jacob Viner, *International Economics* (Glencoe, 1951).

53 J. W. Vandercook, *King Cane: The Story of Sugar in Hawaii* (New York, 1939).

54 Hawaiian Sugar Planters' Association, *Sugar in Hawaii* (Honolulu, 1949).

55 R. A. Duncan, "Status of Cane Cutting in Hawaii," as abstracted in *Sugar*, November 1953.

56 Edgar Anderson, *Plants, Man and Life* (Boston, 1952).

57 *Sugar*, August 1952.

58 E. F. Ferguson, "Cane Cleaning Plant in Hawaii," *International Sugar Journal*, May 1953.

59 Hawaiian Sugar Planters' Association, Experiment Station, *1953 Factory Report* (Special Release 96, Honolulu, 1954).

60 *Sugar*, November 1950.

61 *Sugar*, April 1952.

62 *Sugar*, May 1952.

63 Hawaiian Sugar Planters' Association, *Report of the Engineering Committee 1941*, in *Proceedings of the Sixty-First Annual Meeting 1941* (Honolulu, 1942).

64 A. T. Hassack *et al.*, "Mechanical Planting of Sugar Cane in Hawaii," as abstracted in *Sugar*, July 1950.

65 *Sugar*, April 1951.

66 R. A. Duncan, "Development of a Sugar Cane Harvester," *International Sugar Journal*, August 1953.

67 A. A. Lloyd, *A Survey of Sugar Cane Production Conditions and Methods in the South African Sugar Industry* (Durban, 1949).

68 *International Sugar Journal*, June 1954.

69 E. A. Maier, *A Story of Sugar Cane Machinery* (New Orleans, La., 1952).

70 J. N. Efferson, "Mechanization in the Louisiana Sugar Cane Industry," *Sugar*, December 1947.

71 U.S. Dept. Agr., *Crops in Peace and War: The Yearbook of Agriculture 1950–1951* (1951).

72 U.S. Dept. Agr., Commod. Stab. Serv., Sugar Div., *Agricultural, Manufacturing and Income Statistics for the Domestic Sugar Areas: Volume II of Sugar Statistics* (Statistical Bulletin No. 150, 1954).

73 Gr. Brit., Col. Off., *Digest of Colonial Statistics*, September-October 1954.

74 U.S. Mutual Sec. Admin., Mission to China, *A Survey of the Taiwan Sugar Industry*, Part V (Consultant Rept., No. 1-A, Taipei, 1952).

75 "The Sugar Cane Industry in Fiji," *Tropical Agriculture* (London), October 1954.

76 Mauritius, Cen. Stat. Off., *Year Book of Statistics, 1950*.

77 Gr. Brit., Col. Off., *The Sugar Industry of Fiji*, by C. Y. Shephard, Col. No. 188, 1945.

78 C. van Dillewijn, "Sugar Cane in Venezuela," *Sugar*, June 1954.

79 "Brazilian Sugar Progressing," *Sugar*, December 1951.

PART III
SUGAR POLICY IN PEACE AND WAR

CHAPTER 7

THE SUGAR SUPPLY OF THE UNITED STATES

THE TARIFF SYSTEM

For almost 150 years after 1789, the main instrument of American sugar policy was the tariff (cf. *1; 2,* pp. 52–55). An important contributor to customs revenues, it also provided from the beginning an incidental protection to the Louisiana cane-sugar industry on the mainland, and generally also discriminated in favor of raw-sugar imports in the interests of metropolitan refiners. Only for a four-year period in the 1890's, when domestic sugar received a direct subsidy of two cents per pound but imports were admitted duty-free, did the tariff temporarily lose its pre-eminence.

So simple an instrument was nevertheless consistent with striking changes in the relative fortunes of different supply areas (Table 7). The Louisiana industry was full-grown by 1900, and indeed was

TABLE 7.—ENTRIES AND MARKETINGS OF SUGAR IN CONTINENTAL
UNITED STATES, SELECTED YEARS, 1900–34*

(Million short tons, raw value)

Year[a]	Total	Mainland Beet	Mainland Cane	Hawaii	Puerto Rico	Philip-pines	Cuba	Other foreign
1900.....	2.41	.09	.31	.25	.04	.03	.35	1.34
1905.....	3.11	.34	.39	.42	.14	.04	1.03	.77
1910.....	3.79	.55	.36	.56	.29	.09	1.76	.21
1915.....	4.72	.94	.14	.64	.29	.16	2.39	.16
1920.....	6.34	1.17	.18	.55	.41	.15	2.88	.99
1925.....	6.93	.98	.14	.76	.60	.49	3.92	.03
1929.....	7.59	1.09	.22	.88	.51	.71	4.15	.03
1930.....	6.68	1.29	.22	.87	.81	.79	2.65	.05
1931.....	6.73	1.34	.21	1.00	.80	.87	2.48	.03
1932.....	6.30	1.32	.16	1.05	.94	1.03	1.79	.01
1933.....	6.33	1.37	.32	.99	.79	1.25	1.57	.04
1934.....	6.57	1.56	.27	.95	.81	1.09	1.87	.03

* U.S. Dept. Agr., Prod. and Mkt. Admin., Sugar Branch, *Sugar Statistics,* I (1953), pp. 204–06.
[a] Fiscal year 1900–18; calendar year 1919–34; mainland production on crop-year basis 1900–30.

shortly to enter a period of decline as the result of plant disease that reduced output to a low point of less than 50,000 tons in 1926. The mainland beet industry, which had received a considerable stimulus during the subsidy period, expanded from negligible size in 1900 until it had become a major factor in supply by 1915. The Hawaiian Islands offshore, which were admitted to duty-free status under the terms of the Reciprocity Treaty of 1876, also developed their sugar production most spectacularly in the period before World War I.

Benefits of the United States market were extended to additional territories following the Spanish-American War. Puerto Rico had earned duty-free entry by 1901, and the Philippines by 1914, while a 20 per cent tariff preference became effective for Cuba in 1903. American interests, including refiners anxious to assure themselves an offshore supply of raw sugar, invested large sums to develop output in the Caribbean, particularly in Cuba. By the outbreak of World War I, Cuba had entirely displaced Java, European beet producers, and other foreign areas from the United States market. Except for the year 1920, when skyrocketing prices drew sugar from all parts of the world, the subsequent position of other foreign producers (except the Philippines) remained trivial. Indeed, in 1922, two-thirds of gross American supplies came from Cuba, some of it destined, however, for re-export as refined sugar.

The Emergency Tariff of 1921 and the Fordney-McCumber Tariff of 1922, the aftermath of the postwar collapse of commodity prices and the growing concern for the welfare of agricultural producers, raised the effective duty more than 75 per cent. The culmination of this process came with the Smoot-Hawley Tariff, which brought a specific rate of 2 cents per pound on Cuban raws and represented an *ad valorem* equivalent of 400 per cent at the low prices reached in 1932. Moreover, the Great Depression halted the rapid expansion in the volume of sugar absorbed by the American market, which had climbed from less than 4 million short tons in 1912 to over 7.5 million in 1929. Cuba could sell 4.1 million tons to the United States in 1929, but only 1.6 million four years later. American tariff policy contributed to social unrest and political revolt in Cuba, as well as to a sharp decline in Cuban imports of American farm products.

Nor were the results of that policy satisfactory even to domestic sugar growers. In response to the successive tariff increases, to be sure, domestic beet-sugar production expanded about one-half million tons between 1920 and 1934. But the major benefits of protection were accruing to territories offshore. In Hawaii and Puerto Rico combined, the full rise exceeded one million tons, while production in the Philippines rose almost 1,200,000 tons (*2*, pp. 32–33). At prevailing low levels of consumption, large offshore supplies available for shipment to the mainland in 1934 weighed heavily on the domestic market and jeopardized the competitive position of mainland producers. To the extent that Cuban sugar was not salable in the United States, exports were pushed more aggressively elsewhere, with disastrous effects on the "world" market price. So far did the import price ex-duty fall that domestic prices remained at record lows despite the high tariff. Orthodox sugar protectionism was proving incapable of increasing the returns or improving the competitive position of the mainland industry.

INTRODUCTION OF QUOTAS

To meet the situation, a new system of sugar control was initiated by the Jones-Costigan Act, May 9, 1934, and was formalized without major modification by the Sugar Act of 1937. Total consumption requirements of continental United States are estimated by the Secretary of Agriculture, in rough accordance with certain statutory criteria, and total marketings limited to that figure. Annual marketing quotas, also following statutory prescriptions, are in effect for the various producing regions. Accordingly import quotas, rather than the tariff, become the effective means of protecting the domestic industry, so that the prices in New York and in the "world" market largely go their separate ways. When conditions warrant, quotas are supplemented, without grower referendum, by marketing allotments to individual domestic processors or acreage allotments to growers. Specific quotas on direct-consumption sugar limit the volume that can be imported in refined form. Finally, domestic growers receive "conditional payments," a direct subsidy financed out of an excise tax collected from processors on all sugar marketed. These are recurrent annual expenditures, not emergency disbursements. Though quotas were inoperative during World War II, these basic features have continued to the present without major change.

Of necessity, the quota system bore with unequal weight upon the various supply areas. Mainland beet and (by 1937) cane producers enjoyed quotas in excess of marketings in any previous year, and their further expansion was countenanced. Hawaii did almost as well, with rate of production held barely 10 per cent below the peaks reached immediately before quotas were imposed. Of the strictly domestic areas, Puerto Rico, whose expansion had been the most pronounced, was the most severely limited. Potential output from cane already planted for 1934/35 harvest was an estimated 1,148,000 short tons, but sugar production in fact was cut to 772,000 tons. Some diversion to high-test molasses and syrup eased the adjustment, while restrictions in succeeding years were more moderate. Even this degree of initial restriction was exceeded in the Philippines, where the 1934/35 crop was cut back from a potential 1,578,000 tons to a mere 696,000 tons raw sugar. Only 524,000 tons were for shipment to the United States, as contrasted with 1,241,000 tons in the calendar year 1933 (*3*, pp. 188–89). For the longer haul, however, the Philippine Independence Act of 1934 guaranteed a duty-free outlet in the United States, during an interim period, for 800,000 long tons of raw sugar and 50,000 tons of refined.

Notwithstanding this differential treatment, the original conception and administration of the quota system must be considered enlightened, particularly if one remembers the depression environment of that day. The Presidential Message to Congress proposing the new approach, while endorsing a share of the market for mainland producers, spoke also of providing "against further expansion of this necessarily expensive industry." Secretary of Agriculture Wallace rejected a private agreement on quotas that was overly generous to producers' groups at the expense of the consumer (*4*, pp. 128–29), and he subsequently resisted efforts to have aggregate consumption estimates held too low. To further shield consumers against a price rise, the new processing tax was offset by an equivalent reduction in the effective tariff. Moreover, direct subsidies have one clear fiscal advantage over restrictive tariffs, since they require recurrent Congressional appropriations and are therefore the more likely to attract public attention and periodic review. In fact, annual expenditures amounting to some $60 million have apparently been considered too trivial to invoke much serious concern. If growers are to continue to receive a premium of 20 to 25 per cent above the going price of sugar crops, at least the consumer is better served when that margin is not pyramided into the final cost of refined sugar or of sugar-containing products.

The distribution of quota gains within the industry has not been left entirely to chance. Growers have been favored over processors, relatively small growers over relatively large, and there has been a real concern for the welfare of field labor. The excise-tax-and-subsidy device is itself a carryover from the early days of the Agricultural Adjustment Administration, when the effort was made to increase the farmer's share of the consumer dollar. Additional features introduced in the interwar period were a reduced subsidy rate for larger farm enterprises, official determination of field wages, governmental fixing of the price that grower-processors must pay for purchased sugar crops, and restrictions against the use of child labor in the fields. From 1937 to 1941, the subsidy due the Philippine industry was paid to the Commonwealth Treasury instead of to private parties, to be used as a general developmental fund rather than to promote unmarketable sugar.

The interests of Cuba, as a low-cost foreign supplier, were given overt recognition. The absolute quotas in the 1930's represented a 20 per cent recovery above Smoot-Hawley days, though far below the level of the 1920's. Cuba was also the chief beneficiary of the liberalized tariff; once the effective rate was cut from 2 cents per pound to 0.9 cent, her per-unit receipts automatically doubled. The fact that the quota system has operated smoothly, avoiding in peacetime both temporary gluts and temporary shortages, and without that emergence of government stocks so characteristic of agricultural price-support programs, has in turn owed much to the role of Cuba as a buffer. Excess capacity in Cuba could be called on to meet any sudden increase in American requirements or major quota deficits from other areas of supply. When controls had to be tightened, as a matter of practical politics the responsibility for contraction could be placed more easily on her shoulders.

Sugar legislation involved a number of incidental benefits to the sugar-processing industry, as distinct from sugar-crop farmers. As perishable agricultural products with a peculiarly local market, beets and cane were traded on no commodity markets and did not lend themselves to price support by government-stockholding methods. The quota system held its umbrella over the wholesale price of sugar rather than of sugar crops alone, so that beet-sugar factories and cane millers were quite as sheltered as farmers. As commercial growers themselves, a number of important processors have also directly received—or, as landlords, shared—conditional benefits. Moreover they have enjoyed significant advantages on the cost side. In the absence of direct subsidies to independent growers, factories would have had fewer purchased beets (and mills somewhat less cane) to process at going prices for the refined product. In 1936, the one year since 1934 in which conditional payments were temporarily not forthcoming, processors paid growers a higher price per ton of sugar beets than at any time in the decade 1931–40 (*16*, pp. 91, 95).

Indeed, successive sugar acts were much in the spirit of the industry codes under the National Recovery Administration. The sugar program was an instrument that restricted total supply and divided up the total market but left ownership and management in private hands and did not interfere directly with private contracts between sugar buyers and sugar sellers. It accordingly displayed the distinctive characteristics of a cartel arrangement for a farming and processing industry, in which the scope for freedom of enterprise was narrowly confined. Just as the excise-tax-and-subsidy device perpetuated, in thin disguise, the processing tax that was declared unconstitutional when applied overtly (39), the new sugar control represented Congressional endorsement of a set of industrial practices similar to those denounced by a second Supreme Court decision (15).

Even the seaboard refiners of imported raw sugar, who had been successfully prosecuted on antitrust grounds by the Department of Justice in the very year that the quota system began (32; 33), enjoyed two particular advantages under the sugar-control system. In view of the relatively small margin between raw and refined sugar prices, the risk of inventory loss resulting from declines in the price of raw sugar is of more serious financial concern than the absolute level of price. The quota system has consistently meant a steadier price for raws, as compared either with pre-quota years or with world-market behavior. Price stability as such is less important to beet factories or cane mills, since prices paid for the raw crop and even wage rates are frequently geared to returns from sale of the finished product. Furthermore, the volume of refiners' operations was protected. Expanded production of beet sugar, which required no further processing, had been pre-empting a portion of the refiners' domestic market, an effect concealed in the early 1920's by the heavy volume of re-export business. By 1927, however, the refining industry was estimated to be operating at only 60 per cent of capacity (34, p. 113). Moreover, while Cuba's total exports of raw sugar to the United States fell from 4.2 million short tons in 1926 to 1.1 in 1933, her shipments of refined actually increased from 69,000 tons to 470,000 tons (2, p. 114). The quota system placed a ceiling on competitive beet-sugar supplies and, by means of the specific quota limitations against imports of "direct-consumption" sugar, cut back the shipment of sugar refined offshore.

Under the shield of successive sugar acts, the leading refining companies took steps to correct some of their excess capacity. Their last new plant was built in 1926, and at least four major refineries were dismantled after 1934, some immediately following purchase. But the legislation did not prove an entirely appropriate instrument for reducing capacity. In the process, returns were evidently held sufficiently high to attract into the field several independents, especially innovators supplying liquid sugar to industrial users. The economic basis for very large-scale refining operations in consuming centers appears to have weakened in recent years. Even after the war, several projects for replacing obsolete refining facilities had repeatedly to be postponed because of high building costs (cf. *20,* p. 11), and seaboard refiners have found it difficult to compete for labor in periods of rising wage rates.

The direct-consumption quotas remain perhaps the least defensible feature of the entire sugar program. They open the United States to the serious charge—one not without its own illustrious counterpart in American colonial history—that a metropolitan country is denying to its raw-material supplying areas a kind of industrial processing entirely appropriate to their resources. Not only shipments from a foreign country, Cuba, are handicapped by this device, but also those from Puerto Rico. So convinced were Puerto Rican refiners of their ability to provide refined sugar on a competitive basis that they challenged the constitutionality of the limitation before the United States Supreme Court. Their effort failed. The Court ruled that it was not "a tribunal for relief from the crudities and inequities of complicated experimental economic legislation" (*36,* p. 618), but the experiment is now almost a quarter-century old.

IMPACT OF WAR IN EUROPE

Though Cuba indeed performed the mechanical function of a buffer, the peculiarities of sugar supply make the quota system a more complex mechanism than one might reasonably expect. The extent and ramifications of administrative discretion are well illustrated by the actions taken shortly before and immediately after the invasion of Poland.

For several reasons, domestic sugar production had been unusually heavy in 1938. With agricultural prospects generally bleak that spring, there was strong inducement to plant sugar beets, a crop assured a steady market and a firm price under government policy. The same considerations, augmented by a larger statutory quota under the Sugar Act of 1937 and the further spread of high-yield varieties, promoted expansion of sugar-cane acreage in Louisiana and Florida. Puerto Rico, prevented from grinding at full capacity in earlier years, took advantage of Section 301 (e) of the Act, which prohibited production restrictions on the first crop harvested from cane planted prior to enactment of the new law. As a result, mainland beet and cane production exceeded, and Puerto Rican output approached, the highest levels ever previously attained. In each case, production was greater than marketing quotas, with a considerable increase in year-end carryover resulting.

A series of steps was taken by the Sugar Division, United States Department of Agriculture, to correct the situation in 1939. To reduce the protective margin that had promoted higher domestic production, the total quota figure was raised moderately. Marketing allotments to processors, which had been in force for Puerto Rico continuously since 1934, were now extended to Florida, Louisiana, and the beet companies. Production was also restricted directly. Output in Puerto Rico was cut back some 20 per cent. An acreage reduction of 25 per cent was planned for mainland cane, but the adjustment was to take place over a two-year period, because of the grower investment in cane ratoons. Louisiana could not carry over unharvested cane in the field until the next season, as was practiced in Puerto Rico, so that growers could effect a 10 per cent reduction in acreage only by actually plowing under considerable stubble. While acreage allotments were assigned to beet areas on a generous scale sufficient to allow plantings equal to those in the previous record season, drought and flood in the growing regions somewhat reduced yields.

Together, these production controls were not adequate to avoid the prospect of heavy over-quota stocks on January 1, 1940, except in Hawaii, which had typically failed to meet its generous quota. The burden of storage was by no means the entire cost at issue. Under Section 301 (c) of the Sugar Act, stocks in excess of normal carryover would have involved an offsetting reduction in permissible 1940 acreages, with the likelihood therefore of heavier restrictions than in 1939. The Department of Agriculture was frankly and understandably "subjected to continuous pressure, direct and indirect, to utilize whatever powers . . . interested parties believe it to possess to enhance the value of their stocks of sugar" (5, p. 108).

An avenue of escape from these pressures was afforded when the outbreak of war in Europe brought a quick rise in sugar prices. As an emergency measure, President Roosevelt on September 11, 1939 suspended the statutory quotas on sugar marketings. The effect of this action was to make some 750,000 additional tons of sugar available to domestic purchasers before the end of the year, a full 10 per cent of the total quota previously in force. The domestic price of raw sugar consequently eased and by October 31 was actually below the prewar level. So effective was the measure that on December 26 it was felt necessary to reimpose the quota system in order once more to support the domestic price.

Of the increased marketings permitted by suspending quotas, by far the major part came from Puerto Rican cane (319,000 tons), and mainland beet sugar (243,000 tons) and cane (162,000 tons). Only a small portion came from Hawaii (18,000 tons), none at all from the Philippines, which had little of its current crop remaining for sale, and none from Cuba. Beet processors received little immediate benefit from the high September prices, since much of their old-crop stock was covered under long-term contracts negotiated earlier (*3*, pp. 228 ff.). Not only were domestic interests able to dispose of potential carryover stocks during the 1939 quota year; beet producers and Louisiana cane producers also managed to sell a portion of their 1939/40 crop, which was being harvested during the period of quota suspension, sales of which would otherwise have been charged against their 1940 quotas, and some of which was actually delivered after January 1, 1940 (*7*, p. 314). Despite Presidential advice to the contrary, Louisiana growers gambled that the 1940 harvest would also be unrestricted and planted additional acreage in the fall of 1939 for harvest a year later.

This episode was officially alleged to demonstrate "the value to the national welfare of a reserve supply of sugar built up in accordance with the ever-normal-granary principle" (*6*, p. 130). In fact it represented a dissipation of reserve stocks actually available on the mainland and, through perverse effects on the Cuban crop, limited the supply of sugar available to the United States after Pearl Harbor.

Although negotiations had been proceeding for some months to cut the Cuban tariff to 0.75 cent, the 0.90-cent rate itself was conditional upon the existence of quotas. With the suspension of quotas, the tariff automatically reverted to 1.50 cents. Cuban sellers at the beginning of September had hoped to unload on a rising market; they now held back in hope of restoration of the former tariff. Nor did sales to other countries come up to expectations. The United Kingdom had immediately put sugar purchasing under the control of the Ministry of Food (see chapter 8) and was in the process of harvesting the domestic beet crop. Despite slightly heavier sales on the European Continent, Cuba actually sold less sugar outside the United

States in 1939 than it had in 1938. Even the reimposition of quotas on December 26, 1939 was something of a mixed blessing. To be sure, the lower tariff was restored, and some 200,000 tons of sugar, previously held in bond, could now be passed through customs in time to qualify as 1939 entries instead of being charged against Cuba's 1940 quota. But the windfall from the tariff reduction accrued more particularly to current holders, especially seaboard refiners, than to Cuban producers.

Not only was Cuban sugar displaced by quota-free domestic supplies in the last quarter of 1939, but there were automatic repercussions on the administration of quotas in 1940 and 1941. Since sugar stocks on January 1, 1940 were known to be large, including those that had already passed from processors and refiners to households, to wholesalers, and to sugar-using industries, the aggregate quota set for 1940 was relatively low. It was soon reduced further, to levels at which statutory minima for all domestic supply areas and the Philippines became effective. Further restrictions were almost exclusively at Cuba's expense. Transfer of surplus stocks from beet processors to other hands in September-December 1939 spared domestic beet growers the imposition of acreage allotments in 1940. Production reached a new record figure, some 300,000 tons in excess of beet sugar's marketing quota. Puerto Rico, despite new pressures to limit its access to the mainland market, produced a higher crop of almost one million tons. While the two-year crop adjustment program for the mainland cane regions, begun in 1939, was carried to completion, special legislative dispensation was given producers who had increased their plantings during the period of quota suspension. However, hurricane, flood, and freeze combined to reduce the Louisiana crop to little more than half the previous season's.

Cuba's allowable entries in 1940 were accordingly below her actual shipments at any time in the preceding thirty years with the single exception of 1933. The reductions came at a particularly unfortunate moment. While Cuban sales to the European Continent held up well in the early months of 1940, the invasion of the Low Countries and France in May quickly cut off that market. The United Kingdom, by introducing rationing and arranging to purchase the exportable surpluses of British Dominions and colonies, was little dependent on dollar sources in the Caribbean. Whatever sales could be made to the United Kingdom were at prices barely above the lowest level reached in the Great Depression. Tighter United States quotas were one means used to insulate American producers against these same external factors. The degree of protection is reflected by price differentials: late in the year, United States companies were offering refined cane sugar (based on imported raws) for export at 1.60 cents per pound, at the same time that the wholesale domestic price was 4.35 cents (*8*, p. 444).

Simultaneous difficulties both in world and United States markets brought serious unrest in Cuba. The direct counterpart of "ever-normal granary" operations in 1939 was economic, social, and political instability on that island in 1940. A real threat was posed to the military security of the United States in the Caribbean area (*6*, p. 132), not to mention its future sugar supply, at a time when warfare was already on a large scale and spreading. Yet throughout the year measures were before Congress that would have reduced Cuba's quota even further. One proposed a tax on high-grade molasses, to which product Cuba was turning substantial amounts of its sugar cane, but which competed with grain as a base for industrial alcohol. The Administration succeeded, however, in getting a one-year extension of the existing Sugar Act, largely in its old form.

Substantial domestic harvests of sugar crops in 1940, superimposed upon the large sugar carryover in all hands on January 1, 1940, meant a high year-end carryover in spite of the reduction of imports from Cuba. The initial consumption determination for 1941 was accordingly set even lower than it had been for 1940. This time a sharp reduction in beet acreage was imposed, the first since the control system had been initiated in 1934, to a figure some 15 per cent below the record level of the immediately preceding three years. But the cumulative effects of the past sixteen months continued to weigh most heavily on Cuba. Burdened with stocks which she was prevented from selling to the United States and was little able to market elsewhere during 1940, with a low initial quota in the United States for 1941 and little prospect of improved world-market conditions, Cuba saw no alternative to severe crop restriction. Less sugar was produced in 1941 than at any time since 1912, with the exception again of depression lows in 1932 and 1933.

Few could have anticipated the precise sugar problems that were shortly to come. Yet it seems undeniable that an avoidable strategic mistake was made by giving so low a priority to Cuba's place in the American supply system. The mistake was continued beyond the time when the danger signals had become conspicuous. The Japanese were diverting vessels from commercial routes in the Far East early in 1941, placing Philippine shipments in jeopardy. By virtue of the

Lend-Lease Act of March 11, 1941, the United Kingdom could again consider purchases of dollar sugar from the Caribbean and begin shortening lines of sugar supply that extended all the way to Australia. These developments, both in the Far East and in the Caribbean, occurred before Cuba had completed its grinding season. Yet the general fear of agricultural surpluses was such that Cuba was given little incentive to relax its severe restrictions on output. The United States accordingly lost the opportunity of building up reserve stocks of Cuban sugar on the mainland while peacetime shipping conditions still prevailed. The indirect and almost automatic effects of quota administration in 1939 contributed to subsequent sugar shortages after the United States entered the war.

WARTIME SUGAR MANAGEMENT: SUPPLIES AND PRICES

Changing circumstances of wartime supply were hardly reflected at all in the domestic price of sugar. Effective August 14, 1941, four months before Pearl Harbor, the Office of Price Administration and Civilian Supply imposed a price ceiling of 3.50 cents per pound on Cuban raw sugar of 96 degrees polarization duty-paid at New York (equivalent to 2.60 cents c. and f.). Sugar was accordingly not only the first commodity subject to price control, but the ceiling involved a rollback of 0.30 cent below the prices formerly prevailing. The ceiling was subsequently raised to 3.74 cents on January 5, 1942. Except for an adjustment to 3.75 cents effective September 3, 1944, it remained rigid at that level for more than four years. Maximum prices set at succeeding levels of processing and distribution at a somewhat later date were held almost equally inflexible. Indeed a constant price for sugar was a key feature of wartime sugar policy.

Success of the stabilization policy, all the more notable because this standardized commodity was not subject to serious deterioration in quality, depended on several extremely favorable factors. Heavy reliance on offshore supplies weakened the political pressure of domestic producers for higher prices, at the same time that it put the public authorities in command of raw sugar at a strategic point in the distribution system. Subsequent control over primary supply could be administered through regulation of a limited number of refining companies, mainland cane mills, and beet-sugar enterprises, with little risk of evasion at that stage. The considerable degree of governmental intervention in peacetime provided personnel, agencies, relationships, and practices that had served to protect the industry against price decline before the war and could rather readily be converted into instrumentalities for preventing price movements in the reverse direction. Finally, members of various elements in the industry, particularly the mainland refiners, well remembered the disastrous effects of abandoning sugar controls too hastily after World War I, and were anxious to avoid repeating past mistakes.

Prices aside, certain general measures served to stimulate domestic sugar production. The mere removal of crop restrictions paved the way for successive million-ton crops in Puerto Rico in 1942 and 1943, the former a record high. The mainland cane area had also been held back by peacetime controls, and advantage was taken of their relaxation, especially in Louisiana. In beet regions, alternative crops did not look particularly attractive in the spring of 1942, so that beet plantings were heavy and the crop a large one. Conditional payments, originally intended as compensation to growers for conforming to output restrictions, now became an outright production subsidy to encourage maximum output. The base rate itself was raised one-third by the extension of the Sugar Act in December 1941. Moreover, while Congress never classified beets or cane as "war crops," and accordingly denied them the incidental postwar guarantees attached to that category, their long-term prospects were not entirely neglected. The 1941 extension of the Sugar Act for a three-year period gave reasonable assurances of a continuing sugar program, and a further two-year extension was voted in 1944.

The kingpin of the stable price policy was the purchase of successive Cuban crops by an official agency of the United States government. Government purchase of Cuban sugar had demonstrated its usefulness as a price stabilizer during World War I, and the sugar inflation of 1919–20 could have been largely avoided had the device not been given up so early. Under World War II conditions, the crop-purchase mechanism also facilitated systematic loading in Cuba, economy in use of available shipping, and the sharing of raws among refiners on an equitable basis (45). Shipping advantages alone were substantial, inasmuch as raw sugar was the largest ocean-borne cargo entering the United States during the war. The Commodity Credit Corporation (CCC) also handled some Puerto Rico sugars, but in view of the close proprietary links between Hawaiian plantations and refineries in the San Francisco Bay area, the West Coast trade was left entirely in private hands. Purchases of Dominican Republic and Haiti crops from 1942 to 1944 were for United Kingdom account rather than for the American market.

Except for assurances that their position in the postwar American market would be at least as favorable as under the Sugar Act of 1937 (*41*, p. 16), Cuban producers had no incentive beyond a rigid rate of 2.65 cents per pound throughout the war years 1942–44. A fixed rate fully one cent below the domestic ceiling on raw sugar might appear to have been an unwarranted exercise of bulk purchasing power superimposed upon control of wartime ocean shipping. Actually, it was initially an extremely attractive price from the Cuban standpoint. Available milling capacity, far in excess of recent production levels, could enjoy lower overhead costs in processing the large 1942 and 1944 crops. All exports now received the United States price, whereas in peacetime the portion sold to the world market went at a considerable price discount. The island's economy was spared many special costs and losses associated with wartime shipping difficulties: the price was specified f.o.b., and it became customary to make payment at 90 per cent of the full rate on sugar remaining unshipped for lack of available tonnage. Besides, the contract price paid on those portions of the crop delivered in the form of invert brought practically the same return as sugar itself, as compared to the peacetime practice of processing surplus cane into invert for sale at roughly one-half the world-market rate for sugar. Under contract, by-product molasses brought 13.6 cents per gallon. While lower than the ceiling on the mainland product, this was also considerably better than the customary price of about four cents. At a rigid price of 2.65 cents for sugar, Cuba's shipments of sugar and by-products had a higher export value in 1943 and in 1944 than at any time since the middle 1920's (*40*, p. 104).

In a territory so large as the continental United States, intermittent regional stringency of supplies posed a serious problem with which the wartime control machinery had specifically to deal. Shortages were most intractable in the metropolitan areas of the northeastern states, which had depended almost exclusively on cane refineries operating in Baltimore, Philadelphia, New York, and Boston on the basis of imported raws. These were the centers especially inconvenienced when rampant submarine activity and diversion of ocean-going vessels to military cargo created a transportation bottleneck in 1942 and 1943. Only by taking advantage of the short distance between Havana and Florida was a complete breakdown of distribution prevented. A "shuttle service," employing vessels as small as 500 to 1,000 tons in capacity, hauled sugar to the mainland, and raws were subsequently railed to northeastern refiners (and, to a lesser degree, to New Orleans). The Port of New York, the customary destination for almost one-third of all ocean-borne sugar arriving in the United States (*41*, p. 15), received less of Cuba's 1942 crop than either Port Everglades or Tampa, neither of which had any significance as sugar ports before 1942 or after 1944 (*38*, pp. 105–11, 281). While the shuttle service overcame the basic difficulty, cane refiners continued to be plagued with irregular raw-sugar arrivals and interrupted factory operations. The consequence was not merely higher costs but a loss of labor supply attracted by more continuous employment in other industries. Inasmuch as the Atlantic seaboard was also the main source of supply for overseas sugar shipments to the Armed Forces or under Lend-Lease, the sugar deficit confronting consumers from New England to Philadelphia was perpetuated even after ocean transportation had ceased to be a problem. Despite all difficulties, however, refiners melted more sugar in 1944 than ever before (*53*, p. 910).

Several devices were relied on for rapid adjustment to the changing pattern of local deficits. Early in 1942, with the western states much better supplied than the eastern, beet-sugar processors were ordered to set aside 15 per cent of their stocks and the same portion of their subsequent monthly production for shipment to areas to be officially designated (*42*). These sugars became a vital if unusual source of sugar for the New England and Middle Atlantic States throughout 1942, although the use of scarce freight cars for the long haul was not looked upon with favor by the Office of Defense Transportation. The set-aside order remained in effect until October 26, 1943, by which date beet-sugar surpluses no longer existed and shortages were beginning to occur in certain beet areas of the Middle West. By 1944, it had even become necessary to forbid any transshipment of beet sugar east of Chicago.

Even more important was the Zoning Order introduced in March 1942 (*43*). Primary distributors, i.e., refiners, mainland cane mills, beet factories, and sugar importers, were permitted to make deliveries within specified territories only. The zones reserved for Atlantic Coast refiners were extremely small when submarine activity was at its height, with correspondingly larger territories designated for beet and western cane sugars. Each major change in circumstances—the subsequent decline in beet-sugar production, restoration of ocean shipments to the Atlantic seaboard, heavier receipts at New Orleans, and diversion of New York and Philadelphia refineries to production for overseas shipment—brought a redrafting of the individual zones. Zoning had the clear advantage of husbanding scarce local supplies, promoting movement of sugar from areas of relative surplus to deficit regions, and economizing on transport facilities by eliminating cross-hauling. Under wartime conditions the fact that competitive forces in any particular region were damped seemed not particularly important. Though the system may have inhibited price concessions at times when local conditions may temporarily have warranted them, official price ceilings provided a prime restraint against exercise of monopoly power. The possibility of concentrating operations at a limited number of refineries was explored, but no such program was ever put into operation.

A third device had also significant functions to perform in equalizing the regional supply situation. Throughout the war, the Sugar Act quotas against imports of direct-consumption sugar remained in effect. Even at their restricted level, over 400,000 tons of refined sugar was available for disposal on the mainland. Under the zoning system, these sugars were mainly employed, at the discretion of the Office of Price Administration, for bridging local gaps in supplies. The fact that they originated offshore permitted shifts to changing destinations to take place more readily than was the case with refined sugar processed from imported raws. Indeed, it can be argued that relaxation of the direct-consumption quotas would have been warranted by the wartime emergency and economy in transportation. Even after mainland refineries had become seriously overburdened, raws continued to be imported for refining and re-export, and only at a late date were limited purchases made of Cuban refined for direct shipment overseas.

That the over-all supply situation had nevertheless seriously deteriorated by 1945 is clearly indicated in Table 8 (p. 177). Setbacks came in different regions at different dates, and price was not necessarily the primary consideration. Philippine sugar was completely lost by conquest, and the Japanese were soon stripping island sugar mills of vital equipment and converting cane lands to rice and cotton. The Hawaiian crop suffered throughout the war from a combination of adversities: diversion of cane acreage to military use, shortages of fertilizer and labor, the wearing out of heavy field equipment, and the inconvenience of various security measures. But as a favorable factor, vessels returning from delivery of war material in the Pacific Theater permitted regular sugar shipments of some 800,000 tons a year throughout the war period. Production in Puerto Rico fell off more than one-third between 1942 and 1944 through shortage of fertilizers (upon which the island's yields heavily depended) as well as by compulsory diversion of some cane acreage to local food crops in lieu of customary imports (*3*, p. 264). If Florida and Louisiana had to contend with labor scarcity and interrupted processing schedules, at least their production record barely suffered. The United States Department of Labor helped bring field workers from Jamaica and Barbados to harvest the Everglades crop, while Louisiana was able to field some 200 mechanical harvesters by 1944, in spite of the national shortage of equipment.

Beet-sugar production in wartime was disappointing. Cultivating and harvesting the large crop of beets planted in 1942 by labor-intensive methods became a grower's nightmare. Growers in succeeding years preferred to make use of scarce field labor on such crops as potatoes, field beans, or tomatoes, which also had a significant role to play in the war food program but which did not suffer from so inhibiting a price ceiling. If official incentives to sugar-beet plantings were inadequate, this crop nevertheless received more attention than cane, and for some good reasons. At the very time when beets' fall from grower favor had become evident, the difficulties of shipping raw sugar by water from the Caribbean were at their peak and continental supplies correspondingly vital. Production of beet sugar, even after a decline of more than 40 per cent, was double that of cane. As an annual crop, beets enjoyed further advantages as policy moved from crisis to crisis. The short six- to eight-month gap between planting and harvesting lured officials eager for prompt returns, while cane's longer crop cycle looked less appealing and involved the additional risk of unwanted ratoons at a later date. There was also a political factor in the background. Short beet crops meant idle beet-factory capacity. This was a situation that the large number of legislators representing beet states wished to remedy.

The drastic decline in beet acreages that occurred in 1943 could hardly have been avoided by any reasonable price increase. Labor supply was the critical factor. Resort to such temporary expedients as Japanese relocated from the West Coast, prisoners of war, Indians off New Mexico reservations, and migrants from Jamaica, Puerto Rico, and Mexico (*37*, pp. 38–40) were merely holding actions awaiting the technological revolution that subsequently occurred in the beet fields (see chapter 5). To be sure, a program of producer subsidies was introduced. On the 1943 crop, the CCC, through contracts with processors, assured growers $1.50 per ton of beets (equivalent to about one-half cent per pound of sugar) above their 1942 receipts, which had been about $7.00 from the processors together with a

Sugar Act payment of about $2.50. The new gross figure represented some 120 per cent of statutory parity. But the program failed, less because it was announced relatively late than because a still better support program was subsequently announced for beans, a closely competing crop. Instead of the two million tons of beet sugar that had been hoped for, less than one million were in fact produced. The following year the support payment was raised to $3.00, assuring growers the highest price they had ever received. The program was this time announced sufficiently in advance of the planting season, but results were little better than in 1943. To compensate for reduced volume of operations and higher production costs, beet processors also received direct subsidies in 1943 and in 1944.

The CCC was responsible for similar subsidy programs to other domestic areas, though on a somewhat less generous scale. This government agency, and earlier the Defense Supplies Corporation, also absorbed many of the expenses involved in transporting sugar to unusual destinations: beet or West Coast cane railed to the East (cf. 44); direct-consumption imports shipped to deficit areas; and the additional costs involved in operating the "shuttle service" via Florida. Wartime shipping costs, whether due to ocean freight surcharges, war risk insurance rates, or exceptional handling and wharfage charges, were also levied against the CCC instead of being reflected in a higher landed price. The total government outlay in these directions is extremely difficult to estimate. Part of the cost was compensated out of tariff revenue on Cuban sugar accruing for a time to the CCC, while expenditures by the CCC on sugar transport provided the War Shipping Administration with its most remunerative cargo (38, p. 108).

Under a more aggressive pricing policy, domestic areas might have made a more substantial contribution to American sugar supplies. But for the kind of expansion that emergency requirements were shortly to make necessary, the reliance had necessarily to be on Cuba. Indeed, that island's potential for spectacular response to moderate price stimulus was itself a source of official concern. The risk of building up excess postwar capacity inhibited full use of this latent resource, and policy plodded from expediency to expediency. Throughout the war period, crop purchases were negotiated at late dates, and only from year to year. Though price clauses in successive contracts were necessarily debated, they did not represent the major stumbling block, particularly after Cuba received protection against higher prices on the goods it imported from the United States. When a reduced 1945 crop meant that low sugar prices could no longer be compensated by a large export volume, a moderate increase from 2.65 cents to 3.10 cents per pound was sufficient to close the contract.

What Cuba had most to contend with was lack of continuity in official policy. Cuba's desire for the assurances involved in a multiple-crop purchase ran up against CCC concern lest stocks remain in government hands after the shortage was over. Sharp reversals in American wartime policy became demoralizing to Cuban producers and contributed to the postwar shortage. Although Cuba had been encouraged to extend its plantings in 1942, under the contract for the following year the Cuban exportable crop was limited to 3 million tons, and about one-third of the cane acreage was left unharvested. With the advantage of hindsight, one must not underestimate the large Caribbean stocks and the transportation difficulties which served to justify the crop decision. But ocean shipping had improved sufficiently by March 1943 that the director of the CCC's Sugar Division obtained concurrence of Cuba in the production of an additional half-million tons, only to be overruled by the Corporation's president (*38*, p. 36). Carryover cane available for milling in 1944 contributed to the enormous crop of that year but stored up trouble for the years immediately thereafter. Producers unable to cut their standing cane in 1943 saw little reason to undertake new plantings that would have brought their first full harvest in 1945 and subsequent ratoon crops. The risk of new reversals also gave every incentive to grind the entire crop in 1944, rather than to allow immature plants to mature fully. The reduction in Cuba's 1945 crop was accordingly a by-product of American policy, unfortunately intensified by a Caribbean drought.

WARTIME SUGAR MANAGEMENT: UTILIZATION

While greater wisdom and longer-range planning could have brought forward somewhat larger sugar crops, common prudence in disposing of available supplies would have considerably alleviated the pending sugar shortage. So far as distribution policy was concerned, the failings were both in broad conception and in specific execution. Supplies likely to be forthcoming in the early postwar years were considerably overestimated, while the extent of postwar requirements was not fully appreciated. The notion that large stocks of sugar would be discovered in Java or the Philippines died hard, though such sugar would have had to survive a three-year interval of occupation, social disorganization, and hostilities. In early 1944, the prevalent attitude was that the need for price and rationing controls would disappear in the fall of 1945, regardless of the outcome in the Pacific, on the basis of the formula that abundance would be restored with "the first full harvest after the first full European planting" (*56*, p. 28). That attitude could not have taken into account the German offensive that delayed victory in Europe. But it also ignored two facts: that the severest shortages in the period of World War I had come after the hostilities were over (cf. *57*, p. 72) and that recovery of beet-sugar production in Europe depended not merely upon getting the crop planted but upon normal operations of transportation and processing in the economy at large. American authorities recognized too late the special role of sugar in emergency food policy, wherein caloric adequacy rather than quality of the diet becomes the uppermost consideration: its cheapness as a concentrated form of food energy, its relative economy of transport and ease of storage, and its acceptability to countries of widely different food habits. No caloric reserve was built in anticipation of world needs, but instead sugar stocks in the United States were dissipated.

To be sure, consumer rationing of sugar in the United States began earlier and survived later than control over any other commodity. But the program violated many of the most obvious canons of wartime food management. Consumers were alerted in January 1942 to official plans for introducing rationing, though the rationing machinery did not go into effect until the end of April. Expectations of an early end to control were allowed to arise prematurely, as the result of a Presidential announcement in July 1943, only to go unfulfilled. Public pressures for easing restrictions in 1944 were not sufficiently resisted, although the Department of Agriculture was aware that the underlying sugar situation was deteriorating in spite of Cuba's huge current crop (*54*, p. 702). No organizational unity existed to assure that the allocations laid down by the War Food Administration would in fact be honored by the Office of Price Administration, which had to deal directly with the consuming public. Moreover, for reasons that had nothing to do with the sugar outlook, the practice of validating ration stamps without terminal date was begun in January 1944 (*59*, p. 41). By December, when the supply situation could no longer be ignored, the float of outstanding ration currency had abruptly to be canceled, despite OPA's promise that due notice would first be given. Since sugar reserves had not been built up in wartime, the rationing system had to be tightened after the war was over, when consumers were in a mood for some easing of restraints.

The general distribution of available sugar among competing claimants, as indicated in Table 8 (*47*, p. 240) has one serious omis-

TABLE 8.—DISTRIBUTION OF UNITED STATES SUGAR SUPPLIES, 1937–48*

(Thousand tons, raw value)

Year	Total supply	Domestic civilian	Domestic civilian (adjusted)[a]	Military	Commercial exports[b]	USDA net purchases for export	Visible stocks end of period[c]	Civilian consumption per cap. (Lbs. refined)
1937.....	8,474	6,651	6,643	—	93	—	1,730	95.1
1938.....	8,820	6,437	6,650	—	78	—	2,305	94.5
1939.....	9,472	6,859	7,215	—	153	—	2,460	101.6
1940.....	9,580	7,031	6,765	—	193	—	2,356	94.4
1941.....	10,295	7,960	7,354	95	91	—	2,149	102.8
1942.....	7,835	5,034	6,113	424	55	184	2,138	85.7
1943.....	8,610	5,623	5,566	730	18	474	1,765	79.7
1944.....	8,752	6,170	6,157	1,014	24	317	1,227	88.3
1945.....	7,766	5,045	5,106	1,094[d]	34	175	1,418	72.9
1946.....	7,519	5,551	5,558	119[d]	258	139	1,452	74.0
1947.....	9,641	7,357	7,001	114[d]	232	—	1,938	90.5
1948.....	8,912	7,263	7,518	76[d]	77	—	1,496	95.5

* U.S. Dept. Agr., Bur. Agr. Econ., *Consumption of Food in the United States 1909–52*, Agricultural Handbook No. 62 (Washington, 1953), p. 240, except as otherwise noted.

[a] Adjusted for "invisible" stocks held by wholesalers, retailers, and industrial users, rough approximation only.

[b] For 1937–41, data from U.S. Dept. Comm., *Foreign Commerce and Navigation;* for 1942–48, quantities delivered for export as compiled by U.S. Dept. Agr., Sugar Branch.

[c] Holdings of primary distributors; beginning 1939, includes mainland cane-sugar stocks; beginning 1940, includes raw sugar for processing held by importers other than refiners.

[d] Includes small net shipments for civilian feeding.

sion. Sugar was also diverted to nonfood uses, with repercussions suggested in chapter 1. On the basis of the poor market outlet for sugar at the beginning of its 1941 season, Cuba had processed fully one-third of its crop in the form of high-test molasses. In that year, molasses was the basis for two-thirds of the industrial alcohol required by a high level of general economic activity and rising ordnance requirements (*46*, p. 130). Emergency shortages required a priority system on alcohol for defense purposes even before Pearl Harbor (*50*), while supplementary restrictions on end-use of molasses also became necessary on December 31, 1941 (*51*). Indeed, negotiations for the purchase of Cuba's 1942 crop were initially motivated more by the desire to ensure a continued supply of invert syrup than out of concern for prospective food shortages. The contract initially specified that one-third of the crop should go into invert syrup. But sugar shortages on the East Coast, and the availability of surplus grain as an alternative (though more expensive) raw material for the manufacture of industrial alcohol, permitted a halt to be called after the equivalent of about 700,000 tons of sugar had been diverted. No invert was produced in 1943. By 1944, the synthetic-rubber program required 50 per cent more industrial alcohol than the United States consumed in all uses during 1941 (*46*, p. 128), and cane sufficient to produce 900,000 tons of sugar had to be used for invert syrup. Unless these tonnages in 1942 and 1944 are kept in mind, Cuba's contribution to American supplies is seriously underestimated. But invert syrup was not sugar.

Before 1945, military requirements expanded without putting civilian supplies in serious jeopardy. Until the fighting in Europe was over, civilian authorities were reluctant to assume responsibility for cutting back on Armed Forces' claims, so that "military requirements were seldom reduced" (*49*, p. 11). With armed service personnel numbering over 11 million, consuming at about double the civilian rate per capita, and with the practice of building a nine months' stock to supply overseas pipelines, distribution to the forces exceeded one million tons by 1944. Yet per capita consumption by the civilian population that year was only 7 per cent below the 1937–38 average.

Figures for civilian consumption conceal an important distinction between industrial users and final households. Whereas the small number of enterprises producing refined sugar made administrative control over primary supplies no particular problem, the large number of relatively small firms in food-processing industries made regulation of sugar distribution more difficult. This was owing partly to the attentiveness of congressmen to the needs of their small-scale industrial constituents such as confectioners and bakers, and the corresponding restraint on the severity with which controls could be applied. Various industries could press their claims with special zeal. Bakery products promoted the consumption of cheap grain calories; in canneries, adequate sugar prevented waste of perishable fruits or fresh produce; fluid milk and even eggs could be converted to more storable form in the dairy industry; while the soft-beverage firm could argue that in the absence of adequate supplies of its major raw material, the only alternative was to shut down the entire plant. The blunt fact is that total industrial use exceeded prewar levels throughout the war years, even without taking full account of set-asides of end products for the Armed Forces (*52*, p. 214). Despite an ostensible restriction of industrial users to some 70–80 per cent of their 1941 usage, aggregate flow of sugar to the food-processing sector for civilian purposes evidently exceeded that base level in 1944, although industrial consumption in the base period itself stood one-quarter above 1939. The expansion in industry's takings from a prewar one-third to a wartime one-half of total civilian usage was in part a by-product of the rationing controls, but it also reflected a change in consumption patterns that persisted in the postwar period. It is somewhat ironic that industrial consumption, which operates in favor of more rational (because higher) peacetime sugar quotas, complicated the administration of the wartime sugar program.

WORLD SUGAR SHORTAGE, 1945–46

American supplies in the entire period 1940–44 would have been less ample but for the possibility of drawing on quantities of Cuban sugar that would normally have moved to Europe. The distribution of Cuba's sales between the United States and all other destinations during the war period was as follows, in thousands of short tons, raw value (*69*, p. 53):

Destination	1935–39 average	1942	1943	1944	1945	1946
United States:						
Raw	1,668	1,482	2,987	3,508	2,764	2,321
Refined	366	373	370	427	427	364
Total	2,034	1,855	3,357	3,935	3,191	2,685
Others	911	154	809	441	993	1,504
Total	2,945	2,009	4,166	4,376	4,184	4,189

As against a peacetime share amounting to more than one-third, countries other than the United States received only about 8 percent of Cuba's shipments in 1942, and 10 per cent in 1944. Part of the decline was due to the fact that customary outlets in Western Europe were under German control. But the major European outlet, the United Kingdom, was cut off from Cuban supplies during 1942 by difficulties of ocean shipping. A more normal flow, about 650,000 tons, was restored the following year. In 1944, however, the United States could take so much of Cuba's exports as the result of circumstances that would not repeat themselves. A legacy of the shipping problem had been a year-end carryover in the Dominican Republic on January 1, 1944, equivalent to a full year's crop. The United Kingdom received about 700,000 short tons from that source alone in the course of the year (*70*, p. 16).

Requirements in other countries overseas became increasingly urgent with the passing of time. German conquest of the Ukraine in 1941 required unusual shipments to the USSR. Though the target rate of 70,000 tons a month exceeded the available shipping tonnage, over 500,000 tons of refined sugar were exported to the USSR from the United States in the five years beginning July 1, 1942 (*48*, p. 52), and some 200,000 tons were shipped from Cuba as well (*38*, pp. 281–82). Almost 200,000 tons were drawn down from United States supplies to fill the sugar gap in liberated North Africa. Victories in Western Europe created additional deficiencies, met in part by unrecorded civilian feeding out of military supplies. Insufficient delivery of fuel to sugar factories, the result of a disrupted transportation system, crippled the French harvest in 1944, but in large parts of Western Europe economic disorganization was more serious the following year.

While sugar had been short in wartime mainly for reasons of transport, now lack of physical supplies became the intractable factor. World sugar crops in 1944/45 and 1945/46 amounted to only 27.6 million tons raw value and 26.3 million tons, respectively, as compared with a figure of 34.7 million in the years 1935–39. Despite the fact that French output was already on the upgrade, the European campaign in the fall of 1945 outside the USSR produced only 3.9 million tons, some 3.5 million less than prewar. Unfortunately, Cuba was in no position to duplicate its 1944 efforts, which had brought a sugar crop of 4.7 million tons and the equivalent of an additional 900,000 tons in the form of invert. In the face of heavy new demands, drought and lack of holdover cane unfortunately held production to 3.9 million tons in 1945. Damage to young cane continued to be a factor even in 1946, when output was only a half-million tons higher.

Unsatisfied consumers in the United States, short crops in Cuba, and the postwar disorganization of European beet-sugar production placed a heavy burden on the international machinery for sugar allocations. Essentially, the Combined Food Board perpetuated the two nationalist sugar systems, American and British. Both the United States and the United Kingdom accordingly had jurisdiction over indigenous supplies, while American control over Puerto Rican and Hawaiian sugar was paralleled by the Ministry of Food's purchases of the exportable surpluses of the Commonwealth. By custom, sugar from the Dominican Republic and Haiti was also assigned to the United Kingdom, whose wartime supply commitments extended to Canada, New Zealand, the Middle East, Malaya, and Ceylon. Distribution of the Cuban crop was the crux of the entire system. Although the CCC was the ostensible purchaser, Cuba was concerned that her customary outlets be protected, and both parties considered that the American agency was acting on behalf of the Allies.

Decision did not come easily. By September 1944, the Combined Food Board officially recognized that the sugar position was "extremely difficult" (*18*, pp. 249–51). In the face of prospective requirements about one million tons in excess of available supplies, deadlock continued within the Board's Sugar Committee until the second quarter of 1945. The needs of liberated areas were finally provided for by the decision to hold consumers in the United States, the United Kingdom, and Canada to 70.8 pounds per capita in the last 9 months of 1945. The sense of this formula was to reduce the three parties to roughly the level of rations in effect in the United Kingdom, a severe drop from the 89-pound figure enjoyed by the United States during the preceding year. Nor were supplies sufficient to support any increase in rations the following year when the more representative International Emergency Food Council took over the task of recommending allocations. It was only in 1947 that the allocations for the United States, the United Kingdom, and Canada could be raised. The new per capita level became 93 pounds, about 90 per cent of the prewar United States figure as compared with 75 per cent for Council members generally (*61*, pp. 13, 65). This was a sacrifice that foreign countries were prepared to make in order to encourage the United States to continue its rationing controls. At that, the requirements of Western European liberated and neutral countries were dealt with more liberally than the postwar deterioration of world supplies might have implied, mainly by ignoring the import requirements of defeated Germany and Japan. Latin America, however, remained outside the allocation system.

Under circumstances of world stringency, there was no place for the kind of laxity that had characterized United States sugar rationing in 1944. That year, civilian disappearance had run some 800,000 tons higher than the allotments specified by the War Food Administration. By a series of steps, control over distribution was tightened. Cancellation of ration currency outstanding in consumer hands in December 1944 was paralleled by similar action with respect to industrial users. After sugar-using firms had taken an inventory of their sugar stocks and ration evidence, the OPA recaptured all accumulations in excess of a 60-days' supply (*66*). The 20 per cent underestimation of industrial consumption in the base period, which had by itself resulted in overissuing some 480,000 tons at the allocation rates in effect in 1944, was uncovered and corrected (*58*, p. 251). Of almost equal importance in draining off sugar in 1944 had been the home canning program. Ration evidence representing 1.1 million tons had been issued to the public, compared with an allocation of only 750,000 tons. While local boards were ultimately instructed to issue in 1945 only 75 per cent of the previous year's amount, this information arrived after many applications had already been processed. Consequently total civilian distribution in the first five months of 1945 actually exceeded the level of the preceding year despite the attempts at stricter regulation.

Only after the Anderson Report on *The Sugar Situation* (*58*) was filed in May was the lid finally clamped tight. Mr. Anderson became Secretary of Agriculture, with ultimate responsibility for sugar rationing, allocations, and procurement finally vested in a single authority. Effective control over distribution was imposed at the source in June (*67*), when primary distributors were for the first time assigned specific quotas for United States civilian consumption and other major categories of end use. Household rationing, which had been running at .461 pounds per capita per week through most of 1944, was cut to .393 pounds in the first quarter of 1945 and then to .285 pounds. At this rate, it continued into early 1947, moderated only by a 10-pound allotment for home canning that raised the total household allotment to 25 pounds for the year 1946. For industrial users, the standard rate of 80 per cent of base use for most categories during 1944 had been cut to 50 per cent by July 1945 and was still running no higher than 60 per cent in March 1947 (*64*, pp. 40–42). A few favored categories did somewhat better: the bakery groups that promoted consumption of cheap grains, and the canners of perishable commodities. Despite the decline in absolute level, industrial use absorbed its peak proportion of total civilian consumption in the year 1946 (*52*, p. 214) though it no longer enjoyed the heavy business in end products formerly supplied to the United States Armed Forces.

At low rates of postwar rations, questions of equity in distribution of the burden became extremely important. Rationing sugar in the United States involved some 5,500 local OPA boards, 125,000 registered industrial users, 550,000 institutional users and restaurants, as well as the entire civilian population. Administering and policing the sugar program were an immense task, in a hostile public environment, with few persons in a position to consider particular defects in proper perspective. Widespread publicity surrounded cases of evading rationing controls (*38*, pp. 87–89); denial of small quantities to new industrials users evoked criticism and ultimately even Congressional action. It also became necessary to counteract import of close sugar substitutes. Sugar syrups were serviceable to the soft beverages, and crude hard candy was a useful raw material for the confectionery trade. But for introduction of import and rationing controls on sugar in these disguised forms (cf. *68*), particular industrial users would have enjoyed unwarranted advantages.

The Cuban counterpart of these difficulties was the withholding of successively larger amounts of sugar from sale under contract. Leakages occurred in part through raising the set-aside for "local" consumption, from 200,000 tons in 1942 to 350,000 by 1947. Much of the increase was intended to be shipped to the United States, but in disguised form and at correspondingly higher prices. More serious was the drain to the attractive Latin-American market, where money incomes and sugar consumption were on the rise. Only 65,000 tons had been considered necessary to satisfy this outlet in 1942, but the figure had reached 300,000 tons in the contract for 1947. Even at this level, the drain did not represent as much as five pounds per capita per year on the United States ration. But further sacrifice in favor of Latin-American consumers, who were spared rationing, could not easily be justified. As a final legacy of the crop-restriction program in 1943, the problem of leakages also extended to molasses. Cuba had had every incentive to develop local facilities for processing cane products, whether as spirituous beverages or as industrial alcohol. In the face of postwar feed shortages and international allocations of blackstrap, molasses continued to be diverted to these uses. The American government was required to purchase unwanted alcohol in 1946 and 1947 as a condition of purchasing sugar.

Such considerations contributed to the increasing difficulty with which negotiations over successive Cuban crops were carried through. Purchase of the 1944 crop had proceeded smoothly enough, since a huge crop was in prospect and all limitations on harvesting had been eliminated. The 1945 discussions involved more protracted negotiations over price. But the principle of maintaining customary relationships between returns to Cuban and to Puerto Rican producers brought a settlement just prior to the end of the war in Europe. The bargaining sessions that commenced after V-E Day continued for more than thirteen months before a contract could be consummated. An impasse was reached with a Cuban threat to stop loading sugar ships and United States preparations to retaliate by halting export of essential foodstuffs to Cuba. Only the direct intervention of Secretary of Agriculture Anderson and Cuban President Grau in July 1946 produced a contract for the two crops, 1946 and 1947.

Part of the difficulty was an underlying difference in the points of view of United States and Cuban negotiators. While short-run considerations dictated Cuba's stand on such questions as alcohol sales and exempted exports, her representatives did not press their full bargaining advantage in a time of world shortage. They were prepared to make considerable concessions on the price of sugar in return for longer-term assurances. That meant resisting any adverse change in peacetime quotas under the Sugar Act, a concern for Cuba's position vis-à-vis an independent Philippine Republic within the United States supply system, and a strong preference for a multicrop contract that included a year or more of normal sugar markets in addition to the period of immediate stringency. The American negotiators were in no position to grant guarantees on matters involving future Congressional action. They were not averse to a multiyear purchase and had made such a suggestion as early as 1944 (*38*, p. 45). But the CCC preferred to deal in sugar only so long as was necessary to prevent a runaway price, and wished to avoid any accumulation of government stocks that would have to be disposed of on a falling market.

Negotiations over base contract price were less heated than those over provisions permitting prescribed adjustments in price. There was a general understanding that the CCC ought not to profit from resale of contract sugar, and that Cuba was entitled to additional compensation in the event of a reduction of tariff or increase in the American ceiling price. Indeed, that contractual provision gave additional incentive for perpetuating a stable domestic price. There was also agreement in principle that if Cuba was prepared to accept a tight lid on the price of its major export, the price level of its essential imports from the United States could not be left entirely to the mercy of current inflationary forces. Beginning in 1943, a number of concessions were made to implement that purpose, whether by stabilizing the price of its imported rice, lard, and wheat flour; by guaranteeing that minimum quantities of specified foodstuffs would be made available; or by inserting escalator clauses relating the price of sugar to general indexes of United States commodity prices.

Difficulties in arriving at agreement on specifics arose both because the relevant commodities tended to be in short supply elsewhere, and because these provisions involved domestic political complications in both countries. While Cuban authorities were concerned with the national terms of trade, the United States Department of State took the danger of domestic inflation in Cuba more seriously than did Cuba itself. This was partly because the rising political power of the *colonos* and labor, which led to improvement in their share of the industry's earnings, had its inflationary side. Within the United States, the hiatus in domestic price ceilings after June 1946 preceded the signing of the final contract. With United States prices now on the rise, the escalator clause became a pivotal matter, and it was liberalized as a final gesture in obtaining the new contract.

The Cuban contract rate rose to the point where it could no longer be absorbed by the CCC. The settlement price, a rigid 3.10 cents per pound in 1945, turned out to be 4.1816 cents in 1946, and 4.9625 cents the year following. Since every tenth-of-a-cent rise represented $9 million on a crop of 4.5 million short tons, large additional revenues accrued to the Cuban economy. But token sales were transacted outside the United Nations' pool at prices three times as high (*60*, p. 15), while the rate that would have resulted from competitive bidding in the circumstances of 1945–46 can only be imagined. In response to the contract rate on raw sugar, the price ceiling on refined sugar was raised by stages from 5.50 cents in February 1946 to 8.00 cents in November and to a high of 8.40 cents by the following August. Even these delayed adjustments failed to match the increase in food prices generally. Some wartime subsidy programs were continued into the postwar years, but the increases in price ceilings, together with improved availability of fertilizers and equipment, were the major elements in recovery of domestic sugar production. By the fall of 1947, the beet harvest on the mainland was back to near-record proportions, though acreages remained considerably below prewar.

DECONTROL

Despite the general rush to remove all wartime controls, sugar was recognized as an exception to the rule. Only 5.6 million tons of sugar were available to the United States in 1946, as compared with the 8 million or so tons that consumers could be expected to absorb in the absence of controls. When price controls were removed from foods generally on October 24, 1946, ceilings were retained only on sugar and syrups, corn sugar and corn syrups, and rice. While production prospects improved after the turn of the year, the 6.8-million-ton allocation to the United States announced by the International Emergency Food Committee (IEFC) in February 1947 was still far short of the requirements for a free market. Mindful of the disturbance to the entire trade that resulted from premature decontrol in 1920, major sugar producers and users were almost unanimously in favor of continuing controls as of March 1947 (*61*). By special legislation, the power to ration sugar and to retain price ceilings on sugar and syrups was accordingly extended until October 31 (*63*), and responsibility for administration was vested in the Secretary of Agriculture. President Truman expressed concern that October 31 was too early a cut-off date, particularly in view of the fact that stocks are seasonally low at that time of year.

Replenishment of stocks in all hands was a prerequisite of decontrol. Visible stocks had declined from 2.5 million tons on January 1, 1940 to 1.2 million on January 1, 1945, with little rebuilding during the following two years (Table 8, p. 177). In view of strong demand, sugar moved quickly from refiners to consumers, so that part of the decline in effect represented better efficiency in sugar distribution. But year-end stocks in Cuba had also to be sufficient to maintain refinery activity in the period before new-crop sugar began to move in volume. In fact, Cuba's carryover on January 1, 1946 was so low that housewives had difficulty cashing ration evidence, and some allowance had to be made for a sugar carryover in the future. Invisible stocks, though exceedingly difficult to estimate, were also far below normal. The best industry estimates suggested a prewar level of 820,000 tons, which had risen to some 1,500,000 tons during the year 1941. As of December 31, 1946, they were down to 350,000 tons (*61*, p. 111). The Sugar Branch took the reasonable position that 1947 supplies had best be conservatively estimated and rations held firm so long as there was a risk of a run on limited stocks.

The process of decontrol was one of the smoothest features of the war sugar experience. Despite the larger crops in prospect, consumer and industrial rationing in the first quarter of 1947 was continued at the very low 1946 level (*64*, pp. 40–42). On the basis of the IEFC allotments announced February 21, the basic allocation for industrial users was raised from 60 per cent in the first quarter to 75 per cent in the second. Inventory controls over distributors and industrial users were gradually relaxed. By May 26, households were fully assured of at least a 35-pound ration for the year, 10 pounds higher than in the previous year. Events moved rapidly thereafter. On June 12, three days after IEFC raised the United States allocation by 350,000 tons, consumer and institutional rationing was brought to an end. In the same month, the basic industrial allocation for the third quarter was moved up to 85 per cent. On July 28, industrial rationing was also abandoned. Although sugar distribution rose sharply in the weeks immediately after consumer rationing came to an end, there was little rush for supplies following the termination of industrial controls (*65*, p. 51). Controls over inventories were retained until August 30, price ceilings until October 31. (By September 22 the IEFC had also suspended its recommended sugar allocations, and the international machinery came to a halt on December 31.)

In contrast with decontrol of butter and meat, removal of sugar ceilings on October 31, 1947 was followed by no abrupt rise in price. Refined sugar continued to sell at $8.40 per 100 pounds until the end of 1947 and was to fall as low as $7.50 after the turn of the year. An easier market had been anticipated by the price history of the May 1948 future on the New York Coffee and Sugar Exchange. Futures trading had been permitted in June 1947, in part as a method of providing the best private judgment on market prospects. So much had tightness disappeared by August 1947 that Congressional hearings were being held on re-enactment of the peacetime quota system (*28*). With restoration of import quotas in store after December 31, the risk was not one of deficient stocks but of possible excess. To prevent Cuba from being penalized by late shipment of 1947-crop sugar, the CCC undertook to complete import under the crop-purchase agreement before the year's end. The CCC in turn had a ready purchaser for surplus supplies, the United States Army, which agreed to buy almost half a million tons for use in occupied areas. Any possible embarrassment from disposing of government stocks on a falling market was accordingly avoided. However, the price guarantee to the beet industry, which covered the 1947/48 marketing year, made the current price level a matter of government concern. Processors had later to be reimbursed to the extent of $16,505,000 (*26*, p. 15).

RESTORATION OF THE QUOTA SYSTEM

With the rebuilding of stocks on the mainland and the recovery of mainland production on the basis of new agricultural techniques (see chapters 5 and 6), the prewar sugar controls were restored effective January 1, 1948 (*9*). The new legislation had several important merits, and some aspects of administration continued to be highly enlightened (cf. *10*). The Act was for a five-year term, with the specific intention of granting Cuba a reasonable period for downward adjustment from postwar production peaks. By a temporary expedient, and at new sacrifice to domestic producers, Cuba's quotas were inflated, for she was allocated 95 per cent (later 96 per cent) of the amount by which the devastated Philippine industry fell short of filling its official quota. Cuban raw sugar paid a tariff of only one-half cent per pound, and for some years was to be the almost exclusive beneficiary of any increase in consumption requirements. Moreover, continued purchases under the auspices of the United States government, to meet needs overseas, bolstered Cuban dollar sales for a further period. On April 7, 1948, the CCC purchased one million tons, and on August 11 a further 238,000 tons, for civilian feeding in occupied areas in Europe and the Orient as well as for other countries receiving special aid (62, p. 19). For the first time in the twentieth century, Cuba's exports to the United States fell short of its exports to other countries.

The control system has continued to operate smoothly and with considerable moderation. Aggregate quotas have been set high enough, owing in no small part to strong pressure from a substantial group of sugar-using industries, to permit the price of refined sugar to decline relative to the Bureau of Labor Statistics index of wholesale food prices (cf. 11, pp. 6 ff.). Under all conditions short of fullscale war, there has been sufficient flexibility to counteract even such sudden bursts of consumer hoarding as occurred in 1950.

At the agricultural end, also, successive sugar programs have typically provided domestic growers with a floor rather than an umbrella, protecting their crop from heavy market pressure but not precluding all adjustment of output or every reduction in receipts. As a consequence, sugar beets in particular have served as something of an income hedge. Acreage tends to increase when other farm prices are unpromising (as when production of wheat and cotton was being restricted in 1950), but then to decline when farm conditions are generally favorable. Furthermore, the quota system has not frozen the regional pattern of mainland production. The relative positions of California and Colorado as sugar-beet producers were reversed between 1933 and 1950, California increasing from 15 per cent to 28 per cent of the national total, and Colorado declining from 24 per cent to 16 per cent (12, p. 83; 13, p. 440). The sugar program did not inhibit, though it can hardly take full credit for, a rapid rate of technological improvement. Bulk ocean shipment of raw sugar, bulk shipment of refined sugar, the advent of liquid sugar for industrial users, as well as the mechanical transformation of beet agriculture, the successful mechanization of the Louisiana cane harvest, and the advance of mechanization in cane production in Hawaii, have all taken place in recent years. Although innovations in transportation have been influenced by high current costs of handling, and agricultural mechanization was in considerable part a response to wartime labor shortage, at least the market security represented by sugar policy has been an important conditioning factor.

Particularly as compared with the period before 1939, however, there has been a noticeable tightening in the severity of the sugar controls. The custom is now to set initial determination of sugar requirements lower than expected consumption, as a frankly price-supporting device. Domestic groups have demonstrated the political strength to obtain successive increases in marketing quota as their productive capacity expands. Basic quotas established for domestic areas as a group in the postwar Act exceeded their combined output for any previous year. The figures for Hawaii and for mainland beet and cane were more permissive than restrictive, and only in Puerto Rico were producers under any particular restraint. Minimum quotas for domestic areas were fixed in absolute terms, rather than as a proportion of total consumption, so that Cuba's status as buffer and residual claimant was accentuated.[1] Moreover, these statutory domestic quotas, which are presumably one of the obligations to be met as a condition for receiving subsidy payments, came to be regarded as minima as well as maxima. When growers shift away from sugar beets because other crops are more remunerative, a claim is made for reducing the consumption estimates in order to raise sugar prices and sustain beet production.

The most serious defects of the sugar-control system, however, are less economic than political. Quotas and direct restrictions, with their shifts in the rules of the game and their windfall gains and losses, place a premium on winning political favors rather than on performing economic services. Indeed, the legislative function has in part been delegated to private industry groups. Whereas the Secretary of Agriculture in 1933 played a primary role in drafting the initial sugar legislation, the main details of the postwar acts have been formulated primarily by producer agreement. At public hearings in 1947 and again in 1951, all major domestic producers were represented by a single agent, and it was to him that the Secretary turned, at one point, for an explanation of "what the drafters have in mind" (*28*, p. 24). Congressional testimony in 1951 by counsel for a smaller firm in the

[1] The 1956 amendment to the Sugar Act improved the position of domestic producers and foreign suppliers other than Cuba by revising the provisions for meeting future increases in United States consumption (*27*).

industry is revealing: "The Department of Agriculture told us that it could do nothing for us unless we got the approval of the sugar industry. The attitude we met in the industry was that there would be a lot less trouble in getting the act passed if everyone went along with the majority and no one rocked the boat . . . We cannot believe that this committee wants to delegate to a closely knit group in the sugar industry the power to close down the plants of an independent business" (*14*, p. 220).

The agency responsible for administering the program exercises a degree of discretion that is quite on a par with the powers in the hands of a public utilities commission. The administrative functions could hardly be performed without reasonably cordial relations with producer groups. But for that very reason there is always a danger that the welfare of the industry may be confused with the welfare of the general public, and that the agency may find itself in the position of industry advocate. It is an open question whether generous quota allotments to the domestic industry are accurately described as "fair shares" (*25*, p. 2), or whether a government-sponsored cartel arrangement for a farming and processing industry is well characterized as "competitive" enterprise.

Rational appraisal of sugar policy by the Congress is inhibited by certain questionable interpretations of the program's effects (cf. *14*, pp. 3–20):

a) *That the narrowing of the "world" sugar market has been due to nationalistic policies abroad.* But world-market difficulties are more the result than the cause of American sugar policies. The United States consumes each year over 8 million short tons, raw value, about 20–25 per cent of the entire world production of centrifugal sugar. Relatively small changes in United States imports of foreign sugar can have important effects on external prices. The extreme depression of sugar prices in the early thirties was due as much to the displacement of Cuban sugar from the United States market by successive increases in tariff as to any other single factor. Similarly, high world prices in 1950–51 reflected the heavy absorption of Cuban sugar by the United States, with a consequent shortage in sugar available to other destinations. It is, of course, true that prices to consumers are frequently higher and per capita consumption usually lower outside the United States, and that total sugar consumption within the United States has been rising. But it is also true that industrial users of sugar are not permitted to purchase this commodity in the volume or on the terms that would promote faster growth. The degree of protection enjoyed by the domestic industry is reflected by the lower prices at which the United States could obtain imported sugar, if it so chose. Comparisons with the prices or consumption prevailing within poorer countries, some of whom still rely heavily on excise taxes for public revenues, are largely irrelevant.

b) *That the Sugar Act deserves credit for maintaining the United States price below the "world" price for some months during 1950– 51.* But such a quota discount could exist only because Cuba was prepared to meet increases in its United States quota, out of obvious long-run interest, even at a price below that temporarily prevailing in other markets. Domestic producers were less cooperative and were restrained from taking advantage of the higher external prices only by imposition of export controls, a device appropriate to the Korean war period but not itself a feature of the Sugar Act. By far the more normal situation, and the very purpose of the Sugar Act, is a substantial quota premium. By USDA calculations, the premium had recovered to more than 60 per cent over the world price by August 1953 (*29*, p. 25).

c) *That the sugar program is "self-financing."* In recent years, Sugar Act payments to growers have been running at the rate of about $60 million, as against sugar-tax collections (excluding the tariff) of about $80 million. In this sense, the program has earned a net revenue for the Treasury. But the domestic industry is taxed on the basis of .5 cent per pound raw sugar, while growers receive subsidies at a basic rate of .8 cent per pound. Cuba and the Philippines, which contribute pro rata to tax collections, receive no subsidy. From the standpoint of international trade and public revenues alike, the excise on imported sugar has all the characteristics of a tariff, a large part of which is transferred to the account of domestic producers instead of to the United States Customs. Moreover, even *if* domestic sugar paid the full equivalent of the subsidy, the question would remain whether it is an appropriate exercise of the tax function to impose a levy on a commodity for redistribution to producers of that commodity.

d) *That sugar policy can be justified on grounds of the need for continental supplies in wartime.* But in two world wars, the United States in fact came to depend overwhelmingly on Cuba for overcoming wartime and immediate postwar sugar shortages. During 1943– 46, despite good cane crops, mainland sugar output was 20 per cent less than prewar and fell 30 per cent short of wartime production targets—a larger percentage deficit than for any other major agricultural commodity (*21*, p. 512). Wartime shipments from Cuba never fell as low as they had been held by quotas or tariffs before America entered the war. While the island's contribution to the total United States supply was less than 30 per cent in the 1930's, it exceeded 50 per cent in 1944 and in 1947 (*30*, p. 205). Moreover, Cuba, though a foreign country, is physically closer to the deficit regions of the Atlantic seaboard than are many of the western beet-sugar states and also has good access to the Middle West via New Orleans. Transcontinental rail haulage is itself a bottleneck under wartime conditions. Had the importance of Cuba as wartime supplier been recognized earlier in World War II, the wartime sugar shortage could have been considerably alleviated. While government agencies have of late been stockpiling strategic commodities and the CCC has been accumulating domestic crop surpluses, no effort has been made to raise the regular level of sugar stocks in the United States. Stocks and excess capacity located in Cuba remain America's insurance against any unusual or emergency increase in requirements, as events in late 1956 and early 1957 were again to demonstrate.

POSSIBLE REFORMS

Whatever its merits or defects, sugar policy remains a bipartisan affair. Extension of the Sugar Act in 1951 was sponsored by 31 senators, attacked by only one, and passed by a vote of 72 to 4. The basic pattern of a quota system combined with direct payments to growers seems hardly open to serious question. Yet within the quota-subsidy framework much might be done to check the increase in domestic subsidization, rechannel benefits within the sugar industry, and encourage the supply of sugar to the United States on a more economic basis. Change might intelligently proceed along the following lines:

1. *A ceiling on the total direct subsidy.*—Between 1937 and 1949, annual expenditure for conditional payments to domestic growers rose from $36 million to $61 million (*16*, p. 661; *17*, p. 720). Congress might at least consider freezing the figure at existing levels, or lowering it, or, better still, scheduling further reductions at specified dates. Growers as well as processors enjoy the price-supporting effects of the quota system in any case.

2. *A ceiling on the total subsidized production.*—Legal quotas for domestic producers have also increased during the life of the program. Reducing the total subsidy might by itself prevent any further

displacement of foreign sugar. But the matter might also be approached frontally. A binding limitation on the total amount of subsidized production could reasonably be laid down, and at figures less than previous historical maxima. The rate of subsidy, as contrasted with its aggregate amount, might be inversely related to total domestic sugar production, in order to positively discourage undue expansion of output. A special quota might be set aside for nonsubsidized domestic production.

3. *Revised scale of benefits to individual growers.*—In so far as subsidization of sugar beets and sugar cane is part of American "farm policy," benefits have been distributed rather differently than might reasonably have been expected. The average conditional payment to all beet growers in 1949 was a moderate $600, but it ranged from $250 to $300 in Michigan, Utah, and Ohio, to $3,000 in California (*17*, p. 720). More detail is available for some states in 1948. In Utah, 61 per cent of the total was paid in amounts of under $500, with 89 per cent of the payees in that category (*22*, p. 68). In California, while 43 per cent of the payees received less than $1,000 each, almost 50 per cent of the total expended was in amounts of over $5,000 (*23*, Table 5).

Regional disparities among cane growers are even more striking. While Louisiana ranks roughly with the national average for beet, payments to individual Florida growers have averaged as much as $36,000; to Hawaii growers, $8,000; and even to Puerto Rico growers (where small-scale cane growing has been promoted by auxiliary measures), $1,200. By virtue of the importance of company farming in cane areas, 39 cane-sugar enterprises received payments in excess of $100,000 each on their 1948/49 crop (*31*, p. 10514). Clearly, subsidies benefit large-scale enterprises quite as much as "the small family-size farm" in whose name the program may be defended before Congress (*11*, p. 4). The data seem conspicuously to substantiate Professor Theodore Schultz's view that "government payments do not effectively get at the income problem" in agriculture (*24*, p. 156).

To be sure, deductions from the base rate of 80 cents per 100 pounds of raw-sugar content are made at higher individual outputs. Even for sugar beets, the scale of deductions, ranging from zero at 350 tons to about one-third at 3,000 tons and under one-half at 30,000 tons, seems rather mild. Taking the national average yield as two short tons of equivalent raw sugar per harvested acre, a figure somewhat low for Pacific Coast states, the limits correspond to 175, 1,500, and 15,000 acres. These are to be compared with estimates of average acreage planted to sugar beets per farm in 1953: 12 acres in Utah, 16 in Michigan, and only 108 even in California, where individual operations are typically on a larger scale (*26*, p. 26).

Three revisions would seem entirely appropriate. Deductions might well begin at a much lower sugar output, one that properly distinguishes large-sized beet operations; the scale might be given a steeper graduation; and an absolute ceiling might then be placed on the amount payable to any one recipient, whether as landlord or as grower. Large-scale farm enterprise would prosper according to its economic merit, but with less government help.

4. *Separate treatment for sugar cane and sugar beets.*—Though Puerto Rico and Hawaii producers have few alternatives to their present dependence on cane production, sugar-beet growers in the western states can switch fairly readily from crop to crop. Size of the operating unit, sugar yield per acre, importance to a regional economy, and the very justification for subsidization are quite different for beets and cane. Already domestic producing regions are given differential treatment by the allotment of quotas. The program's officials have the additional responsibility of setting, for various localities, the minimum price that grower-processors must pay for purchased cane or beets, and minimum field wages. Little further administrative difficulty would be introduced if benefit payment per 100 pounds of sugar were now differentiated as between cane and beets. The maximum subsidy per payee and the graduation of deductions might also be specified separately for the two crops. In establishing the new scales, some attention would of necessity be paid to the economic realities of sugar-crop production.

5. *A new price target.*—No specific legal or economic content attaches to the legislative prescription to seek prices "that will not be excessive to consumers and which will fairly and equitably maintain and protect the welfare of the domestic sugar industry" (Section 201). The setting of total consumption requirements is a discretionary process, and might as well be frankly recognized as such.

Certain guides to administrative decision can be meaningful. Continental consumers were formerly given a minimum assurance, not re-enacted in the 1948 law, of "a per capita consumption equal to the average of the two-year period 1937–38" (*19*). However, protection of consumers still springs from the bargaining strength of sugar-using industries, which now absorb a large share of total United States consumption and naturally press continually for higher quotas on a major raw material. Nevertheless, estimated sugar consumption in the United States surpassed the per-capita peaks of the 1920's only in 1941, when hoarding was the dominant factor. The Sugar Act of 1948 replaced a consumer-protection clause with a producer-oriented parity formula, taking as its target the base period January–October 1947 when sugar was as its highest price of the inflated 1940's.

An entirely new criterion could be economically justified. The height of the duty-paid New York price above the equivalent world price for raw sugar reflects the protective margin under a quota system. With a tariff law, that margin is precisely specified. The maximum price premium to be maintained under quotas could be similarly designated. During the 1930's the price of raw sugar per pound in New York ex-duty was held as much as 1.50 cents (or over 100 per cent) above the London price. Postwar reduction of the tariff on raws to one-half cent might now be complemented by a moderate price target in enforcement of quotas.

Several desirable results could be expected to follow. The government's commitment to the industry would become more precise, and would invite revision downward in due course. No longer would the United States be largely insulated from world-market developments. If world sugar prices should fall because of improved efficiency abroad, United States consumers would benefit and United States producers would be exposed to an appropriate degree of competitive pressure. Should world prices be temporarily depressed for other reasons, the United States would become directly interested in correcting a situation that is not in any case independent of United States sugar policies. A new international sugar agreement that merely acquiesces in and reconciles restrictive national policies is clearly a less satisfactory alternative, partly because the burden of adjustment is placed on the countries least able to bear it.

CONCLUDING OBSERVATIONS

The complex issues involved in sugar policy afford no simple solution but cannot properly be ignored. Domestically, the extent to which freedom of enterprise has here given way to government price-fixing and allocating of markets, in the industrial and the agricultural sectors alike, had best be openly recognized. One is entitled to ask whether continuous subsidization of plantation agriculture and sugar-crop processors has exceeded reasonable limits. Yet much direct intervention by the governmental sugar agency into the internal affairs of the industry may have been less effective than the "penetrative power of the price system," which has operated relentlessly via the relative profitability of alternative crops or the relative scarcity of labor. Important regional conflicts are also at stake, between those agricultural sections that benefit from Cuba's imports of American lard, cotton, rice, or wheat as compared with those that are damaged by the import of sugar, or between the refining interests on the Atlantic and Gulf coasts as compared with the manufacturers of beet sugar in the interior. In the semiarid West itself, logic does not necessarily support heavy investment in federal reclamation projects for the purpose of applying scarce local water to the production of subsidized sugar beets.

But legislative action on this highly political commodity is at least as important from an international as from a domestic point of view. One of the few major agricultural commodities produced domestically yet supplied in substantial volume from foreign sources, sugar, like wool, is a sensitive indicator of United States foreign trade practice. By preferential trading arrangements, private foreign investment, and ties of economic and political history, sugar links the United States closely to Cuba in the Caribbean, and to the Philippines in the Orient. The military defeat of Japan brought still another major sugar exporter, Formosa, within the orbit of American policy, though not into the United States sugar market. Is the United States prepared to use its political power for the further expansion of domestic sugar production, at the direct expense of a weak Cuban economy, upon which the United States inevitably depends for wartime sugar supplies? Are the advantages of an expanding volume of international trade so trivial that the United States will be satisfied to pay a higher and higher subsidy for a larger and larger volume of domestic sugar? Can we continue to urge the reduction of trade barriers in other countries at the same time that the margin of protection afforded the domestic sugar industry increases? Is our concern for the economic development of less advanced countries consistent with our blocking appropriate industrial processing in exporting countries? Questions such as these dig deep to the roots of the foreign economic policy of the United States. By congressional action on measures like the Sugar Act, United States commercial intentions are judged and her international leadership is tested.

CITATIONS

1 F. W. Taussig, *Some Aspects of the Tariff Question* (3d ed., Cambridge, Mass., 1931).

2 United States Cuban Sugar Council, *Sugar Facts and Figures 1952* (Washington, D.C.).

3 J. Bernhardt, *The Sugar Industry and the Federal Government, A Thirty Year Record (1917–1947)* (Washington, D.C., 1948).

4 M. Lynsky, *Sugar Economics, Statistics and Documents* (United States Cane Sugar Refiners' Assn., New York, 1938).

5 U.S. Dept. Agr., *Report of the Secretary, 1939.*

6 U.S. Dept. Agr., *Report of the Secretary, 1940.*

7 *Weekly Statistical Sugar Trade Journal,* Sept. 5, 1940.

8 *Weekly Statistical Sugar Trade Journal,* Dec. 19, 1940.

9 Sugar Act of 1948, Public Law 388, 80th Cong., 1st Sess., Aug. 8, 1947; as amended by Public Law 140, 82d Cong., 1st Sess., Sept. 1, 1951, and Public Law 545, 84th Cong., Second Sess., May 29, 1956.

10 B. C. Swerling, "A Sugar Policy for the United States," *American Economic Review,* June 1952.

11 U.S. Dept. Agr., Production and Marketing Administration (PMA), Sugar Branch, *Sugar Reports No. 11,* June 27, 1951.

12 U.S. Dept. Agr., Bureau of Agricultural Economics (BAE), Crop Rept. Bd., *Crop Production (Annual Summary 1950).*

13 U.S. Dept. Agr., *Yearbook of Agriculture, 1935.*

14 U.S. Cong., House, Com. on Agr., *Extension of Sugar Act of 1948, Hearings on H.R. 4521,* 82d Cong., 1st Sess.

15 Schechter *vs.* United States, 295 U.S. 495 (1935).

16 U.S. Dept. Agr., *Agricultural Statistics, 1946.*

17 U.S. Dept. Agr., *Agricultural Statistics, 1950.*

18 R. J. Hammond, *Food: The Growth of Policy* (History of the Second World War, U.K. Civil Series, H.M.S.O., 1951.)

19 U.S. Cong., Public Resolution No. 104, 76th Cong., approved Oct. 10, 1940.

20 Farr & Co., *Manual of Sugar Companies 1949/50* (New York, 1950).

21 U.S. Dept. Agr., *Agricultural Statistics, 1947.*

22 U.S. Dept. Agr., PMA, Utah State Off., *Annual Report, Farm Programs, 1948.*

23 U.S. Dept. Agr., PMA, California State Off., *Annual Report, Sugar Beet Program, 1948.*

24 T. W. Schultz, *Production and Welfare of Agriculture* (New York, 1949).

25 U.S. Dept. Agr., Press Release 464–51, Feb. 26, 1951.

26 U.S. Dept. Agr., Commod. Stab. Serv., Sugar Div., *Agricultural, Manufacturing and Income Statistics for the Domestic Sugar Areas: Volume II of Sugar Statistics* (Stat. Bull. No. 150, 1954).

27 U.S. Cong., Public Law 545, 84th Cong., 2d Sess.

28 U.S. Cong., House, Com. on Agr., *Sugar Act of 1948, Hearings,* 80th Cong., 1st Sess.

29 U.S. Dept. Agr., PMA, Sugar Branch, *Sugar Reports No. 20*, September 1953.

30 U.S. Dept. Agr., PMA, Sugar Branch, *Sugar Statistics*, I (1953).

31 U.S. Cong., Senate, *Congressional Record*, Vol. 97, Part 8, 82d Cong., 1st Sess., Aug. 22, 1951.

32 United States *vs.* Sugar Institute, 15 Fed. Suppl. 817 (1934).

33 Sugar Institute, Inc., *et al. vs.* United States, 297 U.S. 553 (1936).

34 U.S. Cong., Temp. Nat. Econ. Com., *Trade Association Survey*, Monograph No. 18, 76th Cong., 3d Sess.

35 Andrew van Hook, *Sugar: Its Production, Technology, and Uses* (New York, 1949).

36 Secretary of Agriculture *vs.* Central Roig Refining Company *et al.*, 338 U.S. 604 (1950).

37 U.S. President's Commission on Migratory Labor, *Migratory Labor in American Agriculture* (1951).

38 E. B. Wilson, *Sugar and Its Wartime Controls 1941–1947*, 4 vols. (New York, n.d.).

39 United States *vs.* Butler, 297 U.S. 1 (1936).

40 *Anuario Azucarero de Cuba 1951* (Havana).

41 *Weekly Statistical Sugar Trade Journal*, Jan. 15, 1942.

42 U.S., War Production Board (WPB), Supplementary Order M-55-C, Mar. 27, 1942.

43 U.S., WPB, Supplementary Order M-55-D, Mar. 27, 1942.

44 U.S., Defense Supplies Corp., Circular Letter to Beet Sugar Refiners No. 1, Mar. 31, 1942, quoted in *38*, pp. 1106–08.

45 U.S., WPB, General Preference Order M-98, Feb. 14, 1942.

46 U.S., Civilian Prod. Admin., Bur. of Demobilization, *Alcohol Policies of the War Production Board and Predecessor Agencies May 1940 to January 1945* (by Virginia Turrell, Historical Reports on War Administration: WPB, Special Study No. 16, 1946).

47 U.S. Dept. Agr., BAE, *Consumption of Food in the United States 1909–52* (Agricultural Handbook No. 62, 1953).

48 U.S. Dept. Agr., BAE, *The World Sugar Situation*, August 1946.

49 R. A. Ballinger, *Sugar During World War II* (U.S. Dept. Agr., BAE, War Records Monograph No. 3, 1946).

50 U.S., WPB, General Preference Order M-30, Aug. 28, 1941.

51 U.S., WPB, General Preference Order M-54, Dec. 31, 1941.

52 U.S. Dept. Agr., PMA, *Competitive Relationships Between Sugar and Corn Sweeteners* (Agr. Inf. Bull. No. 48, 1951).

53 U.S. Cong., House, Special Committee to Investigate Executive Agencies, *Hearings on Sugar Allocations*, 79th Cong., 1st Sess., Mar. 20–21, 1945.

54 U.S. Cong., House, Special Committee to Investigate Food Shortages, *Hearings Part 2 Sugar*, 79th Cong., 1st Sess., May 8–14, 1945.

55 Food Industry War Committee, *Report by Food Supply and Allocation Committee on the Sugar Situation* (Washington, D.C., Dec. 1, 1944).

56 *Weekly Statistical Sugar Trade Journal*, Jan. 20, 1944.

57 League of Nations, *Food, Famine and Relief* (Geneva, 1946).

58 U.S. Cong., House, *Report of the Special Committee to Investigate*

Food Shortages, The Sugar Situation, 79th Cong., 1st Sess., H. Rep. 602 (as reproduced in *38,* pp. 238–55).

59 J. A. Kershaw and H. Alpert, "The Invalidation of Food Ration Currency," *The Journal of Social Issues,* December 1944.

60 U.S. Cong., House, Com. on Agr., *Hearings, Sugar Situation,* 80th Cong., 1st Sess., January 1947.

61 U.S. Cong., Senate, Com. on Banking and Currency, *Hearings, Sugar Control,* 80th Cong., 1st Sess., February-March 1947.

62 The World Sugar Situation, September 1948.

63 U.S. Cong., *Sugar Control Extension Act of 1947,* Public Law 30, 80th Cong., 1st Sess., Mar. 31, 1947.

64 The World Sugar Situation, September 1947.

65 I. L. Rice, "The Sugar Rationing Administration, Its Operation and Liquidation" (Washington, D.C., 1947, mimeo.).

66 U.S. Dept. Agr., War Food Administration (WFA), Amendment 11 to Second Revised Rationing Order 3, Mar. 15, 1945.

67 U.S. Dept. Agr., WFA, War Food Orders 131 and 131.1, June 12, 1945.

68 U.S. Dept. Agr., WFA, War Food Order 63, July 20, 1945.

69 The World Sugar Situation, September 1949.

70 Commonwealth Economic Committee [Gr. Brit.], *Plantation Crops* (1948).

CHAPTER 8

SUGAR IN THE UNITED KINGDOM[1]

THE RETREAT FROM FREE TRADE, 1919–39

British sugar policy in World War II stems from the experience of World War I in two distinct ways. The overnight interruption of the country's supply of refined beet sugar from Central Europe in 1914 provided the home refiners (and the exporters of raw cane sugar) with an opportunity to capture the British market; the inconveniences of sugar shortage provided an argument, which the need to foster agriculture reinforced, for founding a native sugar-beet industry; the demise of the Brussels Convention furnished the opportunity for doing so. Twenty years of retreat from free trade left the British sugar industry in a state of regulation that was worthy of the heyday of mercantilism and lent itself very well to the establishment of wartime control. (Since 1939 the pattern has been no more than stereotyped and consolidated.) That control, however, inherited more from World War I than a changed industrial organization; its policy of procurement was marked deeply by earlier experience. The British had paid dearly, in cash and short rations, for becoming dependent on Caribbean, United States-dominated supplies in 1919–20. The primary aim of the Ministry of Food's Sugar Division in 1939–45 was to prevent a recurrence of that situation.

Before 1914, British supplies of sugar had consisted of imported raw cane and imported white beet in roughly equal amounts, totaling about 1.8 million tons annually. When war broke out, sugar supplies from Central Europe were abruptly cut off—the supplying countries having become enemy territory—and the United Kingdom government promptly set up a sugar procurement agency, in the odd form of a Royal Commission responsible to the Home Secretary (1, pp. 6–7, 120–26). This body retained its powers throughout the war, subject to the nominal suzerainty of the Food Controller after the Ministry of Food was created in December 1916. For the first two years of war the Commission was able to maintain imports at roughly the prewar level by drawing on Java and Mauritius to replace the lost

[1] This chapter has been prepared by R. J. Hammond.

European supplies; but from mid-1916 onward the shipping shortage became even more acute. Imports in 1917 and 1918 averaged less than 1.4 million tons, despite recourse to the shorter haul from the Caribbean; and at the end of 1917 sugar rationing was belatedly introduced. In spite of divided administrative counsels within and without the Ministry of Food, reflected in abrupt changes of rationing method, it was an immediate success, paving the way for more general food rationing. The corollary of national control of consumption in time of scarcity, however, was inter-Allied cooperation in procurement; this was forthcoming until after the Armistice. But the United States Administration, which had participated in buying the Cuban crop of 1917/18, and had actually bought that of 1918/19, on behalf of the Allies, allowed the crop of 1919/20 to slip through its fingers. The fantastic boom in prices that followed, when Cuban sugar reached 10*d.* (20 cents) a pound f.o.b., only to slump to one-fifth of that amount within six months, made a lasting impression upon the British sugar trade; the more so because the slump was attributed largely to the policy of the Sugar Commission in first forcing down the sugar ration to its lowest level ever—6 ounces per head per week—and then buying the whole of the Mauritius crop. The heavy financial loss that the Commission none the less incurred on its stocks rubbed in the lesson: in another war, the United Kingdom must avoid becoming dependent upon a single source of supply, particularly the Caribbean where it would be at the mercy of changes in United States policy.

The post-1919 atmosphere was propitious in more ways than one for a change in policy toward sugar-beet growing in Britain. Experiments with the crop had been made sporadically during the nineteenth century and more persistently in the first decade of the twentieth; in 1912, a factory was opened by a Dutch company at Cantley, in Norfolk, which produced two or three thousand tons a year—always at a loss—from 1913 to 1915. In 1913 a grant was secured from the government's "Development Fund" in aid of experiments in sugar-beet cultivation; in 1917 and 1919 further grants were made to buy an estate at Kelham, near Newark in Nottinghamshire, for the production of beets on a commercial scale, and to erect a factory there. The factory was completed in 1921, but, like the Cantley factory, which had been reopened in 1920, found itself unable to work at a profit on the amount of protection then provided by the difference—6*s.* 3*d.* a cwt.—between the customs and excise duties on sugar. At the request of both companies, backed by the farming interest, the government agreed to remit the excise duty entirely in 1922, thus giving the industry protection at the rate of 25s. 8d. a cwt. This was possible only because of the steep rise in the sugar duties that had been imposed for wartime revenue purposes, and permissible only because the Brussels Convention had lapsed. It marks the decisive stage in the retreat from free-trade principles, already begun in a small way by the levy of a more favorable rate of excise duty (none, till 1915) than of customs duty, and the introduction in 1919 of a preferential rate of duty on Empire sugar.

The protection was still precarious, in that it depended on import duties remaining at an abnormally high level; and when, in 1924, these were reduced to 11s. 8d., it became necessary to devise another form of assistance if the industry was to survive. In July 1924 the Labour government announced its intention of granting a subsidy for ten years, "to enable the industry to get upon its feet and stand alone" (2, p. 19); and in 1925, under a Conservative government, the British Sugar Subsidy Act was passed, with retroactive effect from the "campaign" of 1924–25, embodying these proposals. There was to be a subsidy on a diminishing scale—19s. 6d. a cwt. of refined sugar for four years, 13s. for the next three, and 6s. 6d. for the final three, plus a smaller subsidy on molasses; but an excise duty would be reimposed at the preferential Empire rate (3).

As a result, a substantial sugar-beet industry became established. The number of factories rose from two, in 1924, to eighteen, of which one was in Scotland (a second there failed to survive); the acreage under beets, from 22,000 to a peak of over 400,000; the output of white sugar, from 23,000 tons to a peak of 602,000. This was no mean achievement, considering that the crop was a complete novelty to most farmers, that the manufacture of beet sugar was likewise novel, and that the world price of sugar was falling. The value of beets in agricultural rotations, and of the residues for animal feed, together with the advantage to the farmer of the contract system under which the crop was grown, made a strong appeal in the arable areas of eastern England where most of the factories were situated. Two-thirds of the total acreage, in the seasons 1932/33 to 1934/35, was in the four counties of Norfolk, Suffolk, Isle of Ely, and Lincolnshire; another 15 per cent in the surrounding areas of Nottinghamshire, Soke of Peterborough, Cambridgeshire, Huntingdonshire, and Essex. Elsewhere the response from farmers was poor, and factories in the West Midlands and Scotland particularly had to haul beets over long distances. The following list shows the location of the factories:

Administrative county	Factory
Essex	Felstead
Fife	Cupar
Isle of Ely	Ely
Lincolnshire (Holland)	Spalding
Lincolnshire (Lindsey)	{ Bardney { Brigg
Norfolk	{ Cantley { King's Lynn { Wissington
Nottingham	{ Colwick { Kelham
Shropshire	Allscott
Soke of Peterborough	Peterborough
Suffolk	{ Ipswich { Bury St. Edmunds
Worcestershire	Kidderminster
Yorkshire (West Riding)	{ Poppleton { Selby

Given the limited amount of arable land suitable for sugar beets, the expense in their transport, and the rather onerous requirements for a factory site—especially proximity to plentiful supplies of water— the scope for a further increase in the industry, within the four corners of any subsidy likely to be granted, was obviously restricted. Indeed, some factories already established, though not inferior to the others in technical efficiency, were clearly marked out as marginal. The fall in world sugar prices, moreover, had destroyed the hope that any part of the industry would become self-supporting in the foreseeable future, even granted a preference equal to the Empire preference. As the ten years for which the subsidy had been granted drew to its end, the government appointed an independent committee to make recommendations for the further conduct of the sugar industry; and when it became clear that the committee would not report in time, extended the subsidy for another year (1934/35).

The inquiries of the Greene Committee (so called after its chairman, Wilfrid Greene, K.C.) ranged over a wide field, for the establishment of the new industry had led to an industrial war between it and the old established refiners. These had prospered on an unprecedented turnover during and after World War I, and their bargaining power vis-à-vis the government had been much strengthened by the amalgamation of the two leading firms in 1921 into the great combine of Tate and Lyle, Ltd., which, in spite of its size, still retains much of the character of a family business. The refiners objected, not to the establishment of a beet-sugar industry—they themselves had taken an interest in a group controlling four factories—but to the trespassing on their territory by certain factories which not only manufactured white sugar from beets, but actually imported raw cane sugar for refining in the off season. At the same time, for good measure, they complained that foreign-refined sugar was being "dumped" on the United Kingdom market. In 1928, therefore, the government manipulated the scale of customs duties (which hitherto had been strictly related to the polarization, i.e., the purity, of sugar) so that raw sugar should pay a duty corresponding to 2s. 4d. a cwt. less than that on fully refined. This had the effect of shutting out refined imports, and of providing the beet-sugar factories with an incentive (of about 1s. a cwt.) to produce raw sugar for sale to refiners, rather than white sugar. The *quid pro quo* given by refiners for this protection was a reduction in prices by the full amount of the duty concession. They nevertheless profited by being able to increase their output (and so reduce their overhead) and for the first time develop a considerable export trade. In 1933 total production of white sugar in refineries was 1¾ million tons, not far from twice the pre-1914 level—a measure of the benefit that war, first, and then government policy had conferred upon the industry.

It might be thought that by comparison the 360,000-odd tons produced by the factories—one-third of it by off-season refining of imported raws—was unimportant. But the refiners did not think so, and professed to fear an increase in this subsidized competition. They reduced their own prices to such an extent that by 1932 some of the factories were in danger of being unable to carry on without further state assistance; a group responsible for about three-fifths of the total production was given a special temporary "advance" (which was never repaid) of 1s. 3d. a cwt. for the season 1931/32. The government thereupon persuaded the parties to enter an industrial agreement to regulate the production of refined sugar by means of quotas. Taking the annual output as 1,900,000 tons, five-nineteenths (500,-000 tons) would be shared by the factories, the remaining fourteen-nineteenths by the refiners; this ratio would be maintained whatever the actual output in any given year. Thereafter 6d. a cwt. was added to the "refining margin," both refiners and Exchequer gaining at the expense of the consumer.

These facts, and many others, were brought out by the Greene Committee in an investigation that is a model of its kind (2). By a majority of two to one the Committee recommended that the whole experiment be wound up; its benefits to agriculture and to employment generally were not commensurate with its cost (some £40 million since 1924; it was reckoned that the industry could just about pay its way if it got its raw material for nothing); the consequences of continuing it would be a minute regulation of the industry such as the majority were reluctant to contemplate. If, however, the government was not willing to take this drastic step, it must embrace the alternative. An independent sugar commission should be set up to supervise and control the industry, with powers to determine the annual subsidy and the refining margin. No more factories should be built; those existing should be unified into a single corporation, in order not only to promote more efficient management, but also to pool receipts of more and less profitable factories to the advantage of the subsidy figure. Finally, the industrial agreement, which had too much favored the refiners, should be abrogated in favor of a new agreement expressly giving the new factory corporation power to refine its total output; the fiscal discrimination against this should be removed.

It was hardly to be expected that the government would adopt the root-and-branch majority recommendation, even had the dissenting member not pointed out that a comparatively small rise in world sugar prices might bring the home-grown industry at least to a level with the colonial producer receiving special preferences. These preferences, related to a limited quota of sugar, had been introduced in 1932, adding a further complication to the already elaborate scale of duties. In July 1935 the government announced (4) that the subsidy would be continued "on agricultural grounds"; at the same time, the recommendations for reorganization were accepted, though modified in two important respects to meet the objections of the refiners. The fiscal discrimination against white sugar was to stay; but the refiners would pay compensation to the factories by means of a device called the "extra quota." The factories would be allowed a quota of 720,000 tons of refined sugar—as against 500,000—on the understanding that the difference would be bought by the refiners at an agreed price. The new sugar commission would not be given power to fix the refining margin; but the refiners undertook not to increase the margin except on account of increased costs.

These arrangements were duly embodied in the Sugar Industry (Reorganisation) Act of 1936 (5); the new corporation came into being in April of that year; after some tough bargaining on details, a new industrial agreement—the Sugar Refining Agreement (6)— was finally approved in March 1937; and, lastly, a financial agreement between the government and the British Sugar Corporation was signed in March 1938 (cf. 7). The principles on which the subsidy was to be administered had been embodied, though rather ambiguously, in the Act, and the purpose of the Agreement was to provide means by which the Corporation should, granted these principles, enjoy some incentive to make economies in operating costs. The "rate of assistance" for each sugar-beet year was to be calculated and announced in advance, in the light of the performance of the previous year; but this was to be compared with what it might have been, had the level of efficiency of the British Sugar Corporation been the same as that of its prede- cessor companies in the year before amalgamation. The difference revealed by this exercise in accountancy was to be divided between government and Corporation; the latter was to retain the whole amount of economies in the year they were first made (i.e., the "rate of assistance" would not be adjusted after the event on their account), and thereafter a proportion diminishing year by year, but diminish- ing less abruptly as the total economies mounted.

The effect of these contrivances was to make the sugar industry into a tight cartel, one member of which—the Corporation—was sub- sidized, partly by the others (through the extra quota payment), that is to say, by the sugar consumer, partly by the duty preference, and partly direct from the Exchequer. Output was regulated by the Sugar Refining Agreement, refining profits per cwt. pegged to the 1936 level, except in so far as they might be enhanced by increased efficiency; the Corporation was barred from quantity production of "speciality" sugars such as caster and cubes. Moreover, although the government had the right to nominate the chairman and two other directors of the Corporation, both Tate and Lyle and the leading firm of sugar brokers—Czarnikow—were also represented on the board by reason of having had an interest in the transferor companies. When one also considers that, by reason of the International Sugar Agreement of 1937, the imported raw material was subject to export-quota restric- tions, the notion of "free enterprise" applied to sugar (as, for instance, during the campaign by Tate and Lyle against the threat of national- ization by the postwar Labour government) seems to have little mean- ing (cf. 22, pp. 317–22).

It was this prospect of quasi-nationalization that had weighed with the majority of the Greene Committee, when they argued that the cost, financial and other, of the sugar-beet subsidy was disproportionate to its benefits. That the subsidy had helped to keep land under cultivation at a time of agricultural depression, and had mitigated rural unemployment, was not denied; nor was the beneficial effect of beet cultivation on the fertility of the land and the yield of subsequent crops in the rotation. But (argued the majority) the money had been ill-spent in that the factories skimmed off much of it before it reached the grower; and as relief to agriculture generally, the subsidy was "haphazard and inequitable," for it benefited only a minority of farmers within easy reach of the factories. A similar sum laid out differently might have been more beneficial. So too with the argument that the sugar might be useful in time of war; it might be better, even then, to import more sugar and grow another crop instead.

These arguments, for all that they were disputed by the dissenter on the Committee, appear valid if—and it is a very big if—one is considering the problem on an academic plane. As arguments against the original decision to establish the industry they might be considered overwhelming. At a time of general, as well as agricultural, depression, with the industry already in being and the government already committed to a variety of projects for assisting other branches of agriculture, they were bound to appear less attractive, particularly from the political angle. In such matters it is the first step that counts; the industry had really won its battle for existence in 1924, and all that was in question by 1935 was the conditions under which it would be allowed to continue. A Labour government would undoubtedly have brought in outright nationalization, probably of the refiners also; whatever advantages or disadvantages that might have had, it could not but have produced a less complex set-up than that created by the Act of 1936.

PREPARATIONS FOR WARTIME CONTROL

From the point of view of the handful of officials in the Board of Trade's Food (Defence Plans) Department, engaged from November 1936 onward in preparing plans for controlling the country's food supplies in time of war, the existence of a regulated sugar industry was an unmixed advantage. The effect was to limit the scope and extent of the discussions with interested parties that took up so much of the Department's time, and to make it possible, after Munich, to devolve detailed preparations upon the Sugar Commission (except for the rationing machinery, which remained the direct responsibility of the Department and perhaps in consequence was not well adapted to the needs of the commodity control). It was an axiom of British food control, derived from the experience of World War I, that government trading in the principal foodstuffs was a *sine qua non*; for sugar, this requirement was met by substituting the Ministry of Food for the Mincing Lane Terminal Market. The Ministry's Sugar Division would buy imported sugar, preferably under bulk long-term contracts with the supplying countries, and allocate it for refining or direct consumption as the case might be. The brokers would act as Ministry agents for handling individual cargoes on arrival, and would, as in prewar times, continue to draw commission from exporting countries; only on the small quantity of "direct consumption" sugars (Demerara, muscovado, etc.) would the Ministry itself pay commission. Refiners would sell sugar at prices to be determined by the Ministry, in quantities prescribed by *documents of entitlement* issued to the buyers in accordance with Ministry regulations; so too with dealers (i.e., agents buying at first hand), wholesalers, retailers and sugar-using manufacturers. Prices of sugar products, however, were not at first controlled, though those of molasses and sugar-beet pulp, and of the allied product, glucose, were. The management of all these activities would devolve on leaders of the trade, seconded for war work in the Ministry by their firms—Tate and Lyle, the British Sugar Corporation, and Czarnikow—and on the staff of the Sugar Commission. Plans to this effect, including the appointment of Area Sugar Officers to supervise distribution, a scheme for providing alternative supplies of refined sugar should the London refineries be knocked out or damaged in air raids, and the drafting of statutory rules and orders for taking over sugar stocks and regulating prices from the outbreak of war, were completed, against time, in the first eight months of 1939.

Another part of the food defense plans, the purchase of a security stock of sugar, was less successful. By April 1938 the Cabinet had finally overcome its fear that disastrous consequences (particularly for the Exchequer) might result from an entry by the government into the commodity markets, and had sanctioned a secret purchase that included 150,000 tons of sugar. The intention was that 50,000 tons should be in the form of Empire raws and the remainder home-grown raws that would otherwise have been sold by the Corporation to the refiners from the output of the 1938/39 beet campaign, and that all should be housed in especially built stores adjacent to the factories. In addition, Tate and Lyle were asked, and informally agreed, to maintain a stock higher by 50,000 tons than was usual at the seasonal trough, i.e., in the late summer. Thus, by the summer of 1939, it was hoped that the country's sugar stocks would be higher than normal by 200,000 tons.

There was, of course, so much sugar in the world that this extra demand need have had no effect on supplies or prices. But the Food (Defence Plans) Department had reckoned without the International Sugar Agreement, under which export and import quotas were being regulated within very close limits; quotas for the sugar year 1938/39 were, in fact, reduced by 5 per cent, or almost the exact amount that the British government was buying for its security reserves. (If these had been transferred to duty-paid warehouses, they would have counted in the statistics as consumption, and so might have prevented quotas from going so low.) Unluckily, also, the United Kingdom beet crop for 1938 was the worst on record; the British Sugar Corporation had only 60,000 tons of raws to spare above its refining quota, and the government had to buy 40,000 tons of Empire raws to make up the security stock, which was complete in March 1939. By that time the market had risen considerably, but it was still not realized that the International Sugar Council might have to increase quotas at short notice. Nor had account been taken of the fact that the Agreement did not provide for the adjustment of quotas to meet shortages or surpluses of raw or refined sugar, as the case might be, but only for a blanket adjustment of both raw and refined in the same proportion. Tate and Lyle had brought the point up in November 1938, but had not pressed it; and nothing effective was done about it at the Council meeting of January 1939.

When, therefore, in April 1939, the refiners suddenly realized that toward the end of the sugar year their stocks would reach the vanishing point if they were not enabled to obtain an extra quota of raws, it was six weeks before the International Sugar Council could be brought to remedy what was now officially an "acute shortage," by the release of 150,000 metric tons of British Empire raws. (In May, 239,000 metric tons had been released, but one-third of this had been refined sugar, and much of the rest was from faraway sources of supply.) There developed a large premium of spot over forward sugar prices, amounting in June 1939 to nearly 2s. 6d. a cwt., to which sugar users responded by running down stocks. Tate and Lyle, seeing that to maintain the extra stocks under their gentlemen's agreement would cost them £200,000, requested to be released from it. Though the International Sugar Council released another 100,000 tons of quota on July 13, the shortage still persisted; when the government decided to buy a further 100,000 tons for its security stock (in September-

October futures) it could only secure 80,000 tons (which did not, of course, arrive until after war had broken out). As a result of these misfortunes, bonded stocks of sugar on the outbreak of war were not appreciably higher than they had been two years earlier before the purchase of security stocks had been embarked upon; stocks at later stages of distribution were probably smaller. The position was embarrassing to the Ministry of Food at the outset of control; had the war begun with the heavy air raids that were expected, sugar might have been dangerously short (*8*, pp. 26–28).

THE STRATEGY OF OVERSEAS PROCUREMENT

Guided by long experience, the Sugar Division of the Ministry, even before it was formally constituted in September 1939, had a clear picture of what it meant to do: to rely for supplies, so far as it might, on Empire and home-grown crops; and to use collective long-term contracts as a means of preventing prices from getting out of hand. Even before the war, informal approaches had been made to Canada to secure coordinated buying, and an agreement to that effect was made shortly after war broke out; a little later, New Zealand, Ceylon, and British Malaya were brought into the arrangement. Broadly, the United Kingdom would contract for all sugar supplies (including any marginal quantities that might be required from outside the Empire), and allow the other partners their shares at the equivalent of the basic price (which was always expressed in terms of full-duty sugar of 96 degrees polarization, c.i.f. the United Kingdom). By September 21, less than three weeks after war had broken out, long-term contracts with all Empire sugar producers had been reached, on the basis of 7*s*. 6*d*. a cwt. (equal with Empire preference to 11*s*. 3*d*.). The United Kingdom undertook to pay any further increase in freight or war risk insurance.

These contracts were to continue in force for the whole of the war period, subject only to rises in the cost of production and/or transport, and without doubt they assisted greatly in stabilizing prices. The hope that they would enable the United Kingdom to do without "foreign" sugar was, however, to be rapidly disappointed. It had been based on two assumptions: that rationing would be introduced simultaneously with food control, thus permitting an immediate reduction in imports equal to the amount normally drawn from foreign sources; and that there would be no shortage of shipping to compel the Ministry of Food to concentrate on nearer sources of supply—on Cuba rather than on Queensland, Mauritius, or Natal. These assumptions were faulty. The Food (Defence Plans) Department's rationing scheme necessarily took some nine or ten weeks to bring into effect, and actually was delayed till January 1940 (though an unofficial restriction of supplies had been put into force four weeks earlier); the government's advisers on shipping had failed to uncover the factors that were bound to produce a fall in imports at the outbreak of a major war, even were the enemy wholly inactive. The autumn of 1939 saw an awkward interregnum between war and peace in which raw sugar stocks fell abruptly; the port refineries had to be kept going by drawing upon the home-grown crop as it was produced (which wasted inland transport and upset the refining quota arrangements); and about 100,000 tons more sugar was consumed than had been bargained for. As, in addition, the West Indies crop was disappointing, there was no alternative but to buy 300,000 tons of "dollar sugar."

Even so, sugar rationing began with a ration of three-quarters of a pound per head per week, instead of the pound that had been hoped for and promised; and the allowance to sugar-using manufacturers averaged only 60 per cent of their datum usage.[2] The hoped-for improvement in the shipping position did not materialize; the West Indian crop turned out even worse than had been expected; and the Treasury was even less willing than before to spend on sugar its dwindling dollar resources. It did agree, toward the end of May, to a purchase of a further 100,000 tons; but this was for stocking up in the face of the German advance, not for consumption. On May 27 the domestic ration was reduced to 8 ounces per head per week, a level at which it was to remain (except for a few weeks of false dawn before Pearl Harbor, and for a few months before the devaluation of the pound sterling in 1947) for more than a decade. Shortly afterward, the allowance of sugar to manufacturers was knocked down another 10 per cent, and the first of a series of sumptuary restrictions on its use introduced—the icing of cakes was forbidden, even by private individuals, out of their rations. This order caused much distress, particularly to brides, and two years later was replaced by one forbidding the adding of edible substances to cakes after baking *in the course of business* (9).

Thenceforward, the Ministry of Food was, broadly speaking, in this position: when it could get the sugar, there were no ships to be had; when it could once again get the ships, sugar was chronically

---

[2] The severity of this restriction on manufacturing users is still more impressive if it is recalled that the base for later controls in the United States was 1941, a year when sugar usage was highly inflated.

scarce all over the world; last, when ships and sugar were more plentiful, dollars were short. During the war, unremitting care and skill were required to maintain the severely restricted level of consumption—about five-eighths of that before the war. The Ministry had always to look beyond the United Kingdom import program, to the management of supplies in the whole British sphere of interest. Its overriding aim was to maintain production, even when the sugar could not be shipped, in the interests both of the producers—many of them colonial natives for whose welfare the United Kingdom government was responsible—and of the postwar position. At no time did the Sugar Division subscribe to the policies of crop restriction that were being mooted in London and Washington from 1940 onward, even in face of forecast surpluses of 2 or 3 million tons. It assented reluctantly and against its better judgment to the schemes for restricting the use of fertilizer on sugar cane, and for promoting the growth of other crops so as to reduce dependence on imports, that were espoused by the Anglo-American Caribbean Commission in 1942–43. It continued to deplore the policy, dictated largely by the shipping authorities, of relying on Cuban supplies and allowing production in the British West Indies and Santo Domingo to fall because the sugar had not been moved; believing, on the contrary, that Cuban sugar should be stored against postwar requirements and the maximum amount shipped from the smaller Caribbean islands. It even deprecated the American decision to allow offshore sugar purchases to rank for Lend-Lease, on the ground that this deprived the British of control over their sources of supply, and was relieved when, in 1944, the policy was reversed. There can be no doubt that its constant advocacy of the dangers of postwar shortage had, in the end, a salutary influence on those responsible for United States sugar policy.

The pattern of procurement after 1942 in the areas of British responsibility (including Canada) was reasonably straightforward. The Middle East and Indian Ocean countries relied on Mauritius and Natal; exports from Queensland and Fiji, both limited by arrangement with the producers, would go mainly to New Zealand and the Pacific Coast of Canada, the remainder (about one-third) being used as return cargo to whatever destinations might be available; finally, supplies for the United Kingdom, "Atlantic Canada," and the European neutrals would be drawn entirely from the Caribbean (except for casual shipments from other sources). As the Caribbean was also the principal source of supply for the United States, it was over this area that the principal debates in the Combined Food Board Sugar Committee arose. Matters went reasonably well so long as supplies were ample and the main problem was the allocation of scarce shipping for the United Kingdom import program; but when the negotiators were faced with the prospect of a supply deficit, from the calendar year 1944 onward, bargaining became tough. The United Kingdom pressed for further reductions in United States consumption; the United States retaliated by pointing to the high level of United Kingdom stocks, ignoring the fact that they had been built up by severe rationing.

For a time both sides were able to maintain their positions intact, largely because the war continued and other claims on sugar supplies remained in abeyance; in 1944 the British (and Canadians) got the marginal Cuban supplies they required; the Americans did not cut down consumption. In 1945, however, both sides had to yield to the pressure of events. By the Memorandum of Understanding of April 1945 (*8*, p. 251 ff.) it was agreed that sugar consumption in the United States, Canada, and the United Kingdom should be henceforth on a par; but also, that the United Kingdom stocks should be brought down, by the end of 1945, to roughly the prewar level. For other sources of supply within the British sphere, the main problems of the war years were weather and shortage of labor and fertilizer. Crops in Natal were badly affected by drought; in 1943, supplies for Indian Ocean countries were so short that the Ministry of Food entered into an elaborate deal whereby it "borrowed" 70,000 tons from Egypt (rather than pay an inflated price that might have affected its dealings with other suppliers). In Mauritius two disastrous cyclones damaged or destroyed much of the 1945 crop.

It is against this background—the decline of Commonwealth supplies, coupled with the expectation, more than realized in the event, of a chronic shortage of dollars—that the United Kingdom government decided to give long-term assurances to overseas sugar producers. Imperial sentiment and United Kingdom self-interest pointed in the same direction, for on the revival of nondollar supplies hinged the prospects of an early removal of sugar rationing. The very policy that wartime events had partly frustrated thus became the means to postwar salvation. As the dollar shortage persisted, and with it rationing, the commitments to colonial and Commonwealth sugar producers were not only continued, but by their mere continuation became more firmly established for the future. Ultimately the Commonwealth

Sugar Agreement of 1951, the most cut-and-dried of all agreements to date, was to impose on a Conservative United Kingdom government the continuation of government purchase of imported sugar, after the end of control, until a new method of implementing the Agreement could be devised by the Sugar Act of 1956.

THE HOME-GROWN CROP

Although the hazards of wartime sugar supply had been the prime cause for establishing a home-grown industry, the plans made from 1936 onward for a plowing-up campaign on the outbreak of hostilities had overlooked the possibilities of increasing the sugar-beet acreage. For its part, the Food (Defence Plans) Department (which was responsible for food control, not food production) had budgeted in 1938 for an output of home-grown sugar in the first year of war higher than that permitted to the United Kingdom under the International Sugar Agreement. When the Ministerial broadcast of September 4, 1939, that launched the plowing-up campaign, omitted all mention of sugar beets as one of the crops to which the newly broken pasture might be sown (*10*, p. 33; *21*, p. 73), the alarm naturally aroused among the beet growers had to be assuaged by hasty government assurances that the British Sugar Corporation was expecting to issue contracts for the normal quantity of beets in 1940. Within a few weeks, however, the Ministry of Food had decided that it would like the acreage raised from the normal 350,000 to 400,000, which, given average yields, would produce about half a million tons of sugar. This request evoked general agreement that beets were, indeed, a suitable crop for planting on newly plowed land—their "cleaning" qualities commended them to farmers, whose leaders foresaw no difficulty in securing the extra acreage within reasonable distance of the factories. Some inducement would, of course, be necessary in the form of a higher price, and the National Farmers' Union asked that the 1939 basic price of 46s. 3d. a ton should be put up to 50s., with an increase in the sugar-content bonus rate from 3d. to 4d. The basic price related to a sugar content of 15.5 per cent, and the bonus was added for each 0.1 per cent in excess. As the *average* sugar content was 16.9 per cent, the changes in basic price would now, even more than before, underestimate the farmers' takings. The Union further asked for a guarantee that the price would be revised if production costs increased during the year.

Although the government agreed broadly to these terms, the response of farmers at large was unenthusiastic. Under war conditions there were other crops (some of them, like malting barley, uncontrolled in price at first) that promised a better return than sugar beets under less irksome conditions; farmers were, and continued to be, worried about labor in particular; the government guarantee offered at best a refund of increased production costs, and so was not calculated to induce the planting of a marginal acreage. In spite of exhortation, including a special mention in a speech by Neville Chamberlain on February 26, 1940, the acreage contracted for 1940 was actually slightly less than that for 1939. (The harvest for 1939, however, was splendid, yielding well over half a million tons of sugar.) The government was not disposed to take this failure tragically; there was plenty of sugar abroad and the full extent of the shipping squeeze that was coming was not realized, even in August 1940, when sugar-beet prices for the 1941 crop were being fixed. The basic price was fixed at 61s. 6d., a figure which was thought to include an element of "inducement" (and did include 3s. 9d., in respect of enhanced agricultural wages, from July 1940). Even so, the contract did not go well. The situation was saved to some extent by the coming of rationing for animal feed, in February 1941. Departments agreed, and were careful to let growers know, that the sugar-beet pulp they were entitled and accustomed to buy back from the factories, *pro rata* to their deliveries of beet, would not be taken into account when assessing their requirements of feedstuffs. This, coupled with intensive propaganda, sufficed to get 350,000 acres grown.

It was not until the summer of 1941, with a long war in prospect, that the official attitude toward the sugar-beet acreage (that is to say, outside the Sugar Division of the Ministry of Food) became something more than lukewarm, and it was agreed to secure the extra acres by issuing compulsory cropping orders where necessary. It was settled policy that a decision to apply compulsion was no substitute for offering sufficient financial inducement to the farmer; on the contrary, it was thought unfair to compel him to grow a crop that showed a low profit compared with others. The competitor *par excellence* of sugar beets, particularly in East Anglia and the Fen Country, was potatoes. For growing this crop, the Ministry of Food once again evinced what F. H. Coller, speaking of World War I, had called "a strange craze" (*11*, p. 119). In pursuance of this policy a large increase in potato prices had been granted in 1940, followed by an acreage subsidy in 1941; with the result that the returns per acre on potatoes were roughly twice those on sugar beets (*12*, esp. pp. 119–22). In these circumstances it was idle to talk of limiting the rise in beet prices to ascertained increases in cost. Toward the end of 1941, after some months of discussion, the basic price was raised to 81*s.*, a rise of almost 25 per cent over that in the preceding year; 6*s.* of this, however, was deemed to be compensation for a further rise in the statutory-minimum adult farm wage.

This level of prices, coupled with some concessions in the matter of transport charges, proved sufficient to call forth an acreage in each of the remaining war years that, given average yields or more, meant as much beets as the factories could cope with. A further 5*s.* a ton was allowed in 1945, at a time when bad weather and shortage of labor had impeded the harvesting of the 1944 crops, both potatoes and beets, and had discouraged growers from making contracts for the following season. There remained, however, problems of production: shortages of fertilizer, which affected the sugar content; shortages of labor for singling and harvesting; and finally, infestation by eelworm, which made it difficult to find suitable land on which to plant beets. As early as July 1943 a Ministry of Agriculture order scheduled four "infected areas" in which the planting of sugar beets was regulated (*13*). As the main beet-growing districts were also those in which the potato eelworm was becoming rampant, the task of the Country War Agricultural Executive Committees in planning cultivations became a complex one.

To these difficulties in the field were added others in the factories. The extra beets to be handled meant a campaign extending into January or even February, so that the roots were liable to lose sugar content, or worse, be ruined by alternate freezing and thawing; the campaign could not *begin* earlier, because of priority given to the harvesting of potatoes. Inferior coal and inferior labor hampered work, and the factory plant suffered from want of skilled maintenance. (One of the earliest jobs the Ministry of Food had to tackle when Europe was liberated was to secure German technicians to attend to this machinery.) For all these reasons the efficiency of the factories, measured in terms of sugar extraction, gradually fell away, so that the full benefit of the extra acreage was not secured.

Nevertheless, the contribution of home-grown supplies to the national total was very substantial, amounting, as the British Sugar Corporation was fond of remarking, to the whole of the domestic household ration of 8 ounces per head per week. The production details for the relevant crop years are given in the following tabulation:

Campaign	Beets bought (1,000 long tons)	Average sugar content (Per cent)	Average extraction rate (Per cent)	Sugar output (1,000 long tons, white equivalent)
1936/37.....	3,448	17.31	87.45	522
1937/38.....	2,583	17.03	85.74	377
1938/39.....	2,191	16.13	81.91	289
1939/40.....	3,513	16.54	83.90	487
1940/41.....	3,176	18.05	83.71	480
1941/42.....	3,226	17.22	83.21	462
1942/43.....	3,923	16.65	78.92	516
1943/44.....	3,760	16.48	79.35	492
1944/45.....	3,248	15.31	77.88	387
1945/46.....	3,886	16.20	80.12	503

After the war, the beet acreage was maintained at a high level, originally on account of the shortage first of sugar and then of dollars to buy it, thereafter on grounds of agricultural policy that stressed the value of beets as a crop. The marginal acres (perhaps 50,000) constituted a strain on factory capacity as well as on the Exchequer. At the world prices then prevailing, it is likely that an industry of the prewar size could have made do without any direct subsidy. From 1950 onward, however, the British Sugar Corporation undertook a massive re-equipment of the factories, made necessary by the obsolescence of the original plant. This greatly increased their efficiency, so that they were able to handle over 5 million tons of beets and produce 720,000 tons of sugar (white equivalent) in 1953/54. Over the last few years even this larger amount of home-grown sugar has not, it is said, cost more than sugar bought under the Commonwealth Sugar Agreement, so that the opinion of the minority of the Greene Committee has been justified in the event.

RATIONING AND CONTROLLED DISTRIBUTION

Rationing of sugar to domestic consumers began in January 1940 and continued until September 1953. The bit of the rationing system that was visible to the public—the ration book, with its counterfoil to be lodged with the consumer's chosen retailer, and its page of sugar coupons that soon became no more than spaces to be canceled by the retailer when the ration was bought—had only a limited function. British food administrators frowned upon the notion of a ration currency, under which the demand, as expressed by coupon "flowback," should automatically regulate the volume of sugar moving through distributive channels. They started from a completely different notion, namely, that every trader should receive an allocation of sugar, fixed beforehand in accordance with his entitlement; and that, for a retailer, the prime factor in determining the entitlement should be the number of his registered customers, bound to buy their rations from him and him alone (*10*, pp. 203–06). For sugar-using manufacturers, the prime factor was their usage in a prewar base period. The entitlement was conferred upon the trader by the issue of a *permit*: to retailers and small manufacturers, by the local office of the Ministry of Food; to large manufacturers, through trade associations such as the Cake and Biscuit Manufacturers' Wartime Alliance. The permit covered a limited period (which eventually was settled at eight weeks); it was lodged by the trader with his wholesaler, who was obliged to prepare summary statements of the permits held by him. At this point the Ministry's Commodity Control, as distinct from its local office, came into play. On the evidence furnished by wholesalers or manufacturers, the Area Sugar Officer issued *vouchers* authorizing a particular refinery or beet-sugar factory to release the sugar to the firsthand *dealer*, or directly to certain large firms that had been allowed to retain their prewar right of buying in this way. Separate vouchers were used in respect of what was called "direct consumption" sugar—high-grade raws such as Demerara and Muscovado, sold against the ration, and others used by manufacturers—which was released by the Ministry through a limited number of brokers. Arrangements were also made for sugar millers, who prepared fine grades of caster and icing sugar needed by specialty manufacturers, to receive their raw material and sell their products against ration documents.

The control of supplies was thus very complete and yet not inflexible; for the Area Sugar Officers, men drawn from the staff of the refiners or the British Sugar Corporation and usually housed with the firm that had lent them to the Ministry, could vary, by means of vouchers, the sources from which supplies were drawn. Sugar Division Headquarters was accustomed to regulate the movements of raw and refined sugar and the production ("melt") of refineries in accordance with, for instance, the arrival of ships, the stock position, fluctuations in demand, or the progress of the beet-sugar campaign. Thanks to the small number of productive units, the expertise of Headquarters staff, and, last but not least, the fact that only a handful of men at the top, who had long worked together in peacetime, were involved in decisions, these arrangements could be made swiftly, informally, and without fuss. When, in the name of transport economy, the call came from "higher authority" for deliveries of refined sugar to be "zoned," the process was carried out by administrative action within the existing control framework, whereas in the case of other foodstuffs complicated inquiries and often recourse to statutory rule and order were required. All vouchers for sugar to be supplied to traders within a particular zone were simply drawn upon British Sugar Corporation factories. The zoning scheme cut off many customers from specialty types of sugar; manufacturing users in the Corporation zone suffered especial hardship.

Coupled with the control of distribution went that of prices. Here the mechanism of rationing, which had been agreed to by the trade before the war without having been fully explained to them, had unforeseen consequences. In peacetime there was no system of margins, or mark-ups, by which the various links in the distributive chain were remunerated; its place was taken by quantity rebates (and discount for prompt payment) granted by the refiners and hence by other traders. The price a man was charged for sugar depended strictly on the amount he paid for at any one time; his profit came from selling in smaller quantities and perhaps from extending credit. Multiple shops, often buying direct from refinery in large quantities, benefited especially from this system, and frequently used sugar as a "loss leader," or bait with which to entice customers for other goods. Consequently, rates of profit on sugar at all stages of distribution were extremely low. The equilibrium of the system clearly depended upon the freedom of traders to purchase sugar in the customary quantities. When, however, quantity rebates were incorporated (with certain modifications) in the wartime price control orders, no one took into account the fact that releases of sugar would not merely be reduced, but made on a strictly weekly basis, even though the order specifically related discounts to the buyer's "weekly quantity." Many wholesalers, when rationing was introduced, found themselves asked to support a diminished trade on a disappearing profit margin; and their position became worse when the ration was reduced from 12 to 8 ounces. In August 1940, therefore, a new order (*14*) adumbrated a system of statutory maximum prices for every class of wholesale transaction, including a system of allowances and rebates in restraint of "double margins": for instance, a dealer buying *ex-refinery* and reselling to a wholesaler got an allowance of 4*d*. a cwt., but if he sold to a retailer, i.e., acted as a wholesaler, he got only the wholesale margin. The new arrangements involved many accounting complications; as the price charged by a refiner to a buyer now depended on the status of the latter and even on the ultimate destination of the sugar, six types of voucher had to be introduced, each with its appropriate price. This, not perhaps wittingly, paved the way for the next change of policy—a decision to subsidize "domestic" and surcharge "manufacturing" sugar, although the two were, as a commodity, identical.

Although the new system of margins worked satisfactorily, the occasion for it ought never to have arisen in the way it did, for the "weekly quantity" as a basis for the allocation of sugar was quite impractical. It had been laid down by people who made no distinction between an accounting transaction and the physical movement of the commodity concerned. Sugar is put up in containers—whether bags of loose sugar or cartons containing packets—that come in recognized quantities: multiples and submultiples of a hundredweight. The food regulations at first took no note of this simple fact; if a permit were issued allowing a trader a hundredweight and a half a week, he and his supplier would be breaking the rules (and, from June 1940 onward, the law) if two and one hundredweight respectively were delivered in successive weeks. The outcry against this absurdity might have been greater had traders generally been aware of it. In fact, the letter of the law was largely ignored, and it had already been decided to alter it when the first prosecution was embarked upon by an overzealous local Food Control Committee—which had hastily to be called off by headquarters. From March 1941 permit quantities were to be "rounded off" to the nearest 28 pounds: in November a "global" permit, allowing eight weeks' supplies to be bought as and when required, was introduced. By a similar relaxation the consumer was no longer obliged to take his ration in the week for which his coupon (or rather his ration-book space) was valid. From July 1941 onward, rationing worked to a four weeks' schedule, and sugar might be bought at any time during a four-week period. Under war conditions weekly delivery of supplies could no longer be guaranteed.

The allocation of sugar to manufacturers was, except for the different basis of entitlement, similar to that for domestic use. A base period was taken—normally the twelve months ending June 1939—and supplies were expressed as a proportion of vouched-for usage during that period. The proportion varied according to the type of product being manufactured, and at first was set by rough rule of thumb: condensed milk, 75 per cent; fancy molasses and chocolate, 70 per cent, cakes and cookies, 60 per cent; mineral waters and candied peel, 25 per cent, to take a few examples. At this stage (January 1940) the Ministry of Food was exercising no sort of control over the use to which sugar so allocated was put. A year later, with most types of food chronically scarce, it was beginning to talk of a "consumption program," under which allocations of sugar and other ingredients of manufactured foods should be correlated one with another. The simple-datum basis for allocations thus became complicated; sugar and fats for bakers, for instance, were allocated in concert between the Ministry's Sugar and its Oils and Fats divisions, the datum-based quantities being adjusted for variations in the population of different districts, compared with those of June 1939. In other words, differences in cake consumption per head between, say, Aberdeen and London, as reflected in the prewar datum, were perpetuated; but no consumer was allowed to gain new advantage (or to suffer) because of changes in the population of his district. Jam manufacturers, whose product was thought to assist the public to make do with a diet consisting heavily of bread, were allowed *more* sugar than before the war; but jam consisting mostly of sugar and fruit pulp proved less popular than the Ministry had hoped, and it was faced with a recurrent problem of disposal. Chocolate and sugar confectionery was rationed in 1942 and the sugar allocation to manufacturers was based on a production program; similar programs were laid down for less important sugar users, such as the makers of pickles and sauces. Last, the use of sugar in cereal breakfast foods, cake and pudding mixtures, table jellies, and ice cream was forbidden; in others, such as cake, it was regulated so that the maximum amount of an austerity product might be made.

The tighter regulation of the use of sugar allocations really closed the only loophole in the control: namely, the misuse of manufacturing supplies lawfully obtained. There was no black market, properly so-called, in sugar, if one ignores the insignificant trade in supplies obtained by theft. There was, however, one possible source of leakage through legitimate channels: the rationing system itself, with its tie to the retailer, had the effect of gradually inflating demand between annual reregistrations of consumers, enabling retailers to build up stocks which lent themselves to misuse; but there is no evidence that this evil ever reached dangerous proportions. Some outlet for human frailty is almost a *sine qua non* of a sound rationing system. After the war one or two of the Ministry's coupon-flowback schemes (not for sugar) became so perfect as to denude retailers of stocks (*10*, p. 208).

SUGAR AND THE COST-OF-LIVING INDEX

Although the first impulse toward the establishment of food control had come from fear of wartime inflation, the attitude of the Chamberlain government toward the rises in food prices that occurred in the first weeks of war was ambivalent. The Treasury seemed unable to make up its mind whether food prices should be kept down to discourage wage demands or be allowed to go up to discourage consumption. Sugar, as one of the few traditionally dutiable foods, and one, moreover, that was likely to cost scarce dollars, tended—as the Minister of Health remarked with surprise in December 1939—to be regarded by ministers as a luxury. So, in Sir John Simon's emergency budget of September 1939, an extra tax equal to 1d. a pound had been clapped upon it. The retail price of granulated sugar, which had already gone up from 3d. or less to 3½d. on account of increased costs, thus became 4½d., of which 2½d. was tax.

During the first year of war, the Ministry of Food had managed to make a small profit on its trading operations in sugar. But by the summer of 1940 it was clear that this would be eroded in 1940/41 by increased prices under bulk contracts related to production costs, higher freights, and higher costs of refining; moreover, the margins allowed to cover costs of distribution were demonstrably too low. The question arose whether a sugar subsidy should be introduced, or the tax lowered so that the current price might absorb higher costs. But the Treasury, though already alarmed lest rises in the cost-of-living index should set off an inflationary spiral of wage demands, as yet preferred the small rise in the index that ½d. a pound on sugar would represent to the £4 millions that price stabilization would cost. In August 1940, therefore, the price of granulated sugar went up to 5d., on the introduction of the new system of wholesale margins instead of quantity rebates.

By December, however, this policy had gone into reverse. Alarm at the steady rise in the cost-of-living index caused the Treasury to embrace a proposal that had been mooted at intervals since mid-1939; that sugar for domestic consumption (which was included in the index) should be subsidized by that used for the manufacture of sugar products (which were not). As the amounts in question were about equal, this could be accomplished without upsetting the Ministry of Food's arrangements for selling raw sugar to refiners; these would henceforward sell refined sugar on "domestic" vouchers at a paper loss of 1d. a pound, which would be roughly recouped by the extra profit of 1d. a pound taken on sugar sold against manufacturing vouchers. It became, of course, extremely important that no mistakes should be made by Area Sugar Officers in specifying the purpose for which sugar was to be used. Other complications arose in the introduction of the scheme. In order that retailers, who mainly, of course, sold "domestic" sugar (other than sales to very small manufacturers and beekeepers) should not have to take a loss on stocks that the Ministry, for reasons of safety, had asked them to maintain, the reduction in the wholesale price was made to anticipate that in the retail price by a fortnight (*15*). There were those who maintained that this was too generous.

In April 1941, the Treasury at length espoused the policy of maintaining an absolutely stable cost-of-living index. The remaining "index" foods—fish and eggs—were brought under control in July, and the chief impediment to a stable figure that remained was clothing prices, which were still rising. Accordingly, in December 1941 the Ministry of Food was constrained, in order to help the Treasury with the index, to reduce domestic sugar prices by a further 1*d*. a pound, i.e., to the prewar level. This time it was not proposed to raise "manufacturing" prices; the notice was short, the task of revising the price schedules for the whole range of sugar products too onerous, a differential of 4*d*. a pound between identical sugars so great as to verge on absurdity. Thereafter sugar (taken as a whole) was to receive an Exchequer subsidy, which continued until the end of control, and to bear a tax at one and the same time. On the occasion of the last price change there had been many complaints from wholesalers dealing mainly in "domestic" sugar that their losses on the swings had not been balanced by gains on the roundabouts, and also from retailers who had held more than a fortnight's stock. On this occasion, with no countervailing price rise, the Ministry undertook to pay compensation, not only to wholesalers and dealers (on a week's and half a week's stock respectively) but to retailers, who could claim that they held more than the two weeks' stocks allowed for in the "staggering" of price changes. This concession to retailers cost upward of £200,000, and though the Ministry was able to make a convincing defense when questioned about it by the House of Commons Public Accounts Committee in 1943 (*16*), its own accounting experts had already advised it that the experiment should not be repeated.

So far as sugar was concerned, however, the question of compensation did not arise again, as the next adjustment for the sake of the index was to be upward. The introduction of the utility clothing scheme had forced the index down lower than the Treasury intended, and the task of modifying the complicated clothes price schedules had become too heavy for the Board of Trade to undertake. In September 1943 the price of granulated sugar for domestic consumption was once again raised to 4*d.* a pound, 2*d.* below that of "manufacturing" sugar; and there it remained until 1947. Later a number of changes occurred, both in price and in tax; but an element of subsidy remained until sugar distribution was decontrolled in September 1953, and for a short period thereafter. It would be true to say, therefore, that from December 1940 onward the price of sugar (whether to domestic consumer or manufacturer) was an artificial price, unrelated to that at which the country got its supplies. But it should also be emphasized that, quite apart from the policy of price stabilization, it would have been necessary on administrative grounds alone to avoid frequent changes in the controlled price of sugar because of their repercussions on other controlled prices (*8; 10*).

FINANCIAL ARRANGEMENTS WITH THE SUGAR INDUSTRY

The coming of control meant that the sugar industry, like other major food industries such as flour milling and oilseed crushing, would be working on government account and on terms fixed by the government. It differed from them, however, in that its peacetime remuneration was fixed by statute or by agreements, formal or informal, to which the government was a party, and which the advent of war had not automatically abrogated. Indeed, with the beet campaign about to start, it was necessary for the Ministry of Agriculture to prescribe the "rate of assistance" to the British Sugar Corporation in the normal way, even though some of the statutory conditions for its prescription had ceased to exist—for instance, an open market price for sugar. Proposals for the industry's wartime remuneration, therefore, involved legal as well as financial considerations.

It was rapidly agreed that the Sugar Refining Agreement, with its quota provisions, could be continued, even though the amounts of sugar refined by the parties to it would now be fixed by the Ministry of Food; and a new agreement, embodying formal changes made necessary by war (such as the suspension of the Sugar Commission and the exclusion from the "refining margin," as defined, of certain charges for which the Ministry of Food was now responsible) formally came into effect on April 1, 1940. In particular, the refiners undertook to continue to pay the British Sugar Corporation in respect of its fictitious "special quota"—a fact that bore on their own remuneration. For the original agreement had been contingent on, indeed a *quid pro quo* for, the maintenance of a turnover that afforded them a profit they considered sufficient, at a margin the refiners had undertaken not to increase except in respect of increased costs. Under wartime conditions their turnover was bound to be reduced; and though the levy payable to the Corporation, being proportionate to the total tonnage invoiced, would also be reduced, it would still constitute a heavier burden because their costs-profits ratio would be less favorable than before. An increase in the nominal refining margin, even if the Ministry would agree to it, would not serve the purpose, because under the Agreement this would increase the quota payments.

To get over these difficulties Philip Lyle produced an ingenious algebraic formula, based on his investigations of refinery costs at Tate and Lyle immediately before the war. He showed that it was possible to split up these costs into three groups—"marginal," "factory fixed," and "overhead"—of which the first varied directly with output and the other two did not. A formula could be built up for total costs per unit of output in a base year, in which these elements were suitably weighted; a similar formula for, say, any particular month of control could then be calculated by applying cost-index numbers where appropriate. The differences should, it was suggested, be paid over to refiners by the Ministry of Food each month as a refining costs adjustment, thus guaranteeing them the same net profit per ton (though not the same total profit) as they had enjoyed before the war. This proposal, though novel and to the nonmathematician alarming, was soon recognized by both Ministry and Treasury as being fair; as Tate and Lyle were not likely to be less efficient than the small refiners, there seemed no danger in applying it to them as well. All that was necessary was for the Ministry's cost accountants to check the detailed calculations and for agreed weights to be inserted in the formula. By May 1940 agreement had been reached for the formula to be applied from February 1, 1940 onward. It was also applied to the refining done by the Corporation in the off season.

One refinement that had been added to Mr. Lyle's original proposal was to cause some trouble later. Departments had jibbed at his implication that even in the event of a long-term reduction in output, no economies in overheads or fixed factory costs were possible; and hypothetical cost/output functions for each had therefore been inserted in the formula, with the effect of slightly diminishing the compensation payable to refiners when three months' notice of a long-term reduction in "melt" had been given by the Ministry. The intention on the official side had been that the whole of the savings so presumed to be possible should accrue to the Ministry; by inadvertence, however, a formula had been accepted by which they were shared by the Ministry and refiners. This came to light in the spring of 1941;

but the refiners, when asked to agree to an amended formula, naturally brought up points on which the existing formula had worked
to their detriment—notably expenditure on air-raid precautions,
which had turned out to be much heavier than had been expected. It
was agreed that this expenditure should henceforth be dealt with outside the formula payments; the excess expenditure already incurred
was allowed as a set-off against the amount that the Ministry had paid
over in error; and a revised formula embodying the Ministry's original intentions was put into force from April 1941 onward. Costings
investigations at intervals of two years showed that the formula was
working well, and it required no further amendment during the war
and immediate postwar period. Eventually, however, the results became out of touch with reality and in 1952 a new formula was adopted.

The settlement with the British Sugar Corporation took longer to
reach, and moreover required legislation. It was agreed that the statutory arrangements, by which the Minister of Agriculture prescribed
a "rate of assistance" based on "standard conditions" determined by
the Sugar Commission, and by which the actual rate of assistance was
worked out subsequently according to rules laid down in the Sugar
Industry (Reorganisation) Act of 1936 and the terms of the Incentive
Agreement of 1938, were incompatible with wartime conditions.
Clearly, the Minister of Food, who controlled the entire sugar industry, should also administer the home-grown sugar subsidy; all that
would be necessary was a simple "deficiency payment" to the Corporation, after the event. The Corporation, however, took exception
to any arrangement that did not perpetuate the Incentive Agreement,
under which it was assured that any economies in operating costs were
shared with the government instead of being plowed back completely
in aid of the subsidy; and this objection was conceded to have merit.
It also had political weight, for the provisions of the 1936 Act could
not be set aside by emergency departmental regulation: Parliament
would have to be asked to legalize not merely what was now proposed
to be done, but the technical irregularities that, unavoidably, had
accompanied the fixing of the rate of assistance in 1939/40. Moreover, it was necessary to assure the Corporation that the suspension
of its "charter" for the duration of war would not be allowed to affect
its statutory standing. The negotiations were protracted, and it was
not until April 1942 that the amending bill finally became law, as the
Sugar Industry Act, 1942 (17). Shortly afterward the Ministry of
Food concluded a wartime Incentive Agreement with the Corporation,
that preserved the principles of the agreement of 1938. As before,

the Corporation was to get a share of any economies it was able to make, on a scale increasing with the amount achieved in any given year, but diminishing with the passage of years; the whole was to be adjusted with reference to the standard quantity of 560,000 tons of white sugar laid down in the Act of 1936. The basis of comparison, however, was no longer to be that of the former beet-sugar companies, but simply that of the Corporation itself in the year previous to that for which its remuneration was being fixed. As in the original agreement, provision was made for "diseconomies," thought probable in wartime, to be subtracted from the accumulated economies already credited to the Corporation.

It should be observed that the whole of this settlement was in the highest degree artificial, designed to perpetuate as much of the peacetime financial arrangements as possible (with an eye to the future), by contrivances set up within the wartime control. For reasons of general policy it had been decided that the price of sugar was not to be allowed to rise—as it would have done if uncontrolled—to a point at which the beet-sugar industry could have made a profit. Therefore, what had hitherto been a subsidy to enable the British Sugar Corporation to stay in business if it sold its sugar at the market price was now simply a subsidy like the other cost-of-living subsidies—to keep the price down but shown in the old form in the public accounts. As the House of Commons Public Accounts Committee, which kept an especially vigilant eye upon the beet-sugar subsidy, was thus enabled to draw invidious comparisons between its cost in peacetime and that during the war, a good deal of ingenuity was spent in ensuring that the figure was not inflated by circumstances beyond the Corporation's control. For instance, a settlement was made separately for the cost of holding abnormally high white-sugar stocks at Ministry behest; and compensation around a million pounds a year—about one-quarter of the rate at which the deficiency payment proper was running at this time—was paid on account of the Corporation's obligatory losses on molasses and sugar-beet pulp, brought about by statutory price fixing.

An element of artifice likewise crept into the working out of the incentive payments—or perhaps one should say a greater element than was already present by reason of the terms of the prewar legislation. In wartime the Corporation was obliged to do many things, great and small, that it would not have done in peacetime; and hardly one of these but reacted on its apparent efficiency (which the incentive payments were supposed to measure and reward): protracted campaigns because of high acreage, potato-drying for animal feed, processing of frosted beets, to take but a few examples. More obvious impediments to efficiency came from inferior coal and labor, and the restrictions upon lighting (the "black-out") imposed by way of air-raid precautions. (One unexpected, almost bizarre, loss of efficiency resulted from the adoption of a longer working week for off-season maintenance workers. They had to be paid at time-and-a-half for the extra five hours; tea intervals amounting to 2¾ hours had to be conceded; and no staff economies materialized. According to how allowance was made for this situation, the Corporation might be deemed to make an economy or a diseconomy in 1942/43.) An almost endless series of adjustments was necessary between Ministry and Corporation before an equitable assessment of incentive allowances could be made. Even so, it became clearer and clearer as time went on that the fortuitous element in the Corporation's affairs was beyond the ingenuity of accountants to compensate. Seasons like 1944/45 and 1945/46, in which bad and good crops following on one another produced swings in costs of the order of £200,000, were aberrations not foreseen by the negotiators in 1936, that resulted in violent fluctuations in the Corporation's income inconvenient both to it and to the government. The Exchequer, it should be noted, was an ultimate legatee of Corporation economies, but derived no benefit—indeed, was out of pocket—from those arising in one year and lost the next. This was because such economies were not merely pocketed by the Corporation in the year of their making, but carried forward into the "accumulated" economies for which it took credit in the subsequent year. The original incentive scheme had been designed on the assumption that economies would invariably be cumulative year by year.

Despite undoubted anomalies, there is no reason to suppose that a fair balance was not struck between the interests of the Corporation's shareholders and those of the Treasury during the years of control. It is worth while adding, moreover, that from first to last the Corporation has been under far more effective financial surveillance than are the industries latterly brought under public ownership. The main purpose of the foregoing account has been to point out the dangers of drawing any but purely budgetary conclusions from the figures of the wartime financial dealings with the home-grown industry.

SUGAR BY-PRODUCTS AND SUBSTITUTES

The control of sugar necessarily had to take account of kindred substances: molasses, glucose, invert and other special sugars, and the substitutes saccharin and dulcin. Cane and beet molasses were controlled by the Ministry of Supply; fancy molasses (syrup and treacle), directly by the Sugar Division of the Ministry of Food. Relations between the two controls were at all times cordial and efficient, and their smooth working was greatly assisted by the interlocking directorates of the principal companies concerned: Tate and Lyle, the United Molasses Company Limited, and the British Sugar Corporation. Fancy molasses, which was mainly a by-product of sugar refining, was used for human consumption, and was ultimately rationed, first along with jam, latterly under the points scheme. Crude molasses was used for industrial purposes, but the British Sugar Corporation was allowed to retain about one-quarter of its output for admixture with sugar-beet pulp for animal feed. A small quantity of high-grade molasses imported by the Ministry of Supply was allocated to food manufacturers in lieu of sugar and on the same datum basis of usage. In 1941 a zoning scheme was introduced for syrup and treacle, under which the southern half of England was supplied exclusively from London, the northern half (and Wales) from Liverpool, and Scotland from Greenock. A committee drawn from the trade operated a quota-*cum*-compensation scheme on the lines of that for sugar refining, and it fixed standards of quality and firsthand prices; Ministry of Food statutory control was limited to retail prices.

Invert sugar, mainly used by brewers, was subjected to a series of price-control orders that maintained a rough parity between its price and that of granulated sugar. The distribution of invert sugar was not controlled by statutory order; allocation to brewers was accomplished by means of excise permits, as was that of other types of sugar used by brewers. Caramel, of which brewers were also the largest consumers, was likewise controlled by restricting supplies of its raw materials (sugar, glucose, molasses, fancy molasses, or starch, as the case might be) to a proportion of datum. Liquid sugar came under the umbrella of the Sugar Refining Agreement, and output was limited by simple instruction to the two firms concerned. In all these cases the small number of manufacturers that had to be dealt with made control relatively easy.

Glucose presented more difficulties. Its raw material is not sugar, but starch (generally cornstarch); and starch was one of the commodities for which no wartime control had been arranged in advance. Within weeks of the outbreak of war there was an acute shortage of both imported starch and of maize (from which some, but by no means all, manufacturers could make glucose); and this wrecked a well-meaning attempt, made with the agreement of the trade, to impose price control in September 1939. By December 1939 the position had become so acute that the largest single manufacturer of liquid glucose had had to close down for lack of raw materials. A Starch Control was thereupon established as part of the Cereals Control; eventually, in June 1940, a system of controlled allocation for liquid glucose was introduced, and later still, statutory control of prices and margins of profit (18). The industry itself, through its trade association, fixed proportionate quotas for the manufacture of liquid glucose; from May 1941, in order to save transport, permits to buy liquid glucose were made out in the name of the association, which nominated the supplier. In 1942–43 four of the smaller manufacturers were closed down under a "concentration" scheme (8, p. 132), but continued to invoice their quota proportions, which were made for them by the "nucleus" firms remaining in production. By 1941 consumption of liquid glucose was down to little more than half of basic usage, and by 1943 it was about 40 per cent. The shortage of glucose monohydrate, about 60 per cent of which was imported from the United States in peacetime, was equally great, though almost entirely confined to the pharmaceutical industry and therefore not presenting the complications posed by that of liquid glucose.

In time of peace, saccharin was mainly important as an ingredient of "mineral waters"; in wartime, it became important as an auxiliary sweetener for beverages, especially tea. Its principal ingredient, toluene, was almost entirely imported and was required for the making of explosives, and the Ministry of Supply would allow only a limited quantity to be used for saccharin. One firm—Boots Pure Drug Company of Nottingham—held a manufacturing monopoly in saccharin, which was marketed partly through the firm's chain of retail drug stores and partly through a selling agency owned jointly by Boots and their licensors, Monsanto Chemical Company of the United States. Imports were negligible—less than one-thousandth of total production. Control at the manufacturing end was therefore easy; the difficulty lay in the fate of the saccharin once it had left the selling agency's hands. A voluntary arrangement by which the agency agreed to limit its monthly sales and to allocate them among buyers on the basis of a datum-period performance was not sufficient to prevent a black market from developing, even though output should reach three times the March 1940 figure. Very reluctantly the Ministry of Food decided that it would have to control saccharin and dulcin prices. An order (19) of December 1941 standardized the strength of saccharin tablets that might be sold, forbade the sale of flours or starches flavored with saccharin and the sale of saccharin or dulcin solutions, and set maximum prices.

During 1942 and 1943 this order had frequently to be amended. In an effort to make the increasingly scarce supplies of toluene go further, the "full-strength" and "half-strength" tablets hitherto authorized were replaced by a standard tablet containing 0.2 grain of saccharin (as against an average of 0.3 grain for the full-strength tablet). The Ministry would have liked to make these tablets the only type authorized, but was baffled by an ingenious firm that contrived a tablet of saccharin and dulcin combined. Dulcin, unlike saccharin, can be synthesized in more than one way; and the firm had access to one of its possible ingredients—the drug phenacetin—by reason of a legitimate trade in pharmaceutical products. There was, practically speaking, no alternative, therefore, but to license this proprietary tablet and control its price; and it appeared, irrepressible, in successive orders (20), even after the Ministry, in May 1943, had applied full control of the manufacture and distribution of saccharin and dulcin. In the opinion of the trade, these extensions of control achieved no more than had been possible through the voluntary cooperation already taking place between the saccharin interests and the government. It was felt that, so far from repressing abuses, the control had been forced to come to terms with them. The black market in saccharin, however, was one that aroused much public comment and it is doubtful whether the Ministry of Food could have stopped short of a patent endeavor to do everything possible to suppress it.

AFTERMATH AND CONCLUSION

With the end of hostilities and of the wartime Coalition government, sugar control entered a period of transition that was to be prolonged for more than a decade. Its dominant features were, first, the shortage of dollars that dictated the continuation of rationing until the autumn of 1953; second, the prevailing political temper that regarded the wartime system of government bulk purchase of sugar, as of other foodstuffs, as a desirable social and economic principle, not a temporary expedient. Accordingly the Labour government had announced, as early as November 1945, that the Ministry of Food was to be made permanent. This recourse to principle made it more, rather than less, difficult for the government to defend its policy to those who did not share its beliefs and who could consequently never be sure whether continued austerity was being imposed on them from necessity or from mere conviction. The effect was to make the policies that were pursued appear more controversial than they in fact were, at any rate so far as imports and rationing were concerned; for it would be a partisan indeed who would suppose that, had the heirs to the coalition been of a different political complexion, the shortage of nondollar sugar would have been remedied earlier.

More genuinely controversial was the Labour party's attitude toward the United Kingdom industry; in its view, the Act of 1936 was a halfway house toward nationalization, and it would be logical, as well as good socialism, to make the end of wartime control coincide with the transfer of both Corporation and refining industry to public ownership. In April 1949 a draft party statement of policy announced this intention, and in due course it was adopted by the party conference. Meanwhile, however, the refiners had made it clear that, unlike some other industries that had been slated for nationalization in the past, they were not going to sit down and wait to be taken over. The directors of Tate and Lyle announced that they would not personally participate in a nationalized industry, and made arrangements to "hive off" to a separate company various assets that might otherwise be acquired by the State as part of the act of nationalization. Moreover, they began a vigorous publicity campaign that gained piquancy from the fact that sugar was still rationed by the Ministry of Food and so the consumer received his statutory allowance in packets printed, for example, TATE NOT STATE. Tate and Lyle, moreover, succeeded in its claim that expenditure of this sort was an allowable deduction for tax purposes.

The resistance was fortunately timed, for in 1949/50 the government's term of office was running out and the balance of public opinion had tilted, if only slightly, against further nationalization. The general election of 1950 put an end—for legal reasons—to the campaign, but the derisory majority the government secured, and the Conservative victory at a second general election in 1951, removed the threat for the time being.

The King's Speech opening the new Parliamentary Session in October 1950 had forecast a bill to transfer the British Sugar Corporation to public ownership, but this never materialized. The subsequent winding up of the wartime controls—culminating in the Sugar Act of 1956—means that a fresh movement toward nationalization would have to start, if not from scratch, at any rate from the compromise, now more than ever firmly established, introduced in 1936. The disappearance of the Sugar Commission and the advent of a Sugar Board, which will administer a levy-subsidy on the lines of that first used in the Wheat Act of 1932, does not mark a breach of continuity in essentials, though it does mark the end of the wartime interregnum.

* * *

At first sight it might seem cynical to remark that sugar control in the United Kingdom was a triumphant success in every respect but one—the maintenance of a sufficient supply to consumers. Cynical or not, it would be no more and no less than the truth; for few but nutrition experts would deny that for almost the whole duration of control, the sugar ration represented a severe, indeed the most severe, curtailment of a civilized amenity. Indeed, but for the low view those experts tend to take of sugar, as being merely a source of calories, it might have enjoyed higher priority in supply programming; and— who knows?—output in, say, the British West Indies might not have been allowed to fall away so much in the later war years. At any rate, the shortage of sugar cannot be laid at the door of the United Kingdom sugar controllers, but for whose persistence and skill it must have been far worse. Their administration combined economy, firmness, and flexibility in unique measure. Nothing could be further from the popular notion of bureaucracy than the easy, informal expertise with which the industry was managed; and if the rationing system (for which the other side, so to speak, of the Ministry of Food was responsible) was inflexible as a matter of principle (*10*, p. 203 ff.) it, like the control of prices, was highly effective. It is pleasant for a historian to reflect that the success of British sugar control was largely due to the thoroughness with which those in charge of it had mastered the lessons of the past, and its success was only qualified to the extent that they were prevented from putting those lessons into practice.

CITATIONS

1 Sir William Beveridge, *British Food Control* (London, 1928).

2 Gr. Brit., *Report of the United Kingdom Sugar Industry Inquiry Committee* (Command Paper, Cmd. 4871, H.M.S.O., 1935).

3 Gr. Brit., Min. Agr. and Fisheries, *The Sugar Beet Industry* (Economic Series 27, H.M.S.O., 1931).

4 Gt. Brit., *Sugar Policy: Proposals of H.M. Government* (Cmd. 4964, H.M.S.O., 1935).

5 Gr. Brit., 26 Geo. 5 & 1 Edw. 8 ch. 18.

6 Gr. Brit., *Statutory Rules and Orders (S.R. & O.) 1937*, No. 221.

7 Gr. Brit., *Sugar Industry (Re-organisation) Bill; Amalgamation of the Beet Sugar Manufacturing Companies* (Cmd. 5139, March 1936).

8 R. J. Hammond, *Food: The Growth of Policy* (History of the Second World War, U.K. Civil Series, H.M.S.O., 1951).

9 Gr. Brit., S. R. & O. 1940, No. 1237; and *1943*, No. 688.

10 R. J. Hammond, *Food and Agriculture in Britain, 1939–45* (Food Research Institute Studies on Food, Agriculture and World War II, Stanford, Calif., 1954).

11 F. H. Coller, *A State Trading Adventure* (London, 1925).

12 R. J. Hammond, *Food: Studies in Administration and Control* (History of the Second World War, U.K. Civil Series, H.M.S.O., 1956).

13 Gr. Brit., *S. R. & O. 1943*, No. 1136.

14 Gr. Brit., *S. R. & O. 1940*, No. 1539.

15 Gr. Brit., *S. R. & O. 1940*, No. 2057.

16 Gr. Brit., Public Accounts Comm., *Minutes of Evidence (etc.), 1943*, Questions 2728–2756 (H.M.S.O.).

17 Gr. Brit., 5 & 6 Geo. 6, ch. 16.

18 Gr. Brit., *S. R. & O. 1940*, Nos. 848, 849, 2198; *1941*, No. 510; and *1943*, No. 1707.

19 Gr. Brit., *S. R. & O. 1941*, No. 2129.

20 Gr. Brit., *S. R. & O., 1942*, Nos. 1106, 2455; and *1943*, Nos. 669, 839.

21 Sir Keith Murray, *Agriculture* (History of the Second World War, U.K. Civil Series, H.M.S.O., 1955).

22 P. E. Hart, "Competition and Control in the British Sugar Industry," *Oxford Economic Papers*, October 1953.

CHAPTER 9

THE BEET-SUGAR ECONOMY OF CONTINENTAL EUROPE

EARLY HISTORY

The beet-sugar industry of Europe in the nineteenth century provides the earliest example of the market for an important tropical product being seriously eroded by the application of modern scientific methods in relatively advanced countries. There were several reasons for the success of beet sugar in Europe. Originally encouraged to replace cane-sugar imports during the Napoleonic Wars and the accompanying Continental blockade, the beet-sugar industry nearly disappeared after the fall of Napoleon, only to re-emerge in the second quarter of the nineteenth century, first in France and then in Germany, Austria, Belgium, and Russia. This time it was the crisis of European agriculture connected with the sharp decline of prices of grain, particularly wheat, that created conditions favorable for an expansion of sugar-beet production. The disparity between beet prices and wheat prices became so great as to give a powerful boost to beets, the more so since they flourish on the rich soils which so far had been used only for wheat (cf. *1*, p. 102; *2*, pp. 441–42; *3*, pp. 92–93).

The importance of sugar beets to agriculture was understood very early, particularly their role in the introduction of rational crop rotations, which replaced the traditional, extensive three-field system dominated by cereals, still common on the Continent of Europe. It became clear very soon that sugar beets, requiring deep plowing and careful cultivation of soil and substantial use of fertilizers, contribute substantially to the increase in yields of other crops in the rotation, particularly of the only cash grain, wheat. The important role of the by-products of sugar beets and of the beet-sugar industry as a supplier of additional valuable cattle feed came to be generally recognized before 1850 (cf. *4*). European farmers understood that sugar beets not only provided them with a new cash crop, replacing other industrial crops, such as oilseeds, that also were suffering at that time from the overseas competition, but created the opportunity for a complete reorganization of farming on an intensive basis. The link with animal husbandry was particularly important at a time when prices of animal products were less depressed than those of cereals.

Other factors also contributed to the beet-sugar industry's revival and strong development during the second quarter of the nineteenth century. Its forerunner and competitor, the colonial cane-sugar industry, was considerably weakened by the abolition of slavery in the British colonies in 1833 and in French colonies in 1848. Furthermore, overseas cane sugar, colonial as well as foreign, paid fairly high import duties, whereas domestic beet sugar in the early stages of its development was free of taxes. This advantage, however, was somewhat temporary since taxation of domestic beet sugar was initiated in France as early as 1837: during the 1840's the excise on domestic beet sugar was equalized with import duties on French colonial cane sugar. In other European countries beet sugar also began to be taxed from the 1840's on, though the excises were usually much below the import duties on foreign sugar.

But tax levies were not necessarily a deterrent. By taxing raw materials or factory equipment rather than the final product, fiscal measures encouraged improvement in both technology of sugar extraction and in quality of sugar beets. Thanks to such lenient and expedient taxation methods, explainable partly by the desire of European governments to favor a depressed agriculture, techniques of cultivation and processing progressed extremely rapidly, as illustrated by the following rates of extraction of sugar from beets in Germany (5, p. 503): from 5.8 per cent in 1836 the rate increased to an average of 8.1 per cent in 1856–65, 12.3 per cent in 1886, 13.9 per cent in 1896, and 15.6 per cent in 1906. This rapid increase in Germany is in sharp contrast with the temporary stagnation in France during 1864–84, when the French system of sugar taxation involved only the final product. During that period in France the rate of extraction of sugar from beets remained at the low level of about 5 or 6 per cent, while before 1864 and after 1884 it increased fairly rapidly under the impact of the tax on the weight of material used. Rapid technological progress made it possible for beet sugar not only to displace cane sugar from domestic markets, but also to move into export both within Europe and for destinations overseas, particularly to Canada, the United States, British India, and Japan (2, p. 446). Production exceeded domestic requirements in Austria-Hungary by 1863/64, in Germany in 1875/76, and in Belgium, France, and Russia during the 1880's.

The sugar policies of Great Britain were of the greatest importance to the further development of the European beet-sugar industry. Beginning in 1845, Great Britain gradually reduced its fairly high sugar duties, which had protected both colonial cane sugar and the domestic refining industry. From 1874 until 1901, entry to the British sugar market was completely free for sugar of all kinds and origins, and when a moderate revenue duty was subsequently imposed on imports, foreign and Empire sugars were treated alike. Under such conditions surplus beet sugar from the Continent moved heavily into what was at that time the largest sugar market in the world. As of 1903, colonial and foreign cane sugar together comprised a mere 12 per cent of the total sugar imports of the United Kingdom. Furthermore, over two-thirds of her total imports of beet sugar consisted of refined sugar in competition with the British refining industry (*6*, pp. 195–96). Martineau supplies the following information on sugar imports in the United Kingdom (in 1,000 long tons; *1*, p. 3):

	1860	1900
Raw sugar	435	662
Refined sugar	13	962

The transformation in the composition of imports between 1860 and 1900 is striking. The competition implicit in the tremendous increase in imports of refined sugar (practically all beet sugar), far from destroying the British refining industry, evidently brought a healthy reorganization into a few highly productive enterprises employing improved technological methods.

During the 1870's and 1880's, European agriculture, particularly in so far as it depended on selling grain, once more faced a severe crisis, caused by competition with imports from overseas. Again, as during the second quarter of the century, wheat prices were severely depressed and European farmers were looking for more profitable crops. Under pressure from agricultural interests, European governments introduced various measures to expedite sugar exports, and thus stimulate further expansion of sugar beets on "wheat soils," in spite of heavy financial sacrifices (cf. *7*). The late 1870's and the 1880's were accordingly the period when sugar export bounties, open or concealed in the form of excess drawbacks, reached their highest level in Continental European countries (*8*, pp. 26–28, 39–41, 58–61; *5*, pp. 501–08).

With each country seeking to outdo its competitors, the burden of export bounties soon became so heavy that European governments tried on several occasions to eliminate this costly device. Efforts at international agreement in the nineteenth century remained unsuccessful, partly because Great Britain was not prepared to introduce a system of countervailing duties on imports of subsidized sugar (5, Vol. II, p. 506), partly because the critical situation of agriculture on the Continent continued into the second half of the 1890's. When high concealed bounties were replaced by smaller direct ones in the early 1890's, European governments somewhat reduced the fiscal burden of export subsidies. However, the task of subsidizing sugar exports was thereby merely shifted to the cartels of sugar industries, which emerged during this period in practically all sugar-exporting countries of Continental Europe. The cartel policy was to maintain domestic sugar prices on the high level permitted by protective duties, and to finance losses on exports out of sales to domestic consumers.

Only when Europe's agricultural crisis was brought to an end before the turn of the century, partly by the changed situation in the world market for agricultural products and partly by introduction of tariff protection for European agriculture, did the urgency for further expansion of sugar-beet cultivation abate. Abolition of export subsidies could accordingly be achieved by the Brussels Convention of 1902. This Convention not only banned all kinds of sugar export bounties, concealed as well as direct, but limited the margin by which import duties could exceed the domestic tax on sugar. The latter provision made it more difficult for cartels to absorb losses arising from export dumping.

The Brussels Convention, though it removed the unequal competition from which cane sugar had been suffering, was not itself responsible for the revival of the cane-sugar industry. Recovery of the cane-sugar industry in the British West Indies had already begun by the middle of the nineteenth century and was not completely arrested even during the years of heaviest beet subsidies (5, Vol. II, p. 377). The growth of the cane-sugar industry during the period of acute competition with subsidized exports of beet sugar was even more pronounced in the areas (Java, Mauritius, Hawaii, Louisiana) that supplied sugar to the Asiatic and North American markets. While some help arose from the fact that British India and the United States introduced countervailing duties against the exports of subsidized sugar even before the Brussels Convention, the main reason unquestionably was that technological reorganization of the cane-sugar industry began earlier in the named territories, as may be judged by the relatively larger size of sugar factories (1, pp. 98, 107; 5, Vol. I, pp. 185, 224). Indeed, the beet-sugar sector not only developed some of the technical equipment that the cane industry came to rely upon, but in some cases provided a competitive incentive for its earlier introduction.

Nor did the Brussels Convention stop the growth of the beet-sugar industry. The United States and some of the new beet-sugar areas of Europe that produced only for their domestic markets experienced rapid expansion, and the growth in Russia, though slower, was still substantial. Even in the old beet-sugar areas of Europe that exported a large part of their output, production of sugar in 1914 was one-fifth greater than at the time of the Brussels Convention. Yet growth of the world's beet-sugar industry as a whole lagged behind that of the cane industry, and beet's share in world sugar production declined from 52 per cent during the three years preceding the Brussels Convention (1900/01–1902/03) to 48 per cent during the three years just before World War I (9, pp. 103–07).

The old beet-sugar producers of Europe—Germany, Austria-Hungary, France, Belgium, the Netherlands, and Russia (problems of which are discussed in chapter 10)—contributed 99 per cent of all the beet sugar produced in the world at the beginning of the 1880's. Before the turn of the century, production began to spread both northward into Denmark and Sweden and southward into Italy, Spain, and Rumania. These areas accounted for more than 5 per cent of the total European sugar-beet acreage by 1900, and on the eve of World War I their share had risen to above 10 per cent.

The entire growth of beet-sugar production after the Brussels Convention came in response to increased domestic consumption within the beet-sugar area. During the three sugar years (1903/04–1905/06) immediately following the conclusion of the Convention, the net export of sugar from the old beet-sugar-producing countries declined to only about four-fifths of its level of the last three years preceding the Convention (1900/01–1902/03). Exports improved somewhat during the following years, but never went much above 90 per cent of the pre-Convention level. The fairly rapid increase of sugar consumption within the older area may be explained partly by the substantial decline of domestic sugar prices that accompanied a considerable reduction in protective tariffs after the Brussels Convention.

While exports from Western Europe fell sharply, particularly those from France, total exports from Central Europe (Germany and Austria-Hungary) declined hardly at all, and exports from Russia even increased substantially. European beet sugar continued to dominate the important British market; refined beet sugar from the Continent comprised half of total British sugar imports in 1911, and with raw made up more than two-thirds of that total. The beet industry accordingly showed surprising strength in freely competitive markets.

The rapid growth of total cane-sugar production during this period (1902–13) to a considerable extent reflects the response to a preferential position in the highly protected market of the United States enjoyed by some cane areas, notably Cuba and Hawaii. Growth in the British and French colonial areas was, by contrast, rather slow, foreign cane sugar profiting more than colonial sugar in the London market. World War I changed radically the relative position of cane and beet sugars in the British market as well as in world markets generally.

INTERWAR DEVELOPMENT

Postwar restoration.—World War I caused great damage to the European beet-sugar industry, reducing the sugar-beet area in the first postwar year (1919) to below three-fifths of the prewar, and sugar production to about two-fifths.[1] By 1924 the sugar-beet area exceeded the prewar figures but the yield per hectare had not quite been restored. Indeed, from 1925/26 to 1935/36 the average yield of sugar beet per hectare for Europe ex-USSR remained 6 to 7 per cent below the average for 1900/01–1913/14 (9, p. 15). This was due to further expansion of beet cultivation in the new areas of Southeastern Europe, the Baltic States, Turkey, and the United Kingdom, where beet yields per hectare were below the European prewar average during the early years of development.

Restoration of the sugar-beet area was rapid in the old sugar-beet belt from which sugar exports had come before the war. That zone was incorporated within the new frontiers of Germany, Czechoslovakia, Hungary, Poland, France, Belgium, and the Netherlands (but not postwar Austria, which had to build an entirely new industry). The leaders of the Central European sugar industry, encouraged by their prewar experience, had confidence in their ability to compete in free world markets and expected exports to recover once the wartime damage had been overcome. In Germany, which lost an impor-

---

[1] Here and in the following discussions the USSR is not included in the European total. Because of the change in the western frontiers of the former Russian Empire, postwar sugar statistics reported by Licht and other agencies for Europe ex-Russia are not quite comparable with the prewar. A substantial portion of the Polish sugar industry and a small portion of the Rumanian were located in territory formerly within the Russian Empire. The sugar industries of Lithuania and Latvia also are located in former territory of the Russian Empire, but there was practically no sugar production within their territories before World War I.

tant portion of its sugar industry in the east to Poland, the stimulus for restoring the remaining portion became particularly strong. The generally high prices of sugar which, except for the short-lived decline in 1920/21, prevailed during the early postwar years, suggested that European beet sugar could once more conquer the world market.

At the same time, dismemberment of the Austro-Hungarian Empire and the severance of several sugar-beet-growing regions from their domestic markets by the new postwar frontiers in Central and Eastern Europe led to exportable surpluses of beet sugar. The main part of the Austro-Hungarian sugar industry was included within the narrow national frontiers of Czechoslovakia, while Poland obtained sugar surplus areas from both Russia and Germany. So reduced was the area of postwar Hungary that even its modest sugar production could not be absorbed at home. Furthermore, these newly created countries, still not stabilized financially, were obliged to stimulate their exports in order to earn foreign exchange.

The result was that by 1924, when the European sugar-beet area had been restored to its prewar level, exports of sugar from Czechoslovakia and Hungary far exceeded the prewar exports from the entire Austro-Hungarian Empire. Czechoslovakia exported more than two-thirds of her large sugar production and Hungary about half of her output. At that time Poland, too, was exporting around a quarter-million tons, about half of her production, and Germany had so far succeeded in restoring her sugar-beet industry as to enter the ranks as a substantial net exporter of sugar. Moreover, Belgium and the Netherlands had already become net exporters in 1919. Sugar was thus moving from the old beet-sugar area across the new international frontiers in quantities closely approaching the prewar level. Of the old beet-sugar producers only France remained a consistent net importer in the postwar period, receiving sugar mainly from her colonies.

However, heavy exports of beet sugar met a completely new situation in the world's sugar markets. During the war the world supply of cane sugar had been enlarged by about 2 million tons, which compensated for the beet sugar that disappeared from British and other markets. The cane-sugar output continued to grow even faster in the early postwar years, increasing by about 3.5 million tons during the period when European beet sugar was gaining its prewar level. Consequently, by the mid-1920's, when the beet-sugar production of Europe (ex-Russia) was nearly back to its pre-World War I figure of around 6.3 million tons, supplies of cane sugar had risen from a prewar average of 9.5 million tons to about 15.1 million. As the result of excessive supplies, sugar prices dropped sharply in 1925 in the London market and averaged during 1925/26 only slightly more than half the average price for 1923/24. This setback temporarily arrested the expansion of beet cultivation in Europe. For two or three years European beet-sugar production remained about stationary and net exports from the old beet-sugar belt declined to slightly over one million tons, or about half the prewar level.

Several factors limited exports. Most important was the basic change in the sugar policies of Great Britain. European beet sugar, notably Czechoslovakian, which was exported mainly as refined (white) sugar, was particularly hard hit. Throughout Europe the expansion of sugar production in deficit (sugar-importing) countries continued vigorously, particularly during the late 1920's. Ireland, Finland, Latvia, Lithuania, and Turkey, all previously dependent entirely on imports for sugar, initiated domestic production during the interwar period. Some countries which had a sugar industry of their own before the war but were net importers expanded their beet production to the point of self-sufficiency or even became net exporters. That group included Denmark, several countries of Southeastern Europe, and temporarily even Italy. By 1930, the European sugar-beet area reached its interwar peak. More than 30 per cent of the total European sugar-beet area (ex-Russia) was then outside the "old" sugar-beet belt as against substantially less than 20 per cent in 1919.

Bleak market prospects induced European beet-sugar exporters to accede to restrictive international sugar agreements. Although the principal exporters, Czechoslovakia, Poland, and Germany, took part in the so-called Tarafa action of 1927/28, this agreement was endangered from the very beginning by the unwillingness of the Netherlands government to join on behalf of Java, and was then wrecked by the British refining tariff in April 1928 (*10*, pp. 47–48). Consequently, between 1927 and 1930 the sugar-beet area was further expanded even in some of the participant countries, such as Germany and Poland. Effective restriction began only in 1931/32, under the ("Chadbourne") International Sugar Agreement. By then, production control had to be extremely severe if accumulated excess stocks were to be disposed of. Unusually high beet yields in 1930 were a further complication, resulting in sugar surpluses in practically all European countries. At the end of the 1930/31 campaign, sugar stocks in the European exporting countries alone were one million tons above normal. Production restrictions became necessary not only among the beet-growing countries of the Chadbourne group (Germany, Czechoslovakia, Poland, Hungary, and Belgium), but also in France and Italy, normally importers of sugar.

Restrictive policies.—Between 1930 and 1932, the sugar-beet area in Europe (ex-USSR) was deliberately reduced by about 30 per cent, and in the Chadbourne group by more than 40 per cent. Germany cut its beet area practically in half, and Czechoslovakia, Poland, and Hungary (all of whom began restrictions before 1930) even more. But within the environment of shrinking employment, retrenchment of nationalistic economic defense measures, and deflation, it took four years to dispose of accumulated stocks, since markets could not absorb exports up to the agreed quotas. The low level of sugar prices in the free world market, which the Chadbourne Agreement failed to raise, limited European exports and, consequently, production in the later years. Passing the depth of the industrial depression in the summer of 1933, Germany with its large domestic market adjusted production to the requirement of that market from the 1932/33 campaign on. Other Central European exporters, however, were forced to continue exporting a substantial portion of their total production at unprofitable prices, even after reducing sugar-beet cultivation to the minimum compatible with maintenance of their intensive agriculture.

The private Continental beet-sugar cartels of the late nineteenth century had stimulated rather than restricted exports. During the years when the Brussels Convention had been operative, cartels became less active and in some cases were even dissolved. After World War I, with the Brussels Convention abandoned, cartel practices again became widespread, and exports were stimulated anew.

Germany, which had regained full control over her foreign trade policy in 1925, raised import duties on sugar above the prewar level in 1928 and introduced a further increase in 1930. The possibility of maintaining domestic sugar prices above the level prevailing in the free world market had encouraged the organization of an export cartel in 1926. At the beginning of each campaign year, the cartel determined the quantity of sugar that had to be exported, and allotted export quotas among the sugar factories in proportion to the historical base of their production. In the late 1920's these obligatory exports could be sold only at prices substantially below the domestic level. From 1929 on, the allotment system was extended to include sugar sales on the domestic market and to regulate the timing of these sales during the sugar campaign. By lifting the proportion of output required to be exported from 8 per cent in 1927/28 to 18–22 per cent in 1930/31 (*11*, pp. 11–15), the German cartel was able to keep the domestic price slightly below the fixed maximum price which the government had begun setting when the tariff was raised in 1928.

Private cartels in other countries operated along similar lines. The Czechoslovakian cartel, established immediately after the government postwar control was relaxed in 1921/22, controlled sales on the domestic market only. The matter of exports was left to be decided by the individual sugar factories. The Czech government strictly controlled the domestic price (*12*, pp. 57–61). The Polish government introduced by law on July 22, 1925 a quota system for control of domestic sugar supplies, but it too left exports of sugar free (*10*, pp. 80–81).

With the limitation of sugar exports under the Chadbourne Agreement, surplus sugar could no longer be dumped on the world market. In each producer country, total production had now to be adjusted to the requirement of the domestic market plus the agreed quotas for export. Although the governments of respective countries were not themselves signatories to the Chadbourne Agreement, they had to intervene to enforce it when sugar producers could not voluntarily agree on production quotas. Such was the case in Germany. A government order was issued on March 27, 1931, incorporating all sugar factories into the Economic Union of the German Sugar Industry. This order empowered the governing body of the Union to allot production quotas and to fix for each factory the amount of sugar it could put on the domestic market and the percentage it would be required to export. Each factory was required to subdivide the total quantity of sugar beets necessary for fulfillment of its sugar quota among the beet growers in its territory in proportion to their previous production. The new regulation encompassed all stages of production, from the sugar-beet growers through to sugar traders distributing sugar on the domestic market and exporting abroad. This legislation made sugar production, processing, and distribution a compulsory state cartel affair and shaped the pattern for the entire Fascist system for agriculture introduced by the Nazis in 1933.

In Czechoslovakia also the strict and comprehensive control of the domestic market by cartel was completed by government licensing of sugar exports, proclaimed by order of the Minister of Industry and Trade on May 5, 1931. In Poland the existing government control of the domestic sugar supplies under the quota system was extended to exports as well (*10*, pp. 80–81; *13*, pp. 155–57, 173).

Outside the Chadbourne group, production quotas for sugar factories as well as for sugar-beet growers were established in France in

1931, by a voluntary agreement between the organization of the sugar factories and the general confederation of sugar-beet growers. This three-year agreement was extended for a similar term in 1934. However, the government was persuaded to enact a law (August 8, 1935) whereby decisions taken under this private agreement on sugar quotas and their regulation became compulsory when accepted by 80 per cent of the membership with 80 per cent of the productive capacity. Another decree, of the same date, prohibited construction of new sugar factories without special permission by the public authority. A further decree of August 25, 1937 authorized the public power to fix the annual quota for sugar production. Inasmuch as the government had authority to license imports of sugar, its control over the production and supply of sugar in the metropolitan market was practically complete, although all government decisions were taken after consultation with organizations of the interested groups which themselves administered the details of regulation (*14; 15*).

In Italy, already under Fascist rule, restriction of sugar production was also agreed upon by organizations of sugar producers and of beet growers in 1930/31. At first sugar-beet growers merely agreed to stop further expansion of the beet area, but later under a quota system the sugar-beet area was reduced by one-third within two years. Subsequently the control was further tightened by fixing growers' quotas in terms of tons of sugar beets produced rather than hectares of beets grown.

Characteristic of all these organizations that controlled sugar production was their vertical structure. They included growers of sugar beets, raw-sugar producers, refiners, and in some cases traders in sugar. In this respect they assumed the forms approaching the principle of the Fascist or Nazi "corporative organizations." In France, in their later evolution, they were called *Groupement Interprofessionnel*. It is significant that the *Hauptvereinigung der deutschen Zuckerwirtschaft* (Principal Association of the German Sugar Economy), established by the Nazi government in 1934, preserved to a great extent the organizational forms of the old Economic Union of the German Sugar Industry.

Several new principles also guided the structure and functioning of the Hauptvereinigung. While quotas for sugar-beet growers and processors had previously been regarded mainly as their rights, based on their past performance in the industry, the new regulation introduced the idea of an "obligation" on the part of growers and processors to perform certain tasks, under the threat of losing their allotments.[2] Establishment of quotas or allotments was now guided not only by historical facts, such as previous production, but also by principles of economic rationalization: better delimitation of the hauling area of a factory district, in which growers belonged contractually to certain factories, or better utilization of fertile sugar-beet land by growers. Beet growers who held stock in sugar factories or who grew or sold beets under contract were no longer treated differently from those who did not. These "new ideas" paved the way for conversion of the sugar-control organization from a means of restricting output into one of expanding the industry at will in the years immediately preceding and during World War II. Thus it was possible to continue the Hauptvereinigung without change for the wartime control of sugar, by the decree of September 7, 1939, on "the public management of sugar beets, sugar, and byproducts" (*16*, pp. 42–46).

Similarly, the prewar French organization was the basis for the Groupement Interprofessionnel of the sugar-beet growers and processors (cf. *15*), established by the Vichy government under the law of August 7, 1941. Indeed, the same organization was continued with only a few changes at the top by the French liberation government, for restoration and expansion of the industry. And again, in the western zones of Germany, the military governments of occupying powers used the sugar organization created under the Nazi government, with only a few modifications.

The national sugar-control organizations created early in the 1930's did not limit their activity to restriction of the production of sugar and sugar beets. Under the political pressure of tightly organized, agricultural interest groups, in an attempt to preserve as much as possible of the sugar-beet cultivation so important for intensification of European agriculture, these control organizations also pursued a policy of diverting surplus sugar beets to other uses than for sugar production, especially for feeding livestock and as raw material for making alcohol. This diversion policy was sometimes applied also to the final product, by stimulating the use of sugar for other purposes than human consumption. In such cases sugar was usually denatured by various processes in order to permit marketing at various prices. Such a system of diversion was practiced, for instance, in Czechoslovakia, where the use of denatured sugar as animal feed increased from less than one thousand tons in 1929/30 to 35,000 tons in

---

[2] During the war similar principles were applied for the control of economic activity in the democratic countries. For instance, British county land committees had such authority.

1936/37 (*17*, pp. 48–55). At the time when world sugar prices fell to their lowest level, prices paid for sugar beets used for this purpose were even higher than prices for beets used to produce export sugar.

Quantitatively, however, diversion of sugar beets before processing into sugar was of much greater importance than the costly diversion of the finished product. Beet diversion to feed played an important role particularly in countries which imported large quantities of concentrated feed for livestock, such as Germany, Denmark, and Czechoslovakia. In Germany this method of diversion was stimulated by cartel regulation accepted for the 1930/31 campaign; sugar contained in various feedstuffs prepared from sugar beets was chargeable against the required export quota (*11*, pp. 14–15). Increased use of sugar beets for feed, originally meant to divert sugar-beet surpluses, contributed to a positive expansion of sugar-beet cultivation after 1934/35, particularly in would-be autarchic Germany. The beets-to-alcohol policy, at least as far as Italy was concerned, was also part of the search for greater self-sufficiency, rather than a surplus-disposal device. Indeed, in 1935 a program for expanding the production of alcohol from sugar in four years to 1.2 million hectoliters (*18*, pp. 240–41) was launched in connection with preparation for the Ethiopian war and the potential blockade of Italy by countries belonging to the League of Nations.

In the case of France, however, the greatly increased use of sugar beets for making alcohol not only emerged as a means of diversion of surpluses but has retained this role.[3] Originally, distillation was intended to absorb *occasional surpluses* of sugar beets. Such a situation existed in the mid-1920's: in the years 1924–28 an average of only about 200,000 hectoliters of alcohol was produced from sugar beets. However, during the next five years, 1929/30–1933/34, the annual average rose to 1.3 million hectoliters. For 1934/35–1938/39 it reached 2.8 million, with a maximum of 5.3 million hectoliters in 1935/36 (*20*, pp. 980–1016). The quota for production of alcohol from sugar beets was established by the law of March 31, 1933 at 2.3 million hectoliters and was raised by several later laws to 2.5 million in 1939. During the last prewar years, 1935–39, the sugar-beet output from 80,000 to 85,000 hectares—one-fourth of the total

---

[3] The war years, 1941–45, were an exception. The Vichy government favored processing of beets into alcohol, even amid a sugar shortage, because the lack of liquid fuel under the blockade was more critical. More than 30 per cent of the total sugar-beet crop was used in the 1943/44 campaign for processing into alcohol (*19*, pp. 240–42).

area—was processed to alcohol. The French governmental policy of guaranteeing the disposal of surplus beets as alcohol, at the same prices as for sugar production, did not solve the problem of temporary beet surpluses, but created instead the more serious problem of permanent surpluses of alcohol, at enormous expense to the French treasury.

PREPARATION FOR WAR AND THE SUGAR ECONOMY DURING THE WAR

The restrictions placed on Europe's beet-sugar production in 1930–32 were soon relaxed, except in Czechoslovakia and Poland, whose sugar industries depended on exports. The European sugar-beet area in 1934 was about 20 per cent above that of 1932,[4] an increase contributed to mainly by countries producing sugar exclusively for domestic consumption, and by Germany, which also adjusted production to domestic need. Germany's sugar policy during the second half of the 1930's is of particular interest, since it was framed in close agreement with the Nazi aim of self-sufficiency in food. Measures in this direction, in the light of later events, appear to be among the earliest steps in the preparation for war.

Even during the restrictionist period of the early 1930's, the diversion of sugar beets for uses other than the production of sugar accounted for more than 20,000 hectares in Germany, or 9 per cent of her total sugar-beet area in 1932/33, as compared with less than 1.5 per cent in 1930/31. The figure declined to 3 per cent in 1933/34 and remained below 5 per cent during 1934–36.

In the following years, however, the direct use of sugar beets for feed had another purpose than that of diverting a crop surplus. It became the means of reducing imports of feed concentrates from abroad. Before the Nazis' accession to power, Germany was a heavy importer of both bread grain and feed grain. In order to reduce imports of bread grain the Nazi government prohibited the use of rye as feed. Since rye was a feed of considerable importance, this prohibition forced the livestock industry to depend even more on imported feed. Greater use of sugar beets as feed helped to fill this gap. Within five years (1934/35–1938/39) the area of sugar beets for feed increased from 16,000 to 80,000 hectares—15 per cent of the total sugar-beet area. Fully three-quarters of the entire expansion of the German sugar-beet area between 1934 and 1937 served to enlarge the supply of feed rather than of sugar. However, the 1938/39 goal to boost feed

[4] Statistical information in this section is from *21* unless otherwise indicated.

sugar beets to 18 per cent of total production (*22; 16*, p. 404) was not reached.

During the six years 1933/34–1938/39 the production of dried unextracted feed sugar beets increased almost eightfold (from 81,500 to 631,000 tons). A special dehydration industry for sugar beets emerged. Every sugar factory was required to install a dehydrating plant, but was granted a state subsidy for the purpose. By 1936/37 the productive capacity of these dehydrating plants had to be equal to 10 per cent of the factory's basic quota for processing sugar, a figure subsequently raised to 15 and then to 18 per cent.

The increased use of unextracted sugar beets for feed meant a price disadvantage for growers. To compete with imported feed concentrates, beets for feed had to be priced much lower than beets for sugar—20–24 RM per metric ton as against 32–36 RM (*22; 16*, p. 123). Growers received the same *average* prices whichever use was made of their beets, but the average was depressed as the proportion of the crop used for feed increased. In 1939/40 a subsidy of up to 44 RM per ton of dried sugar beets was extended to factories incurring losses on this product, (*24*, p. 148). For the 1942/43 and 1943/44 campaigns that subsidy was raised to 52 RM per ton (*25*, p. 115).

Unextracted dehydrated beets as well as high-sugar-content dried pulp (*Steffenschnitzel* and others) were reserved mainly as feed supply for farmers other than beet growers, since the latter had beet tops and leaves and wet pulp at their disposal. Statistics show that nearly 85 per cent of the concentrated sugar-containing feedstuffs was sold in 1937/38 by factories to other than beet growers. The largest portion was used by hog breeders in northern and northwestern regions of Germany, who depended heavily on imported feed concentrates. The use of sugar-containing concentrated feed was particularly profitable for fattening hogs (*16*, pp. 402–08).

Measures were also taken for better utilization of by-products of sugar-beet growing and processing. Sugar-beet tops are equal in nutritive value to better kinds of hay. Only about one-third of the tops could be used fresh during the short harvesting season. Hence, the problem arose of preserving the tops for use in winter months in such a way as to prevent the heavy loss in nutrients to which storage in primitive trench silos was subject. To feed 30 per cent of the tops fresh, 50–55 per cent as silage, and 15–20 per cent dehydrated was regarded as a satisfactory utilization of tops under German climatic conditions (*22*). The government fostered the construction of dehydrators and concrete silos by large subsidies, and there was a rapid increase in the volume of silos, from 650,000 cubic meters in 1933 to 7.5 million in 1939 (26, p. 114). The greatest expansion of concrete pit-silos took place in the principal regions of sugar-beet cultivation, and consequently a greater proportion of tops and leaves was ensiled. Use of dehydrated tops and leaves apparently developed much more slowly. Although the amount of dehydrated tops, according to Woermann, nearly doubled in the 1936/37 campaign, that quantity comprised only about 2 per cent of all tops and leaves harvested. In order to obtain a by-product with higher feed value, the long-standing emphasis on high sugar-extraction rates was reversed. Special processes of sugar extraction were developed (Steffen process, for instance) that left more sugar in the pulp after diffusion. Such pulp, when dehydrated, represents a concentrated feed with 28 or more per cent of sugar.

The policy of the Nazi government was not only to replace feed imports by increased feed use of sugar beets and better utilization of by-products, but also to substitute sugar for some imported food commodities, particularly fats and oils that were produced largely from imported raw materials. Fiscal demands, reflected in high sugar excise taxes, made it difficult to increase sugar consumption by any substantial lowering of prices. But a scheme was developed for raising consumption of jams. To this end sugar was supplied to the jam manufacturers at reduced prices (11, p. 190). Supplies of subsidized fruit jams were planned to reach 100,000 tons in 1936 and were supposed to replace butter and margarine spread on bread. The per capita sugar consumption in Germany, according to Licht, rose from 22.6 kilograms during the depression years 1931/32–1932/33 to 26.7 kilograms in 1936/37, and the total disposal of sugar reached 1,845,-000 metric tons in 1936/37 as against 1,512,000 metric tons in 1931/32.

The German sugar industry had to solve another problem during the years preceding World War II—the accumulation of sugar emergency reserves. When Germany was a large exporter of sugar, necessary sugar reserves had been carried by the factories and the wholesale traders participating in world trade. After World War I this function was fulfilled less satisfactorily by the wholesale trade, and the stocks accumulated in the factory warehouses at the end of the trade year 1930/31 were considered a heavy burden. Under strict regulation of the sugar industry and with the trade being administered by the Hauptvereinigung, the role of the wholesale trade was reduced to fulfilling certain distribution functions rather than speculative carrying of stocks. The function of carrying stocks as a public service without price risk was accordingly imposed on the sugar factories. An order of the chairman of the Hauptvereinigung on February 20, 1936 required that every factory show on October 1, 1936 (beginning of the new campaign) a stock of sugar equal to at least 12 per cent of its basic production quota (*11*, pp. 114–16). It was understood by sugar interests from the beginning that these stocks were created not only for assuring regular supplies for consumers, but also for strategic purposes (cf. *11*, p. 188). All sugar produced above the yearly sales quota, established for the 1935/36 campaign at 75 per cent of the basic quotas, was regarded as a part of this obligatory stock at the disposition of the Hauptvereinigung. For the next campaign, 1936/37, the requirement for stocks at the end of the campaign was raised to 14 per cent of the basic production quotas, and at the end of 1938/39 (October 1, 1939) to 20 per cent (*16*, p. 126).

The central sugar-control organ enforced the compulsory stock requirement by threatening to reduce the sales quotas of factories that failed to meet their obligations. However, Licht's figures on the sugar stocks in Germany on September 1 for 1934–38 are as follows (in thousand tons):[5]

1934	257	1937	313
1935	367	1938	758
1936	360		

It appears that sugar stocks, after increasing by more than 100,000 tons from September 1, 1934, to September 1, 1935, tended to be stable or actually to decline during the two following campaigns. Apparently the rapidly increasing requirement for processing sugar beets into feed, together with the growing consumption of sugar, prevented further accumulation of stocks in spite of the fact that the yearly sugar production quota was raised substantially during these years. Unusually high yields of beets in 1937 on the greatly enlarged planted area finally permitted stocks at the end of the 1937/38 campaign to be brought up to a level that exceeded the minimum requirement of 20 per cent of the basic production quota. Large stocks notwithstanding, the quota for production of sugar beets for the 1939/40 campaign was fixed at 105 per cent of the basic production quota, 90 per cent to be processed into sugar and 15 per cent into feedstuffs.

---

[5] These stocks normally are larger than October 1 stocks for the same years, since consumption usually exceeds production of sugar in Germany during September.

These expansive policies brought the beet area in the old German territory to double its 1932 level even before the beginning of the war, more than restoring the prerestriction situation of 1930. Meanwhile the beet area of all Europe outside the Soviet Union was expanded during 1932–39 by only one-third. Recovery, quite naturally, was least in Czechoslovakia and Poland. But even in such traditional producers as France, Belgium, and the Netherlands, plantings remained more or less stationary during the period 1932–39. Despite some efforts at expansion during the last two prewar years, their combined beet areas in 1939 stood below the 1930 level by a considerable margin. Germany thus increased her share of the total European acreage, even within the old frontiers. Her share in European beet production rose even more, since she succeeded better than other countries in raising yields by greater application of subsidized fertilizer.

However, the German sugar situation during the war was not determined only by its expansive policies on the old territory. Of great importance was its piecemeal territorial expansion preceding the war, as well as its early war success. It is true that the early incorporation into the German Reich of Austria (1937) and of the Sudeten area of Czechoslovakia (1938) did not increase sugar supplies per capita of population. Austria, with a relatively new industry, just reached self-sufficiency in sugar on the eve of its incorporation into the German Reich; the Sudeten area, with only slightly more than one-tenth of the sugar industry of Czechoslovakia and one-fifth of its population, had not much of a surplus. But when Germany established control over all of Czechoslovakia early in 1939 and conquered Poland in September 1939, the two most important sugar-surplus areas of Europe were brought within the orbit of the German economy. These two countries had exported more than 400,000 tons of sugar in 1937/38 in spite of the fact that their production during the last prewar years (1935/36–1937/38) continued to be held to about three-fifths of the prerestriction level (1928/29–1930/31). These exports could increase German supplies within the old territory by about one-fifth.

German control over the sugar industries of these two countries was established in different manners. The Sudeten area of Czechoslovakia, with little surplus sugar, was incorporated into the Reich and directly subordinated under the German administration. The major part of Czechoslovakia was dismembered in 1939 into two parts. Bohemia and Moravia, with about three-fourths of the total sugar production of Czechoslovkia, formed the so-called Protectorate; it was left under the Czech administration but strictly controlled by the German occupation power. Slovakia, with only about 15 per cent of the Czechoslovak sugar industry, formed a nominally independent state under close German control. It had not much exportable sugar, since about one-third of its sugar industry was located in the territory annexed by Hungary.

Poland also was dismembered. The part that was invaded by the Soviet Union included only about one-tenth of the Polish sugar production whereas its population comprised more than one-third of the total population of prewar Poland. The main sugar-producing area of Poland with 90 per cent of factory productive capacity fell under German control. For administrative purposes it also was divided into two parts: (1) the Western, consisting mainly of districts that were part of Germany before World War I, with the addition of a considerable western fringe of formerly Russian Poland, was incorporated into the Reich; and (2) the Central, located between the German- and the Soviet-annexed zones, formed the so-called General Government under the general governor directly responsible to Hitler, and with administration dominated mainly by Germans. The part directly incorporated into the Reich was the *major sugar surplus area of Poland*, since it included about three-fifths of the prewar Polish industry, while its population, after expulsion of about one million Poles to the General Government, composed only a little more than one-fourth of the prewar population of Poland. On the other hand the area that was included in the General Government normally could not have much surplus sugar, since it produced less than one-third of the total Polish production of sugar and its population comprised more than one-third of the total. This was particularly so when to the General Government was added in 1941 a part of the territory (eastern Galicia) formerly annexed by the Soviet Union, which was rather deficient in sugar.

Table 9 presents further details on the prewar and the wartime production of sugar within the area under close immediate control of the German Reich. Statistics of sugar production presented in this table relate to nearly the same beet-sugar area for the entire period, although a small portion of Polish output on the territory annexed by the Soviet Union is not included for 1939/40–1940/41, and a small portion of the sugar industry annexed by Hungary from Slovakia is excluded for the entire war period.

It appears from the table that at the beginning of the war (1939/40) Germany with its annexed territories supplied more than one-third (36 per cent) of the entire sugar production of Europe ex-

TABLE 9.—SUGAR PRODUCTION, GERMAN SPHERE OF CONTROL, 1937/38–1944/45*

(*Thousand metric tons*)

Region	1937/38	1938/39	1939/40	1940/41	1941/42	1942/43	1943/44	1944/45
Old Reich	2,191	1,862	1,992	2,065	1,745	2,037	1,872	1,750
Annexed territories[a]	36	262	590	619	593	597	485	440
Of this:								
Austria	—	—	190	—	—	125	108	—
Sudeten	—	—	68	—	—	69	50	—
Annexed Polish territories plus Danzig	—	—	332	—	—	403	327	—
Austria[b]	157	—	—	—	—	—	—	—
Czechoslovakia[c]	741	517	510	582	536	520	432	459
Poland[d]	562	546	87	109	100	137	140	45
Total, German sphere	3,687	3,187	3,179	3,375	2,974	3,291	2,929	2,694
As per cent of Europe ex-USSR......	51.9	44.9	44.8	48.8	45.6	51.8	46.1	55.5

* F. O. Licht's *International Sugar Statistical Year and Address Book 1950* (Ratzeburg, mimeo.), pp. 3, 7–8, 26–27, of Part I, and pp. 11–13 of Part II.
[a] Annexed territories include: Danzig from 1937/38, Austria and Sudeten area of Czechoslovakia from 1938/39, and Polish annexed districts from 1939/40. Figures are obtained by subtracting data for the Old Reich from data for Greater Germany.
[b] From 1938/39 on, Austrian production is included in that for the annexed territories.
[c] From 1938/39 data cover Protectorate Bohemia-Moravia and Slovakia; the production of the Sudeten area is included under the annexed territories.
[d] From 1939/40 General Government only. Production of western districts, incorporated in the Reich, is included under annexed territories; that of the eastern districts, annexed by the Soviet Union, for 1939/40 and 1940/41 are not included at all; for 1941/42–1944/45 a portion of it may be included under the General Government.

USSR. Together with those parts of Czechoslovakia and Poland that remained outside the administrative borders of the Greater Reich, but under its strict political control, this share reached 45 per cent. It is significant that these shares did not tend to decline in the later years of war. On the contrary, during the second half of the war Greater Germany's share exceeded two-fifths of European production, and together with unannexed territories of Czechoslovakia and Poland its production approached one-half of the total. In the year of the German collapse—1944/45—its share in the European total actually rose to 55 per cent.

Production within the German-controlled area did not decline much absolutely even during the most strained war years. Average annual output in the Greater Reich together with unannexed parts of Czechoslovakia and Poland during three years 1941/42–1943/44 was only 4 per cent below the last prewar years. This was achieved by persistent efforts to expand sugar-beet areas not only within the borders of Greater Germany but also in the Protectorate and the General Government. The table shows that these efforts were rather successful, since sugar production in the remnants of Czechoslovakia and Poland during 1940/41–1942/43 was substantially higher than in 1939/40, in spite of the fact that war conditions affected beet yields per hectare unfavorably. Expansion of the sugar-beet area on territory immediately controlled by the Reich continued until 1942 when it was 7 per cent larger than in the first year of the war (1939) and 13 per cent higher than in 1937.

German success in maintaining production appears even better when compared with developments in the territories of its allies and enemies, as well as in neutral countries, given in Table 10. The area figures in this table include only beets processed into sugar, while we know from earlier discussion that in Germany in the last prewar year (1938/39) the output from more than 80,000 hectares of sugar beets was processed directly into various feeds. This practice continued during the war years, the annual average use for this purpose in 1940/41–1944/45 amounting to more than 100,000 hectares. Feed use reached its peak in 1941/42, when the entire crop from 133,000 hectares, or nearly one-fifth of the total sugar-beet area of Greater Germany, was fed to animals. This helped to a considerable extent to solve the feedstuff crisis experienced by Germany in this particular year, but the use of beets for feeding animals continued on a high level to the last year of the war. Feeding sugar beets was of the greatest importance in the old German territory; in the an-

TABLE 10.—SUGAR BEET AREA AND SUGAR PRODUCTION, EUROPE ex-USSR, BY SPECIFIED AREAS, 1937/38–1944/45*

Region	1937/38	1939/40 Abso-lute	1939/40 % of 1937/38	1941/42	1942/43	1943/44 Abso-lute	1943/44 % of 1939/40	1944/45 Abso-lute	1944/45 % of 1939/40
				SUGAR-BEET AREAS (1,000 hectares)					
German-controlled area[a]	739.7	779.2	105.3	782.1	834.6	818.4	105.0	736.9	94.6
Italy	108.0	150.7	139.5	154.9	150.2	160.1	106.3	105.3	69.9
Southeastern countries[b]	91.8	158.9	173.1	172.2	152.5	216.5	136.2	185.1	116.5
Western occupied countries[c]	361.5	391.1	108.2	320.7	281.3	332.8	85.1	307.3	78.6
Great Britain	126.8	137.7	108.6	138.5	165.5	164.3	119.3	167.9	121.9
European neutrals[d]	166.4	107.3	64.5	140.8	107.4	132.6	123.6	143.5	133.7
Europe ex-USSR	1,624.4	1,763.4	108.6	1,753.9	1,723.6	1,873.6	106.2	1,689.2	95.8
				SUGAR PRODUCTION (1,000 metric tons)					
German-controlled area[a]	3,687	3,179	86.2	2,975	3,290	2,929	92.1	2,694	84.7
Italy	347	480	138.3	462	423	344	71.7	60	12.5
Southeastern countries[b]	268	410	153.0	373	284	379	92.4	281	68.5
Western occupied countries[c]	1,707	1,899	111.2	1,490	1,273	1,273	67.0	730	38.4
Great Britain	426	546	128.2	524	585	555	101.6	438	80.2
European neutrals[d]	599	471	78.6	595	426	542	115.1	540	114.6
Europe ex-USSR	7,106	7,102	99.9	6,520	6,351	6,136	86.4	4,848	68.3

* Data from F. O. Licht's *International Sugar Statistical Year and Address Book, 1950*, Part II, pp. 3–5, 11–13. Only sugar-beet areas used for sugar production are included.

[a] Germany, Austria, Czechoslovakia, Poland.

[b] Hungary, Bulgaria, Rumania, Yugoslavia.

[c] France, Belgium, Netherlands, Denmark.

[d] Ireland, Sweden, Switzerland, Spain. The sharp decline of the sugar-beet area for this group of countries from 1937 to 1939 was mainly in Spain, where it was caused by the Civil War. But Ireland and Sweden also reduced their beet areas just before the war.

nexed areas it was less developed since their sugar industries were not as well provided with the necessary means of processing.

Table 10 indicates that the area producing sugar beets for processing into sugar also expanded in the German-controlled area. In 1943/44 it exceeded the 1937/38 level by more than 10 per cent and was 5 per cent above that of 1939/40. Only in the year of collapse—1944/45—did this area fall slightly below prewar. Italy and countries of Southeastern Europe that had close economic relations with Germany as its allies or were subjugated (Yugoslavia) also expanded their sugar-beet cultivation. During two years preceding the war, 1937–39, their expansion of beet area proceeded more rapidly than Germany's since they had greater arrears to make up. But they continued this expansion up to 1943. Consequently, at the beginning of the war, the central and southeastern portion of Europe closely associated with the German economy included more than three-fifths of the total sugar-beet area of Europe ex-USSR. During the war this share increased further.

In contrast, Western countries of Continental Europe made last-minute efforts to expand their sugar-beet cultivation in 1938/39 and 1939/40, but this expansion was more than offset by disorganization of their sugar industries by the German invasion in 1940. In no year following the invasion could the Western occupied countries recover their prewar (1937) sugar-beet area, while the countries of Central and Southeastern Europe (including Italy) by 1943 had expanded their plantings 27 per cent above the 1937 level (Table 10).

This contrast was due largely to the situation in France. Her main beet area was cut into three parts during the occupation: (1) two departments that led in sugar production—Nord and Pas de Calais—were included in the sphere of control of the German Military Governor of Belgium; (2) a large part of two other departments, also leaders in sugar production—Somme and Aisne—was located within the prohibited zone, from which refugees who escaped during the German invasion were barred; and finally (3) the remaining major sugar-producing departments were located within the German-occupied zone. In unoccupied France sugar beets were cultivated in several departments, but only in small quantities. Most of the output was processed into alcohol since the unoccupied area had only three of France's 100 sugar factories.

The total French sugar-beet area was reduced by invasion to 130,000 hectares in 1940/41 from the peak of 347,000 hectares the preceding year. During occupation, the highest level was 268,000 hectares in 1942/43. It declined to 255,000 hectares in 1943/44 and 196,800 hectares in 1945/46,[6] chiefly perhaps because of the shortage of labor. With some 10 per cent of those gainfully employed in French agriculture before the war held as prisoners, and the number of migrant workers from Belgium and other countries greatly reduced, there was a reduction of some 15 per cent in the total agricultural labor force. But the shortage of coal for sugar factories, in spite of their priority for certain quantities, also limited possible expansion (29, pp. 523, 526). These factors apparently affected all parts of the dismembered French sugar-beet area.

The situation in other Western occupied countries was much more satisfactory. In Belgium and the Netherlands the area was maintained at the 1937 level during the crucial war years (1941–45), and Denmark was even able to expand her area of sugar beets used for sugar, at the cost, however, of beets used for feed. She was therefore in a position to supply limited quantities of sugar to other Scandinavian countries, particularly Norway and Finland, but only by strictly rationing consumption at home.

European neutrals did little better than the Western occupied countries. Even excluding Spain, which was mainly responsible for the sharp decline in the group's total from 1937 to 1939, their combined beet area was only a few thousand hectares—6 per cent—larger in 1943 than in 1937. Western countries of Continental Europe, because they had better access to sugar supplies across the Atlantic, did not pursue autarchic policies as did the Central European countries. But occupied countries and neutrals alike suffered from the wartime blockade. Among the Western Allies, only Great Britain succeeded in expanding acreage, which was one-third higher at the end of the war than in 1937.

Expansion of sugar-beet cultivation within the area directly controlled by Germany and in the territories of her allies permitted their supplies of sugar in the second half of World War II to be maintained on a fairly high level. It would not be correct to take 1937/38 as a representative prewar year within the German-controlled area, since the yield of beets in 1937 was exceptionally high in Central Europe, just as yields in 1939 were exceptionally good in France. The prewar situation may be better represented by the average production

---

[6] These statistics on sugar-beet area include also sugar beets processed into alcohol. Data except for 1940/41 are according to 27, those for 1940/41 according to 28, p. 10.

for 1937/38–1939/40. On that basis, it may be said that during the three most crucial war years (1941/42–1943/44), following Hitler's invasion of the Soviet Union, the production of sugar within the German-controlled area exceeded 90 per cent of the prewar average. Performance was even better in Italy and in Southeastern countries. At the same time, average sugar production in the Western occupied countries during these years fell about 20 per cent below the prewar level.

This large decline in the Western occupied countries took place despite yields per hectare maintained better than in Central European countries, Italy, and particularly the Southeastern countries. This relates especially to Belgium and the Netherlands. Sugar-beet yields in Belgium between 1940 and 1943 averaged above, and in the Netherlands only slightly below, the 1937–39 level. Even in France, wartime yields of sugar beet on the reduced area declined less than within the German-controlled area. A relatively better supply of skilled labor during the war years in the occupied countries, with the exception perhaps of France, was apparently the principal factor in maintaining beet yields, since the consumption of fertilizers, particularly of scarce superphosphate and nitrogen fertilizers, was maintained in Germany rather better than in the occupied countries (*30*, p. 691). In Belgium, consumption of nitrogen and potash evidently averaged above the 1938/39 level during the war years (1941–43), but available manure declined in quantity and quality (*29*, p. 458). Germany was able to maintain sugar production within the directly controlled area on a level exceeding 80 per cent of prewar even in the year of collapse, 1944/45. At that time, war developments in Italy and the Western occupied countries reduced their sugar productions to a very low level (see Table 10), creating still greater temporary shortages in the liberated areas.

The picture of wartime sugar supply in Continental Europe is incomplete without data on the foreign trade in sugar within Europe. Unfortunately, information on this subject is scarce and sometimes contradictory. The International Institute of Agriculture supplies data on international trade in sugar during war years in the usual form (*33*), but emphasizes that these are neither comparable, because of numerous and frequent changes of frontiers, nor complete, since various kinds of movements of commodities across the frontiers during the war are not covered by trade statistics. The best unofficial source (*21*) completely omits data on the sugar movement within

Europe for the war years. The fragmentary information that follows is based mainly on secret German statistics (31), and on official French estimates of German removals of agricultural products from France (28). These sources clarify first of all the sugar supply and disposition within Greater Germany (in the frontiers as of September 1, 1939—that is, before the annexation of the Polish territories), and her dependence on sugar supplies from areas under her closer control. But they present also a wider view of the intra-European movement of sugar. By virtue of her control over the principal sugar-surplus areas of Continental Europe, under blockade conditions from mid-1940 to mid-1944, Germany commanded the sugar supplies available to all sugar-deficient countries of Continental Europe.

For the years 1939/40 to 1943/44 the sugar balance of Greater Germany (within September 1939 borders) was as follows, in thousand metric tons (31, pp. 1H1, 1H4):

Supplies and disposition	1939/40	1940/41	1941/42	1942/43	1943/44[a]
Disposable quantities:					
German production	2,051	2,061	1,778	2,022	1,900
Stocks on Oct. 1	162	202	214	96	265
Imports[b]	287	389	443	572	390
Total	2,500	2,652	2,435	2,690	2,555
Disposition:					
Exports[c]	152	328	317	213	217
Disposed within Reich ...	2,146	2,110	2,022	2,212	2,196
Stocks on Sept. 30	202	214	96	265	142
Total	2,500	2,652	2,435	2,690	2,555

[a] Data as estimated on July 20, 1944.
[b] Details in following tabulation.
[c] Details in tabulation, p. 262–63.

In these statistics the stability of the sugar supply at the disposal of the Reich is most stricking. Only in 1941/42, when German production declined mainly because of unfavorable weather conditions, was the supply of sugar below the level of the first year of war (1939/40). The fact that Germany could in the later war years take more sugar from outside contributed to this stability.

Data on imports of sugar into Greater Germany in the borders of September 1, 1939 (i.e., not including eastern territories annexed from Poland), by countries of origin, are shown in the following tabulation in thousand metric tons (31, pp. 1H2, 1H5):

Countries of origin	1939/40	1940/41	1941/42	1942/43	1943/44[a]
Annexed Polish area	119	124	160	180	160
Protectorate Bohemia-Moravia[b].	166	248	203	183	120
General Government	—	5	—	28	21
Soviet Union	—	—	43	115	36
Yugoslavia-Banat	—	—	2	4	7
Belgium-North France[c]	—	10	20	43	9[d]
Netherlands	—	2	14	15	15
Other countries	2	—	1	4	22[e]
Total	287	389	443	572	390

[a] Data as estimated on July 20, 1944.

[b] Imports from Protectorate Bohemia-Moravia also include exports from the Protectorate to third countries in execution of clearing agreements of the Reich.

[c] These imports are presumably from the French departments Nord and Pas de Calais administered by the German Military Government for Belgium.

[d] In addition to this export of sugar, France assumed obligation to process 300,000 tons of sugar beets into alcohol.

[e] Of this total, Italy supplied 19,000 metric tons and Denmark 2,000 metric tons.

They indicate that the principal sources of these imports were Polish and Czech surplus areas, while imports of sugar from other sources were sporadic and of secondary importance. Imports from Poland and the Protectorate Bohemia-Moravia were technically merely shipments within the common custom frontiers, since the annexed Polish territory was an administrative part of the Reich and the custom border between Germany and the Protectorate was abolished as of October 1, 1940. It appears likewise that imports from the Protectorate, as they are shown in the tabulation, include also exports from the Protectorate to third countries in execution of the clearing agreements of the Reich. The Reich government disposed of Czech sugar surplus as its own, not only for covering domestic needs but also for export to allies and other occupied countries.

Nearly half of the total production of sugar within the annexed Polish territory was shipped to other parts of the Reich, while imports from the Protectorate comprised about 45 per cent of its production. Together these placed nearly as much sugar at Germany's disposal each year throughout the war as had been exported annually during the last prewar years from the entire territories of Poland and Czechoslovakia. In the later war years Germany also exacted substantial quantities of sugar from the General Government by keeping very low sugar rations for the Polish, and especially the Jewish, population. Substantial quantities taken from the Soviet Union, principally from the invaded Ukraine, represented a war loot. This also was not in any real sense an exportable surplus. Production of sugar in the Soviet territory occupied by the Germans declined greatly in 1941/42 because beets from the large area sown before the invasion could not be harvested satisfactorily, while plantings in the following years were hardly adequate to cover the needs of the local population (see chapter 10).

Imports of sugar into Greater Germany from the Western occupied countries were, by contrast, relatively modest, and were at least partly offset in the case of France by German exports to that country, and for Belgium and the Netherlands by exports from France commandeered by the Germans. Official French sources recognize that direct German taking of sugar from France was small (28, pp. 5, 11–12). This was not an act of German generosity, since they took from Western occupied countries, particularly France, very large quantities of other foodstuffs (29, pp. 611–14). Rather, additional supplies of sugar from Polish and Czech surplus areas were sufficient to cover the German needs. Indeed, the amount of sugar distributed within Greater Germany in 1942/43 and 1943/44 was even larger than in the earlier war years. German sugar balance sheets (31, pp. 1H1–1H5) show that the sugar consumption of the army (direct and industrial) increased by more than 20 per cent over 1939/40, while the total consumption by civilians declined during the same years by 9 per cent. This may point to changes in the structure of the population since the number of persons mobilized in the army increased in the second half of the war, rather than to any change in per capita consumption. Human consumption of sugar increased in the later war years as compared with 1939/40 also in the German-occupied Protectorate of Bohemia-Moravia (32, p. 48). Its population enjoyed even better sugar rations than the Germans, though fat rations were substantially lower (29, p. 286).

German supplies of sugar were so abundant that quantities could also be provided to some allies and to suger-deficit occupied countries. Exports of sugar from Greater Germany (boundaries as of September 1, 1939) to various countries during the war years were as follows, in thousand metric tons (31, pp. 1H2, 1H5):

Country of destination	1939/40	1940/41	1941/42	1942/43	1943/44[a]
Norway	.2	29.9	23.8	25.5	23
Finland	.3	8.2	16.7	16.0	16
Baltic States	7.4	.1	—	—	—
France	—	19.3	29.1	10.0	22[b]
Italy	—	34.2	30.4	20.0	—

Country of destination	1939/40	1940/41	1941/42	1942/43	1943/44[a]
Switzerland	10.1	26.8	6.2	—	5
Greece	—	—	—	12.0	12
Turkey	—	—	—	3.0	—
Bulgaria	.3	19.9	15.1	10.0	15
Croatia	—	—	—	12.5	15
Other countries	16.2	15.8	12.4	—	5
Alsace-Lorraine[c]	—	—	33.0	33.0	33
Luxembourg[c]	—	—	7.0	7.0	7
Eastern Upper Silesia[c]	—	20.0	47.0	47.0	47
Styria-Craina[c]	—	—	17.0	17.0	17
Total	34.5	174.2	237.7	213.0	217
Exports from Protectorate					
to third countries	117.0	154.0	79.0	—[d]	—[d]
Grand total	152	328	317	213	217

[a] Data as estimated on July 20, 1944.
[b] Includes 10,000 metric tons according to old contract and 10,000 for alcohol.
[c] Estimates of required additional supplies from outside.
[d] Included in exports to individual countries.

Substantial quantities supplied to Italy, added to her sugar production which, during the crucial war years (1940/41–1943/44), was one-fourth larger than the prewar average, permitted per capita consumption in that country to exceed the prewar level until 1943. Directly or through the intermediary of the Protectorate, sugar was supplied to deficit countries in Scandinavia—Norway and Finland. Occupied Denmark was also permitted to export sugar to these countries. Net imports into Norway during 1941–45 fell to about 70 per cent of the prewar level, and into Finland, 50 per cent (33). Denmark's contribution was most important, and involved deliberate rationing of her own domestic consumption.

Germany also supplied sugar, directly or through the intermediary of the Protectorate Bohemia-Moravia, to Southeastern Europe, allies and occupied countries alike. Among these countries, Bulgaria, at the time a German ally, obtained the largest and most continuous supplies of sugar. Receipts by occupied, dismembered Yugoslavia, particularly Croatia, then an Italian satellite, also were substantial. On the other hand, very little was received by Greece, which had no domestic production at all. Enjoying less than 10 per cent of her normal prewar imports, Greece experienced a great shortage of sugar throughout the later war years (1941–44), although the parts occupied by Bulgaria fared somewhat better. Imports into Southeastern countries are in fact understated in the tabulation above, inasmuch as exports from the Protectorate to third countries (including Bulgaria, Yugoslavia, Greece, and Rumania) in execution of clearing agreements of the Reich, are not distributed among destinations. Information available on exports of sugar from Czechoslovakia during 1936/37–1945/46 by countries of destination (*34*) does not separately indicate exports from the Protectorate. Furthermore sugar shipped from the areas annexed by Germany (Sudeten) and Hungary (south Slovakia) within the annexing country are included in export figures. Distribution by destination may involve some additional errors, since sugar from the Protectorate was shipped by German order and Czech authorities could not have been well informed about final destinations.

Quantities of sugar indicated by the German statistics as exported to France coincide fairly closely in total with the quantities indicated by the French source as imported from Germany (*28*, p. 11). The French source explains, however, that German deliveries of sugar to France were not in compensation for sugar taken from France, but in exchange for other commodities, such as wheat, coffee, or cocoa, considered of greater value by the Germans. But if Germany's direct taking of sugar from France was relatively small, indirect effects of her policies reduced French supplies of sugar more severely. To help meet her increasing requirements for alcohol, Germany took from France the alcohol equivalent of 240,000 tons of beet sugar during the four years 1940/41–1943/44 (*28*, p. 14). Including the Vichy government's own requirements for alcohol distillation, about 6.5 million tons of sugar beets were diverted during 1940/41–1943/44. This was 29 per cent of the total quantity harvested, as against the prewar average of 27–28 per cent. Thirty per cent of the alcohol produced in this way was for German use. What had been a means of diverting surplus beets before the war became a matter of sheer necessity under wartime conditions.

German insistence upon relatively higher rations for the population of the two French departments under the control of the German Military Government for Belgium imposed a further burden on other parts of France. But the principal factors that reduced the total supply of sugar available to the French population were: (1) the reduction of French output during 1940/41–1943/44 by about 35 per cent below the prewar level (1934/35–1938/39), and (2) the disappearance after 1939/40 of net imports, which averaged 113,000 tons a year during 1934/35–1938/39, mainly from overseas possessions. Since France continued to supply sugar to French North Africa until the landing of the Allied forces there in November 1942, France found herself a net sugar exporter during 1941–43, despite a nearly 50 per cent drop in domestic consumption (*28*, p. 17).

Disappearance of net imports of sugar during the war years in Belgium and the Netherlands was mainly responsible for the reduction of sugar supplies in these countries. This was particularly serious for the Netherlands, which derived nearly one-fourth of its total supply of sugar from imports in 1934–38, as compared with around 10 per cent for Belgium. Belgium had the additional advantage that sugar which she previously supplied to Luxembourg within a common custom border came, after 1941/42, from Germany instead. Generally speaking, sugar supplies in Belgium and the Netherlands held up better than in France, because domestic sugar production in these two countries was better maintained. All three succeeded in accumulating certain reserves of sugar before the German occupation by expanding imports and reducing exports in the last years before the blockade was imposed.

THE POSTWAR PERIOD

Territorial changes in Continental Europe after World War II considerably affected the structure of the European sugar market. Although frontiers in the West were largely restored to their preinvasion state, that was not true in Central and Eastern Europe. West Germany, created by unification of the three Western zones of occupation, comprised a sugar-deficit region, cut off from normal sources of sugar supply in the central and eastern provinces of Germany. Hence, West Germany chose to develop an expansionist sugar policy which is hardly likely to be reversed in the future. Central Germany, the former Soviet zone of occupation, became by contrast an important sugar-surplus area. The ample sugar supply of Poland was further enlarged by addition of the substantial sugar surpluses of the eastern German provinces.

These eastern surpluses were mainly absorbed by the Soviet Union, particularly during the earlier years of occupation when German sugar was taken as reparations. From 1949/50 on, absorption of sugar surpluses by the Soviet Union was temporarily at least a stabilizing factor in the world sugar market, though Soviet imports would be unnecessary if levels of output planned for its own sugar industry were actually attained. In view of the fact that sugar-surplus Czechoslovakia came also to be included among the Soviet satellites, concerted disposal of Eastern European sugar introduces an element of uncertainty so far as longer-run developments in the world sugar market are concerned. Certainly, trends in sugar-beet area and sugar production in Western Europe must henceforth be considered separately from those in Eastern Europe.

The main feature of the postwar development of the Continental European beet-sugar industry is its rapid restoration to the prewar level. This particularly relates to the sugar-beet area, which was restored to the prewar (1938/39) level in the second postwar year (1947/48). This is because expansion of the sugar-beet area continued during the war in some countries, particularly in Central and Eastern Europe, so that the total for Continental Europe ex-USSR in 1943/44 had substantially exceeded that in 1938/39. The subsequent decline, though sharp, lasted only two years, and took place mostly in the regions directly affected by military action. Furthermore, expansion continued without interruption until 1952/53, by which time the beet area of Continental Europe ex-USSR was two-thirds larger than in 1937/38, the last normal prewar year. In this expansion Western and Eastern Europe participated equally. From then on, a certain stabilization of the sugar-beet area took place in response to the decline of sugar prices in the world market.

The data on sugar-beet areas in Table 11, which includes only sugar beets processed into sugar, somewhat overstate the extent of the postwar expansion. The use of beets for preparation of various feedstuffs was greatly reduced during the period. In West Germany, where from 1946 to 1948 food supplies were at a critically low ebb, the use of sugar beets delivered to factories for purposes other than production of sugar was prohibited after the war. Only in Denmark and, to some extent, Belgium and the Netherlands is direct feeding of beets still of considerable importance. In France, however, the use of sugar beets for processing into alcohol has continued to absorb about one-quarter of the crop, as it did in the latter 1930's (cf. *20*, pp. 986–1016; *35*). Not until 1954 were serious efforts made to limit this practice.

It is not yet quite clear whether Europe's sugar-beet area has definitely been stabilized. While acreages in Western Europe declined in 1953, expansion continued in the East. Furthermore, preliminary information indicates that even in Western Europe the sugar-beet area expanded once more in 1954 and this tendency continued in 1955. Such countries as West Germany and Italy, not yet self-sufficient in sugar, are particularly eager for further expansion. Spain, on the other hand, had been expanding her sugar-beet cultivation

TABLE 11.—PRODUCTION OF SUGAR, SUGAR-BEET AREA, AND YIELD OF SUGAR PER HECTARE, CONTINENTAL EUROPE ex-USSR ex-TURKEY, 1937/38, 1945/46–1953/54*

Region	1937/38[a]	1945/46	1947/48	1949/50	1950/51	1951/52	1952/53	1953/54	Average 1950/51–1953/54 Actual	Average 1950/51–1953/54 Percent of 1937/38
	SUGAR PRODUCTION (1,000 metric tons)									
Western Continental Europe[b]	3,348	1,603	2,304	3,650	4,936	4,821	4,616	6,012	5,096	152.2
Eastern Continental Europe ex-USSR ex-Turkey[c]	3,126	1,205	1,659	2,568	3,109	3,139	2,562	3,684	3,124	99.9
Total	6,474	2,808	3,963	6,218	8,045	7,960	7,178	9,696	8,220	127.0
	SUGAR-BEET AREA (1,000 hectares)									
Western Continental Europe[b]	775	475	766	958	1,070	1,190	1,309	1,212	1,195	154.2
Eastern Continental Europe ex-USSR ex-Turkey[c]	648	480	825	942	994	1,041	1,092	1,104	1,058	163.3
Total	1,423	955	1,591	1,900	2,064	2,231	2,401	2,316	2,253	158.3
	YIELD OF SUGAR (metric tons per hectare)									
Western Continental Europe[b]	4.32	3.37	3.01	3.81	4.61	4.05	3.53	4.96	4.26	98.6
Eastern Continental Europe ex-USSR ex-Turkey[c]	4.82	2.51	2.01	2.73	3.13	3.02	2.35	3.34	2.95	61.2
Weighted average	4.55	2.94	2.49	3.27	3.90	3.57	2.99	4.19	3.65	80.2

* Data based for all years except 1945/46 on F. O. Licht's *Internationales Zuckerwirtschaftliches Jahr-und Adressbuch 1954/55*, Part II, Table 2, p. 364, and Part III, Table 2, p. 370; for 1945/46, on F. O. Licht's *International Sugar Statistical Year and Address Book 1950*, Part II, pp. 3, 11.

[a] Data for Germany for 1937/38 are divided between western and eastern Germany on the basis of information in F. O. Licht's *International Sugar Statistical Year and Address Book 1950*, Part I, p. 20. Licht's data for Poland are corrected for loss of territory to the Soviet Union on the basis of information in *Petit Annuaire Statistique de la Pologne 1939*, pp. 78–79.

[b] Includes West Germany, Austria, France, Belgium, Netherlands, Denmark, Sweden, Italy, Spain, Switzerland, and Finland.

[c] Includes East Germany, Czechoslovakia, Hungary, Poland, Rumania, Bulgaria, and Yugoslavia.

rapidly up to 1953, but reversed this policy when a combination of improved yields per hectare and an enlarged beet area resulted in a considerable sugar surplus which could be disposed of only at low export prices. Similarly, France passed special legislation limiting her sugar-beet area, since both processing of sugar beets into alcohol and export of sugar to foreign markets require heavy state subsidies. Certainly further expansion would be entirely feasible on the Continent if the end product could find a ready market (36, p. 27).

Expansion during the early postwar years was stimulated not only by the sugar-market situation, but also by such organizations as the Food and Agriculture Organization (FAO) and the Organization for European Economic Co-operation (OEEC). In their programs of European agricultural reconstruction and development, the goal for principal grains was the prewar area by 1950 and 5–6 per cent beyond that by 1952. The sugar-beet area, however, was to exceed the prewar figure in 1950 by more than one-fifth and in 1952 by one-third (37, p. 8; 38, p. 13). Clearly, these programs aimed at achieving greater self-sufficiency in food as a contribution to the closing of the postwar dollar gap (39, pp. 163–67).

While postwar expansion of the sugar-beet area was equally strong in Eastern and in Western Europe, actual production results were quite different in the two regions. In the West, sugar production regained its prewar level as early as 1948/49 and during more recent years, 1950/51–1953/54, on the average exceeded 1937/38 production by more than one-half, or about in proportion to the increased area. In the East production remained below the prewar level until 1951/52, and the average for the period 1950/51–1953/54 hardly reached the 1937/38 figure. Table 11 indicates that yields of sugar per hectare in the East in 1950/51–1953/54 were nearly 40 per cent below 1937/38, whereas in the West they were close to the level of that year. If the comparison is made with the longer period 1937/38–1939/40, yields show a few per cent increase in the West but almost a 30 per cent decline in the East.

Several factors contributed to the low postwar yields in Eastern Europe. Of primary importance was the radical reorganization of the agrarian systems in satellite countries, with its adverse effects on the efficiency of their agriculture. Especially in eastern and central Germany, Poland, and Hungary, the rapid mass parceling of large estates, not well prepared or planned, hit sugar-beet production particularly hard. In this region the proportion of large sugar-beet farms had been high, while the newly created farms were usually too small and were poorly equipped with draft power and machinery. The eastern provinces of Germany, administered by Poland, had also to undergo a nearly total expulsion of German farmers and their replacement by new settlers unfamiliar with local conditions. The new settlers could not replace the expelled farm population even numerically, and shortage of machinery was aggravated by shortage of agricultural labor. This feature has become characteristic of the region in recent years (*40*, pp. 141–46), and of Czechoslovakia as well (*41*, p. 11). The size to which agricultural enterprises were reduced is indicated by Poland as an extreme case. There were more than 800,000 beet growers in 1950, nearly half of them with less than 5 hectares of *land* per farm, and averaging less than one-third of a hectare of *sugar beets* per farm (*42; 43*, p. 1). Under such conditions, various incentives to growers succeeded in raising the beet acreage, but could not restore sugar production to the prewar level.

Moreover, Western Europe on one side and the satellite countries on the other were very differently situated with respect to the supply of commercial fertilizers. In all OEEC countries of Europe taken together, the consumption of commercial fertilizers was, as early as 1949/50, about 40 per cent higher than before the war (*44*, pp. 37–39), and West Germany had about restored its prewar consumption. In the Soviet zone of Germany, economically the most advanced among the satellite countries, the use of fertilizers that year remained substantially (about 20 per cent) below the prewar level, as the following figures indicate (*45*, p. 543; *46*, p. 20):

Fertilizer	Total use (1,000 tons)		Kilograms per hectare of agricultural land	
	1938/39	1949/50	1938/39	1949/50
N	218	186	32.6	29.2
P_2O_5	182	95	27.3	14.9
K_2O	323	275	48.4	43.1

The supply of fertilizers in other satellite countries unquestionably was still less satisfactory, especially in the Southeast, where yields of sugar beets remained particularly low compared with prewar (*47*, pp. 71–76).

Official Polish statistics, which compare the use of commercial fertilizers per hectare of cropped area in 1948/49 within the present boundaries with prewar use in Old Poland, give the erroneous impression that the postwar use of fertilizers in Poland was two to three times as great as prewar (*48*, p. 65). However, about one-third of the present agricultural area of Poland, and more than one-half of its sugar-beet acreage, consists of land that was within the German frontiers before the war. German farm lands at that time received fifteen to sixteen times as much fertilizer per hectare as did Polish farm lands. Official estimates of fertilizers used in Poland in 1948/49 in fact represent only about one-half the total quantities used *in the same territory* in 1937/38.

Sugar production in Eastern Europe would greatly increase if prewar yields could be obtained from the present sugar-beet area. Exportable surpluses, however, could not be expected to rise proportionally. Postwar consumption of sugar increased considerably in all Eastern European countries, as it did in other countries with low levels of income. This tendency is likely to continue, particularly in the Southeast. Net imports of sugar into Western Europe (including the United Kingdom) from the East European countries (including the USSR and Finland) appear to have declined from the prewar average of 775,000 short tons to below 190,000 tons during 1949–53 (*49*, pp. 156–59). During 1954–55, the sugar supply position of the East European countries deteriorated so badly, partly because of unfavorable weather conditions, that such traditional exporters as Czechoslovakia, East Germany, and Hungary actually imported considerable quantities of sugar from the Western Hemisphere during the early months of 1955 (*50*, p. 2). Although they probably preserved their net export positions over the year as a whole, this episode indicates that sugar surpluses of East European countries are now relatively small, and net positions may fluctuate rather widely in the future.

SUGAR BEETS IN EUROPEAN AGRICULTURE

Despite the postwar expansion, the share of sugar beets in the total crop area of Continental Europe remains extremely small. In eleven countries of Western Europe (Greece, Norway, and Portugal grow no sugar beets), the area in beets amounted to 1,233,000 hectares in 1952/53, or 1.6 per cent of the total arable area of about 78.8 million hectares including tree crops (*51*, pp. 3, 40). This percentage is raised slightly, to 1.7 per cent, when sugar beets processed into alcohol in France are taken into consideration. In Eastern Europe, where the proportion is somewhat higher, the area figures for 1952 and 1953 averaged 1,110,000 hectares, or about 2 per cent of the total arable land roughly estimated for the earlier period (chiefly 1947–49) at 54.7 million hectares (*51*, p. 3).

That over-all percentages conceal substantial variations among individual countries is indicated by the following data on sugar-beet areas as percentages of total arable areas:

Western Europe		Eastern Europe	
Netherlands	6.2	Soviet Zone of Germany	4.3
Belgium	6.0	Czechoslovakia	3.9
West Germany	2.6	Hungary	3.0
Denmark	2.4	Poland	2.1
Austria	2.2	Rumania	1.0
France	2.0[a]	Yugoslavia	1.0
Italy	1.4	Bulgaria	.8
Sweden	1.4		
Switzerland	1.3		
Spain	.8		
Finland	.4		

[a] Or 1.6 if sugar beets processed into alcohol are excluded.

One can infer that sugar beets continue to be of greater importance in the traditional zone of their cultivation, which occupies the middle latitudes in Europe from Belgium, the Netherlands, and northern France in the west to Poland in the east, than in countries either to the north (Sweden, Finland) or to the south (Italy, Spain, and the Balkan countries). One northern country (Norway) and three southern ones (Albania, Greece, and Portugal) continue to depend entirely on imports for their sugar needs. Portugal can rely on a plentiful supply of cane sugar from its colonies. Greece recently planned to introduce sugar beets and to build sugar factories, but the project was postponed when objections were raised by the country that assists her financially. Apparently the rapid expansion of the beet-sugar industry in neighboring Turkey, though mainly in her Asiatic territory, served as an example for the Greek government in its plan for a domestic sugar industry.

Even greater territorial concentration is characteristic of the distribution of beet acreages within individual countries. Sugar beets have rather exacting requirements, not only as to climate but also as to quality of soil. In Europe sugar beets are cultivated mainly on the so-called "wheat soils." Deep loess soils, rich in humus and lime, characteristic of relatively dry regions, are very favorable for sugar beets because they are easy to work and root penetration is good. But sugar beets require rather abundant moisture, and such soils are not common in humid areas. In those areas beets are grown on various grades of loamy soils, not too heavy and not too light. One authority, Lüdecke, argues that the belief that sugar beets may be successfully cultivated only on wheat soils is false (*52*, pp. 23–30). Good crops of sugar beets are possible on various grades of loamy soils, so long as they are not stony and are deep enough. With adequate moisture, says Lüdecke, sugar beets may be grown even in sandy soil. This is probably correct in present-day Europe, with its high cultural condition of soils, but location of sugar-beet cultivation was determined many decades ago, when conditions were different. Furthermore, another authority, Woermann, shows that sugar beets compete successfully with potatoes only when cultivated on wheat lands, where yields may exceed 30 tons per hectare, while yields of potatoes remain the same as on sandy soils (*53*, pp. 238–44).

France may serve as a good example of a great concentration of sugar-beet cultivation within a relatively limited area. In 1951 two-thirds of her total sugar-beet area was located in 6 departments of the north and in the so-called Parisian region. In each of the 6, sugar beets composed more than 10 per cent of the arable area; they averaged 12.4 per cent for all and went as high as 16 per cent in one, Aisne (*54*). Such a large proportion of sugar beets in the arable area of these departments may indicate that the expansion of sugar-beet cultivation reached something of a natural limit. Replanting sugar beets on the same land more frequently than once in 5 or 6 years is not advisable for various reasons connected with rational crop rotations, while a substantial portion of the arable land in these departments is, of course, not suited to this crop. Four more neighboring departments, with a proportion exceeding 5 per cent, plant more than 10,000 hectares each. Together the 10 form a compact area comprising 85 per cent of the total sugar-beet area of France, and claiming 91 of her 105 sugar factories, half of which also have distillation facilities for processing sugar beets into industrial alcohol. The remaining 15 per cent of the French sugar-beet area is dispersed in more than 30 other departments throughout France, but only in 5 of these, adjacent to the compact sugar-beet area, do sugar plantings exceed 5,000 hectares. In the dominant region, beets are a leading crop in the rotations, to which other crops are adjusted.

The high level of agricultural technique applied to sugar beets considerably raises the yields of other crops that follow it in rotation. This is particularly true of wheat, which yields far higher per hectare in the sugar-beet belt than in France as a whole. Some leading agronomists even argue that intensive forms of agriculture in France generally derive from the introduction of beets.[7] That may be too strong a claim. It is the high proportion not only of sugar beets, but of all hoe crops (*plantes sarclées*) in the arable area that is the best indicator of intensive crop rotations. In less than one-third of the departments in which the proportion of hoe crops exceeds 20 per cent of the arable area (1952) are sugar beets primarily responsible for the degree of intensity. Elsewhere, such crops as potatoes and fodder roots are involved. Nevertheless, the 7 leading beet departments do consume 45 per cent of the total quantity of nitrogen fertilizer used in France and the major portion of potash (56, pp. 880–90). Moreover, the northern region, including 8 leading beet departments, is evidently characterized by the highest agricultural output per male worker engaged in French agriculture, by fairly large size of farms, and by high productivity per hectare of cultivated land—30 per cent higher than the average for the entire country (57, pp. 172–95, especially p. 184). For a relatively limited area these statistics reflect an impressive performance in which the role of sugar beets is certainly pivotal.

In *Italy*, sugar-beet cultivation is concentrated within a still narrower area, so narrow, indeed, that the crop could hardly be expected to influence national agriculture as a whole. Some 80 per cent of the entire Italian sugar-beet area is concentrated (1952–54) in the delta of the Po River, divided between the two provinces Emilia and Veneto in which it comprises 7 to 8 per cent of the total arable land. Nearly half the sugar-beet area is concentrated within only two districts, Ferrara in Emilia and Rovigo in Veneto, where its share in arable land reaches 20 to 25 per cent. The remaining 20 per cent of the sugar-beet area is dispersed throughout many provinces of Italy, in none of which does it reach one per cent of the entire crop area. Yields in these provinces are on the average substantially below those in the Po Delta, the differential being 15 per cent during 1952–54 (58, p. 80; 59, pp. 135–36), in spite of the fact that sugar beets in southern Italy are also grown under irrigation. Indeed, some authorities recommend against any beet planting in Italy outside the Po Delta, because unfavorable climatic conditions have had to be offset by excessive protectionism and the policy of multiple prices (60, pp. 62–63).

In the delta of the Po River, sugar beets are cultivated in recently

---

[7] "Toutes les exploitations rurales qui font de la culture intensive ont comme tête d'assolement la betterave" (55, p. 90).

drained marshy land, to the reclamation of which beet cultivation made a major contribution. Even here, however, climatic conditions are not particularly favorable for sugar beets because of the summer heat. Summer heat shortens the growing season in Italy, since harvesting of beets must be done during August–September, before heat wilts the leaves. Heat also shortens the sugar campaign since it speeds the spoilage of harvested beets (*61*, pp. 215–18). Considerable efforts have recently been made to develop special varieties suited to these local climatic conditions. In southern Italy, where new sugar factories are now being built, the plan is to rely on winter-grown beets as a means of avoiding losses from summer heat and at the same time extending the growing season (*62*, p. 448).

In *Spain* sugar beets occupy less than one per cent of the entire arable area, and their cultivation is dispersed throughout the entire country. However, about 90 per cent of their area is irrigated, and as the irrigated land of Spain is *estimated* at about one million hectares (*63*, p. 29), the percentage of sugar beets among irrigated crops must be as high as 10 to 15 per cent. In this respect, the situation in Spain is similar to that in the Mountain and Pacific states of the United States.

In *West Germany* cultivation of sugar beets is fairly well dispersed, although the density is greater in the provinces (*Laender*) of the middle latitude (Lower Saxony and North Rhine-Westphalia). In 1952–53, sugar beets comprised more than 6 per cent of the arable land in Lower Saxony, and less than one per cent in Bavaria (*45*, pp. 153, 157). Until recently, expansion proceeded mainly in the older sugar-beet regions, but with the construction of new sugar factories and enlargement of old ones in Schleswig-Holstein and Bavaria the importance of sugar beets in the northern and southern parts of West Germany is increasing. While beets were expanded recently to some extent in competition with vegetables and oilseeds, the principal expansion of sugar beets as well as of fodder beets took place at the expense of grains.

Detailed computations available for West Germany on gross crop production, net food production, and monetary receipts from sales of agricultural products make it possible to determine more precisely the role of sugar beets in German agriculture during recent years. Gross production of sugar beets (expressed in grain units) averaged 6.2 per cent during 1950/51 on only 2.6 per cent of West Germany's total crop area; in Lower Saxony, the figures were 13 and 6 per cent respectively (*64*, p. 434). As a cash crop par excellence, beets contributed 11.2 per cent of the total monetary receipts from sales of all crops in West Germany during 1950/51–1952/53, although inclusion of animal products reduces the average to 3.6 per cent (*65*, pp. 2–24; *66*, pp. 360–69).

Such percentages do not measure fully the importance of sugar beets for German agriculture. The role of sugar-beet by-products in feeding livestock on the farm without going through the intermediary of the market place must not be overlooked. Favorable effects on yields of other crops in rotations have already been noted in the French case. Historians of German agriculture mention still other indirect benefits. For instance, Krzymowski says that the improved agricultural technique required by sugar beets forced farmers to pay more attention to selection of better varieties of other plants, particularly grains, as well (*67*, pp. 179, 220). The better cultivation and fertilization of land for sugar beets caused lodging of succeeding grain crops grown from common seed. It became necessary to import improved seed from England and Denmark and, later, to organize seed selection at home in order to prevent the lodging. Beets also appear to have stimulated the development of agricultural machinery in Germany.

In *Belgium* and *the Netherlands* the share of beets in the total arable area is the largest in Europe. In the Netherlands sugar beets are cultivated mainly on the marine clay soils of the western coastal region. There they occupy 13 per cent of the crop area, while on the poorer sandy soils in the east their share does not exceed 2 per cent. According to official estimates, sugar beets comprised 6.7 per cent of the total value of crops produced in the Netherlands in 1953, as against 3.4 per cent in 1938/39 and 5.5 per cent on the average for 1951–53 (*68*, pp. 406–07, 488). Estimates for the decade 1932–41 indicate that 12.5 per cent of all Belgian farms cultivated sugar beets, and crop rotations with sugar beets were in effect on 30 per cent of *the better land* (wheat land) of Belgium (*69*, pp. 33–35). Beet acreages have since increased from about 4 per cent to about 6 per cent of total arable land.

Beets are particularly important in the better agricultural land in *Denmark* and *Sweden*. They take up about 7 per cent of the arable area of the Danish island, on which cultivation is concentrated because of better soils. Climate rather than soils restricts beets to the southern portion of Sweden, where as large a proportion of the plowed land is devoted to beets as in West Germany.

Lack of reliable information on the present state of agriculture in the Soviet satellite countries precludes any detailed discussion for Eastern Europe, although the relative importance of beets in the Soviet zone of Germany and in Czechoslovakia has already been indicated. It must be stated, however, that production remains below the 1925 record level for Czechoslovakia and probably also below the 1930 record for the Soviet zone. Only in Poland and in Southeastern countries have previous records definitely been surpassed by postwar expansion.

The great importance of sugar beets in the advancement of European agriculture, on the one hand, and the relatively small share of sugar beets in the total crop area of most European countries, on the other, imply that the limit to further expansion is not so much the shortage of suitable land as it is inadequate sugar consumption. Greater consumption, which might be achieved by reducing the high sugar excises prevailing in many European countries, would induce an increased supply of sugar beets, particularly in countries that do not already cover their sugar requirement from domestic sources. Prevailing low prices of sugar in the world market, however, tend to stop further expansion or even reverse it in countries where supplies exceed domestic requirements. It is noteworthy that, beginning with 1953, crop prices for sugar beets were lowered in most West European countries (23, p. 5).

CITATIONS

1 G. Martineau, *Sugar* (London, 1910).

2 G. Dillner, "Weltzuckerwirtschaft und Marktregulierungen," *Weltwirtschaftliches Archiv* (Kiel), March 1941.

3 K. T. Voblyi, *Essay on the History of the Beet-Sugar Industry of the USSR*, I (Moscow, 1928) [in Russian].

4 M. Lenglen, *Défense et illustration de la betterave* (Paris, 1946).

5 Noel Deerr, *The History of Sugar* (2 vols., London, 1949 and 1950).

6 E. Saillard, *Betterave et sucrerie de betterave* (Paris, 1913).

7 S. von Ciriacy-Wantrup, *Agrarkrisen und Stockungsspannen* (Germany, Reichs- und Pr. Min. für Ernährung u. Landwirtschaft, *Berichte über Landwirtschaft*, N. F. 122 Sonderheft, 1936).

8 F. R. Rutter, *International Sugar Situation* (U.S. Dept. Agr., Bur. Stat. Bull. 30, 1904).

9 F. O. Licht, *World Sugar Statistics, 1937* (Magdeburg, 1937).

10 G. Mikusch, *Geschichte der Internationalen Zuckerkonventionen* (Germany, Reichsministerium für Ernährung u. Landwirtschaft, *Berichte über Landwirtschaft*, N.F. 54 Sonderheft, 1932).

11 K. H. Spielmann, ed., *Jahrbuch der deutschen Zuckerwirtschaft* . . . *1936/37* (Mainz, n.d.).

12 Josef Siegel, *Die tschechoslowakische Zuckerindustrie* (Berlin, 1928).

13 O. W. Willcox, *Can Industry Govern Itself?* (New York, 1936).

14 J. Debordes, article in *Droit Social* (Paris), June 1947.

15 J. Debordes, article in *Revue de Legislation Agricole* (Paris, 1948).

16 K. H. Spielmann, ed., *Jahrbuch der deutschen Zuckerwirtschaft* . . . *1940* (1940).

17 "Possibilities of Reducing Costs and Expanding Use of Sugar Beets," in Czechoslovak Academy of Science, *Agricultural Topics* (Prague), No. 71, October 1938 [in Czech].

18 Italy, Min. Agr., *Notes sur l'Agriculture italienne et sur l'Organisation Corporative agricole* (Rome, 1937).

19 Confédération Générale des Planteurs de Betteraves Industrielles, *Annuaire Betteravier, 1945–46* (Paris).

20 "The Problem of Alcohol," *Etudes et Conjoncture* (France, Ministère des Finances et des Affaires Economiques), September 1953.

21 F. O. Licht's *International Sugar Statistical Year and Address Book 1950* (Ratzeburg, 1951).

22 E. Woermann, "Sugar Beet Cultivation in the Food and Feed Economy," *Mitteilungen für die Landwirtschaft* (Germany), Nov. 12, 1938.

23 F. O. Licht's *Sugar Information Service*, Supplementary Report No. 20, Oct. 20, 1954.

24 K. H. Spielmann, ed., *Jahrbuch der deutschen Zuckerwirtschaft* . . . *1941*.

25 K. H. Spielmann, ed., *Jahrbuch der deutschen Zuckerwirtschaft* . . . *1944*.

26 K. A. Schoeller, *Die Zuckerrübe in der deutschen Ernährungs und Futterwirtschaft* (Berlin, 1940).

27 France, Min. Agr., *Statistique Agricole Annuelle*, various years.

28 Commission Consultative des Dommages et des Reparations, *Prélèvements Allemands de Produits Agricoles, Sucre* (Monographie P. A. 13, Paris, 1947).

29 Karl Brandt et al., *Management of Agriculture and Food in the German-Occupied and Other Areas of Fortress Europe* (Food Research Institute Studies on Food, Agriculture, and World War II, Stanford, Calif., 1953).

30 Mirko Lamer, *The World Fertilizer Economy* (Food Research Institute Studies on Food, Agriculture, and World War II, Stanford, Calif., 1957).

31 Germany, Statistisches Reichsamt, *Zahlen zur Deutschen Kriegsernährungswirtschaft* (Geheime Reichssache [1945]).

32 *Sugar News* (Prague), December 1947.

33 International Institute of Agriculture, *International Yearbook of Agricultural Statistics, 1941/42–1945/46* (Rome, 1947).

34 *Sugar News*, January-February 1947.

35 Siegfried Mielke, "The New French Sugar Policy," F. O. Licht's *Sugar Information Service* (Ratzeburg), Apr. 2, 1955.

36 Hugo Ahlfeld, "On the Sugar Situation in Western Europe," F. O. Licht's *Sugar Information Service: West European Special Sugar Edition* (Ratzeburg), July 1954.

37 FAO, *European Programs of Agricultural Reconstruction and Development* (Washington, D.C., June 1948).

38 OEEC, *Report of the Food and Agriculture Commission,* Sec. I, Vol. III (December 1948).

39 FAO, *The State of Food and Agriculture 1955, Review of a Decade and Outlook* (Rome, September 1955).

40 Germany, Deutsches Institut für Wirtschaftsforschung, *Wochenbericht,* Sept. 3, 1954.

41 "Czechoslovakia Balance Sheet," *News from Behind the Iron Curtain* (New York), May 1955.

42 W. Brykczynska, L. Chrzanowski, and S. Kubas, *Sugar Beets* (Warsaw, 1952) [in Polish].

43 "The Polish Sugar Industry During the First Post-war Period," F. O. Licht's *Sugar Information Service,* June 1, 1953.

44 OEEC, *Second Report, European Recovery Programme* (Paris, 1950).

45 West Germany, Statistisches Bundesamt, *Statistisches Jahrbuch für die Bundesrepublik Deutschland, 1954.*

46 M. Kramer, *Die Landwirtschaft in der Sowjetsichen Besatzungszone* (Bonn, 1951).

47 F. O. Licht's *Sugar Information Service: East European Special Sugar Edition,* July 1955.

48 Poland, Cen. Stat. Off., *Rocznik Statystyczny* [*Statistical Yearbook*] (1949).

49 Lois Bacon, "Europe's East-West Trade in Food," in U.S. Dept. Agr., For. Agr. Serv., *Foreign Agriculture,* August 1955.

50 F. O. Licht's *Sugar Information Service,* Apr. 18, 1955.

51 FAO, *Yearbook of Food and Agricultural Statistics,* VIII, Part 1 (Rome, 1954).

52 Hans Lüdecke, *Zuckerrübenbau* (Hamburg, 1953).

53 E. Woermann, "Betriebswirtschaftliche Fragen des deutschen Zuckerrübenbaus," *Agrarwirtschaft* (Hanover), September 1952.

54 France, Min. Agr., *Statistique Agricole Annuelle 1951* (1953).

55 L. Malpeaux, *La Betterave à sucre* (Encyclopédie des Connaissances Agricoles, 8th ed., Paris, 1945).

56 "Quelques Aspects du Problème des Engrais," *Etudes et Conjoncture,* August 1953.

57 United Nations, *Economic Survey of Europe in 1954* (Geneva, 1955).

58 Italy, Instituto Nazionale di Economia Agraria, *Annuario dell' Agricultura, 1954* (1955).

59 Italy, Instituto Centrale di Statistica, *Annuario Statistico Italiano, 1954* (1955).

60 G. Medici, *Italy: Agricultural Aspects* (Bologna, 1950).

61 Banco di Roma, *Review of the Economic Conditions in Italy,* May 1951.

62 *International Sugar Journal* (London), December 1955.

63 K. Heinrich, *Strukturwandlung und Nachkriegsproblem der Wirtschaft Spaniens* (Kieler Studien 28, Kiel, 1954).

64 G. Thiede, "Überblick über die ernährungswirtschaftlichen Produktions- und Versorgungsleistungen der Landwirtschaft in den westdeutschen Bundesländern," *Berichte über Landwirtschaft,* Heft 4, 1952.

65 O. Thiel and K. Padberg, "Produktion, Verkaufserlöse und Betriebsausgaben der westdeutschen Landwirtschaft," *Berichte über Landwirtschaft,* Heft 1, 1952.

66 *Agrarwirtschaft,* December 1953.

67 R. Krzymowski, in *Geschichte der deutschen Landwirtschaft* (Stuttgart, 1939).

68 Netherlands, Min. Agr., *Report on Agriculture in the Netherlands in 1953* (1955).

69 Louis Decoux, *Dix Années de Recherches à l'Institut Belge pour l'amélioration de la Betterave (1932–1941)* (Brussels, 1945).

CHAPTER 10

THE CHANGING FORTUNES OF THE SOVIET
SUGAR INDUSTRY[1]

BACKGROUND

Sugar production in the USSR is not a new industry created under the Soviet plans of industrialization. It is one of the oldest industries of the Russian Empire, well developed long before the Bolshevik revolution. Now nearly 150 years old, it emerged in the beginning of the nineteenth century at the time of the Napoleonic Wars. There were four beet-sugar plants in Russia prior to Napoleon's invasion in 1812 (1, pp. 78–79). The beginning of the Russian sugar industry was thus contemporaneous with the emergence of beet-sugar production in general. As in other countries of Continental Europe, it was from the outset a highly subsidized industry. Originating in the provinces around Moscow, its center shifted to the Ukraine, where, by the middle of the nineteenth century, three-fourths of the Empire sugar industry was located (1, pp. 154 ff.). This development was in response to favorable natural conditions for sugar beets in the Ukraine rather than to special subsidization. In the meantime, the older center of the subsidized sugar industry in central Russia gradually declined. The Ukraine became the principal center of the sugar industry of the Russian Empire, and so continued until the Soviet revolution and thereafter. A secondary center developed in the central black-soil region immediately north of the Ukraine.

The sugar industry was one of the few industries of the Russian Empire in the development of which foreign capital did not play an important role. It was financed mostly by landed gentry and until the revolution was closely connected with large agricultural estates. Though Russian yields were low relative to the rest of Europe and subject to wide fluctuations, this organization of the industry supported a respectable technological performance. If Russian sugar yields at the beginning of the century were less than half the European average, the fault lay not with the processing industry, whose extraction rates were quite on a par with those elsewhere. Even the low beet tonnages per hectare had been raised some 25 per cent in the period

---

[1] Much of the material in this chapter originally appeared in 89.

immediately preceding World War I, whereas agricultural yields in Europe generally were fairly stationary.

The second-ranking sugar industry in all Europe was to suffer three ruinous blows within the following quarter-century. Loss of a portion of the former beet area to the new Poland was a relatively minor matter, but the agrarian revolution of 1917–20, by completely dismembering the large estates, destroyed the old basis of the industry. Production of sugar, which had exceeded 2 million tons[2] before World War I, had fallen to a mere 57,000 tons in 1921/22 and during the 1920's never recovered to prerevolutionary levels. Acreage, to be sure, reached prewar figures before the end of the decade, but beet yields remained depressed. Since the volume of beets was below the full capacity of the approximately 200 sugar factories carried over from Czarist days, agriculture and not processing stood as the Soviet bottleneck.

The second blow was associated with the mass collectivization of Soviet farming. The state sugar-beet farms, formed by the Soviet government mainly from lands that had belonged before the revolution to the sugar factories or to their stockholders, played an important role in restoration of sugar-beet production; nearly 30 per cent of the sugar-beet hectarage was planted by state farms in 1928–29 (*3*, pp. 176–77). Moreover, collectivization in the early 1930's meant expansion of the sugar-beet area, like that of other technical crops. There were 1.5 million hectares in beets in 1932, double the precollectivization level of 1928–29. But such rapid expansion of sugar-beet cultivation in the chaos of the early years of mass collectivization of farming resulted in a catastrophic decline in sugar-beet yields per hectare. During 1931–32 they averaged only about a half of the low precollectivization level and production of sugar fell below 900,000 tons. Positive reduction of sugar-beet hectarage to a workable size (below 1.2 million hectares) and a considerable increase in the application of mineral fertilizers were necessary before the Soviet government could raise the yields of sugar beets during the last years before World War II (1935–39) to a level somewhat exceeding the precollectivization average (1925–28). But the yields remained below the relatively low level prevailing before the revolution.

The following tabulation, from a Soviet source, shows how rapidly the application of mineral fertilizer on sugar beets increased in the Soviet Union during the later prewar years, and how much it affected the average yield (*2*, pp. 98–108):

---

[2] Metric tons are used in this chapter, unless otherwise indicated.

	1932	1934	1938
Total use of mineral fertilizers (*1,000 tons*)	152	311	826
Average use per hectare of sugar beets (*quintals*)	1.0	2.6	7.0
Per cent of sugar-beet area fertilized	55.0	84.3	100.0
Yield of sugar beets (*tons per hectare*)	6.4	9.6	14.1

In order to indicate the full degree of the Soviet achievement with regard to sugar beets before the war it should be mentioned that the yield per hectare in 1937, the best agricultural year climatically, was 18.3 tons. The average yield of sugar beets in the years preceding World War II may be taken as about 16 metric tons per hectare. It thus required some 15–20 years to restore the agricultural basis of the Soviet sugar industry.

Before 1930 the full capacity of the previously existing sugar factories, in so far as they were maintained in good repair, could not be used because the supply of sugar beets remained below the pre-revolution level. Not until the following decade, 1930–40, when the sugar-beet hectarage was much expanded, was the government obliged to add to its processing capacity. Practically all of the 11 new sugar factories were located in the eastern areas, where there had been no sugar production before the revolution or in the early postrevolution years (*4*).

With the German invasion and occupation of the Ukraine in World War II, the USSR's chief sugar-beet region and 85 per cent of its factory capacity passed under Nazi control. Despite some extension of acreages and new factory construction in unoccupied territories to the east, the effect of this third jolt was to decimate Soviet sugar production. This time, moreover, there was serious destruction of factory buildings and their power installations. Though liberation of the old sugar-beet regions had begun in 1943, the prewar hectarage was not reached until 1948. Moreover, shortages of tractors, agricultural implements, and fertilizers kept yields low; a transportation bottleneck meant great crop loss or damage in prolonged storage as late as 1948; and the full restoration of factory capacity did not come until 1950. Sugar production probably did not reach the preinvasion level until 1952. But, by comparison with the degree of damage to agriculture and manufacturing, mere recovery may be regarded as a notable achievement.

INTERWAR EXPANSION TO THE EAST

Before the collectivization of agriculture the Soviet government did not attempt to relocate the sugar industry. Efforts were limited to restoring sugar-beet culture in the vicinity of existing sugar factories in the Ukraine and the central black-soil area just to the north. In 1928 the Ukraine had 84 per cent of the total sugar-beet hectarage of the USSR, about the same proportion as in the comparable territory before the revolution. But with the drive toward the mass collectivization of farming in the early 1930's, the Soviet government undertook not only to expand greatly the sugar-beet hectarage but also to direct this expansion to a considerable extent to the new areas in the northeast. In 1934, when the sugar-beet area was stabilized near the level of 1.2 million hectares, only 70 per cent of this was in the Ukraine, while the share of the central agricultural region of the Russian Soviet Federated Socialist Republic (RSFSR) increased from below one-sixth in 1928 to nearly one-fourth in 1934. During the 1930's sugar production was also expanded to quite new areas: Western Siberia, Far East, Central Asiatic republics, and Transcaucasia. In 1938 sugar beets in the new regions covered nearly 90,000 hectares, from which level there was practically no change until the invasion of the USSR by the Germans in 1941.

The expansion of the area of sugar beets during the 1930's proceeded in two directions: (1) unirrigated sugar-beet cultivation was expanded in the east to the margins of the climatic zone where this was possible; and (2) sugar-beet cultivation under irrigation was introduced in the warm but dry regions of Central Asia and Transcaucasia, areas more favorable for sugar beets than the nonirrigated areas if the lack of moisture is made up by irrigation.

The factors limiting expansion in the first direction were three: (1) insufficiently high temperatures during the sugar beets' relatively long period of growth; (2) short duration of the frost-free season; and (3) inadequate moisture in certain localities otherwise favorable for sugar beets. A total of 2,200° C. (the sum of the excesses of daily average temperatures over 10° C. during the growing season) is regarded as the minimum amount of heat required for the successful cultivation of sugar beets under Russian conditions. In the eastern regions of the USSR, because of the increasing continentality of climate from west to east, only limited areas satisfy this requirement. One area extends as a narrow band between 52 and 54 degrees latitude in Western Siberia, expanding somewhat in the foothills of the Altai Mountains; another is located along the middle course of the Amur River and the Ussuri River in the Far East (5, pp. 1–15). These areas are much less favorable for sugar beets than the old regions of the Ukraine and the central black-soil region of the RSFSR.

In the Siberian region the frost-free period is too short and the supply of moisture is scarcely sufficient; in the Far East, although rainfall is abundant, its distribution during the growing season can hardly be regarded as favorable for sugar beets. Only in the western portion of the North Caucasus—still another area of unirrigated sugar beets— did this crop find fairly favorable conditions, but even here the excessive heat during the summer months is not quite satisfactory for its cultivation. Moreover, expansion of unirrigated sugar beets was pushed into such submarginal regions as that along the Volga where moisture is insufficient, or in parts of the Urals and Siberia where the growing season is too short.

The expansion of irrigated sugar beets proceeded under more favorable conditions where there was enough water for irrigation. The limiting factor here was shortage of irrigated lands, or rather competition for these lands from more valuable crops, such as cotton. Only in the marginal regions of the irrigated farming areas of Central Asia and Transcaucasia, too cold for cotton, could sugar beets find room for expansion.

The expansion of sugar beets in the submarginal eastern areas was directed by the general Soviet policy of shifting the productive center of the USSR to the less accessible regions of the Urals and Siberia. This was partly a matter of safety, to offset the possible loss of the western sugar-producing areas in the Ukraine in case of war, partly a matter of relieving the permanently strained transport system from the long-distance hauling of sugar. The idea of regional self-sufficiency in food became very popular with planners during the period of the Third Five-Year Plan (1938–42), but certain approaches to it had been considered earlier. Thus, the plan of developing the sugar industry in Western Siberia, particularly in Altai, was closely connected with the development of the Ural-Kuznetsk center of heavy industry (6, pp. 301–10).

The extension of the sugar industry to the Far East, where the cultivation of sugar beets, because of climate and soil conditions, requires special techniques that may not always be applied (7, pp. 44–49), also was motivated by the difficulties of supplying the rapidly increasing population of this region with sugar from the European parts of the Union. The idea of importing cheap sugar from abroad through Vladivostok was, of course, excluded on principle in the Soviet Union.

In both the submarginal eastern area and the Far East, sugar-beet yields are low and very unstable. In Altai, for instance, during the last prewar years, yields averaged below 8 tons per hectare. Moreover, it was frequently necessary to abandon nearly half of the sown area because the young plants died in the early part of the growing season, while the shortness of the season did not permit replanting.

Furthermore, not only were climatic conditions unfavorable for sugar beets in the Siberian areas, but also the labor supply on farms was insufficient for such a labor-intensive crop as sugar beets cultivated without sufficient mechanization, as was the case in the USSR. Compared with the old regions of sugar-beet cultivation in the Ukraine and the central agricultural zone, the farm population in Western Siberia is fairly sparse in relation to the cultivated hectarage. In addition, the larger portion of the disposable labor supply is required for the livestock industry, which has more room for expansion than in the densely populated Ukraine. The labor shortage is particularly critical during the harvesting of sugar beets, because the harvesting of grain crops in Western Siberia is not completed when the harvesting of sugar beets must begin. Furthermore, the harvesting of beets must be completed earlier than in the Ukraine, because of early frosts. Even in the last prewar years, when the hauling of sugar beets from fields was to a considerable extent done with trucks, a substantial portion of the beet crop in Altai remained in the fields, undelivered (*8*, pp. 68–83).

The fact that natural conditions and the density of farm population are more favorable for cultivation of sugar beets in the Ukraine than in the new regions in the east is, of course, known to the Soviet specialists in agricultural economics, and is sometimes mentioned in their publications. For example, Professor Kolesnev, in his book on the organization of socialist agricultural enterprises, gives the following comparison of the labor outlays per hectare of sugar beets and per ton of sugar beets produced in various regions (*23*, p. 76):

Region	Man days per hectare of crop	Man days per ton of crop (1937–39 av.)
Vinnitsa oblast (Ukraine)	128	5
Chernigov oblast (Ukraine)	138	9
Bashkir ASSR (the Urals)	141	20
Altai Krai (W. Siberia)	...	21

It thus appears that, because of the low average yields of sugar beets per hectare in the eastern areas, the outlay of labor per ton of sugar beets there is several times as high as in the Ukraine.

Only in the Soviet economic system, which eliminates any competition among various producing regions, would it be possible to undertake production under such economically adverse conditions as characterize sugar-beet production in Siberia. It is not surprising that the reviewer of Kolesnev's book criticizes the author for revealing information "representing the Soviet reality in a false and perverted light," as well as his idea that the leading principle for the rational geographical distribution of crops and of agricultural products generally should be, in the USSR as in capitalist countries, the principle of least outlays. He reminds the author that in the Soviet Union the planning of a rational geographical distribution of agricultural production must take into consideration such important political and economic factors as the strengthening of the defense capacity of the country, the more equal distribution of industry throughout the country, and the development of productive forces of the borderland (9, pp. 56–63).

Expansion of sugar beets on irrigated lands in Central Asia and Transcaucasia is economically more rational. Here, sugar beets may be regarded as an advantageous cash crop in rotation with alfalfa and cereals, as in the mountain states of the United States, where the climatic conditions of some, such as Colorado, Idaho, and Montana, are somewhat similar to those in the region of irrigated sugar beet in Central Asia. In the cultivation of sugar beets on irrigated lands the Soviet government succeeded in raising the yields per hectare to a level substantially higher than in the Ukraine, though with much heavier applications of mineral fertilizers than are used there.[3] The following tabulation shows the sugar-beet yields (26, p. 454) and hectarages (20, p. 583) in various regions of cultivation for the last prewar years:

Region	Yield (Metric tons per ha.)		Hectarage (1,000 ha.)
	1939	1940	1940
USSR, total	14.3	17.1	1,196
Ukraine	14.5	18.0	824
RSFSR	11.8	12.9	335
Kazakh SSR	30.9	27.8 }	30
Kirgiz SSR	40.6	43.2 }	

The yields of sugar beets on irrigated land in Central Asia were thus two to three times as high as those of nonirrigated beets, but the area

---

[3] The use of mineral fertilizers per hectare of sugar beet in the Kant region of Kirgiz SSR in 1940, according to Pyranishnikov (10, p. 157), was equal to 14.5 quintals per hectare, while for nonirrigated sugar beet in the same year it was 7.8 quintals.

before the war was only 30,000 hectares, or less than 3 per cent of the total sugar-beet hectarage of the USSR. On the other hand the yield of sugar beets in the RSFSR, which includes most of the new expansion of sugar beets in the east, was some 20–25 per cent below that in the Ukraine.

THE SUGAR INDUSTRY ON THE EVE OF WAR

In spite of the expansion of sugar-beet cultivation to the eastern areas, the center of the sugar industry remained in the Ukraine and the central black-soil zone north of the Ukraine. In 1940 less than 10 per cent of the sugar-beet hectarage was located outside these old centers of the sugar industry, and less than 7 per cent of the productive capacity of sugar factories (*11*, pp. 45–49). Consequently, the situation of the sugar industry in these regions characterizes the Soviet sugar industry in general.

The role of state farms in the expansion of sugar-beet cultivation, important during the 1920's, declined somewhat during the 1930's. Practically the whole expansion of the sugar-beet area in that decade was in collective farms. This was natural to expect, since the government had as complete control over the planning of crops in the collective farms as it had had before in state farms. Furthermore, through collective farms the Soviet government could tap more easily, and to a greater advantage to itself, the important reservoir of manpower concentrated in the large villages of the Ukraine and the central chernozem zone where the density of rural population was particularly heavy.

Yet this does not mean that the later role of state farms in the organization of sugar-beet cultivation became negligible; state farms continued to play an important part in the sugar industry of the Soviet Union. The total production of sugar-beet seed, for example, was wholly concentrated in the state farms. The hectarage devoted to the production of sugar-beet seed on sugar-beet state farms was nearly as large as their sowings under factory beets. State farms received elite (selected) seed from the sugar-beet experimental and selection stations, and produced commercial seed for all collective farms as well as for their own use. Achievements in the production of sugar-beet seed were substantial. While before the revolution much of the beet seed was normally received from Germany, the USSR became independent of imported seed even before collectivization. Of course, experience with expanding output of sugar-beet seed during World War I was utilized during the Soviet period. Sugar-beet state farms also played an important role in the extension of sugar-beet cultivation to the new regions. Usually cultivation of sugar beets in the new area was tried first in the state farms organized for this purpose, and only later was it introduced on collective farms in the respective regions. The same was true of experiments with new methods of cultivation of sugar beets, including mechanization.

However, the mass production of sugar beets, requiring large amounts of manual labor, was concentrated in collective farms. It was there that the government had a great source of manpower which it could use without taking too much responsibility for its maintenance. But several years were required before collective farms could achieve in the cultivation of sugar beets as good results as were obtained before collectivization by small farms under contracts with sugar factories.

Beginning with 1934–35, attempts were made to introduce rational crop rotations in collective farms cultivating sugar beets. These were usually six- or seven-year rotations with sugar beets in one field, following normally after winter grain, mainly wheat, sown on fallow. Yakushkin wrote in 1936 that nearly 90 per cent of the sugar beets was sown after winter crops in those collective and state farms in which rational rotations were newly introduced (*12*, p. 485). Sugar beets thus received the best place in the crop rotation. In the western portion of the sugar-beet zone one field in these six- or seven-year rotations should normally be under leguminous grasses, mainly clover, but in practice only a small fraction of the collective farms cultivating sugar beets had actually introduced such rational crop rotations. In the eastern portion of the old sugar zone, as in Voronezh oblast, the hectarage under leguminous grasses was insignificant even in the planned rotations. Of the total crop area of Voronezh oblast in 1936 only half of one per cent was in perennial grasses (*13*, pp. 153–62). In order to obtain satisfactory yields of leguminous grasses in this region, cultivation requires more advanced techniques than were actually applied at that time. But even in the Ukraine the predominant rotations involving sugar beets did not include leguminous grasses, but consisted only of sugar beets and cereals with one field in fallow. Rotations were thus far from rational. The following distribution of the total crop hectarage in collective farms of the sugar-beet zone among various crops, based on the annual reports of collective farms for 1937, shows that in the second half of the 1930's six- or seven-year crop rotations with leguminous grasses were not typical for collective farms cultivating sugar beet (*2*, pp. 98–108):

Crop	Per cent of total crop area
Cereals, including dry legumes[a]	78.4
Technical crops, mainly sugar beet	11.9
Seeded grasses	5.9
Potatoes, vegetables, and fodder crops	3.8

[a] Dry legumes composed about 3.5 per cent of the total crop area in the sugar-beet zone in 1934 (*15*, pp. 322–25).

From this distribution it may be inferred that six- or seven-year rotations with seeded grasses in one field did not make up more than 30–40 per cent of the total crop area. The fact that cereals constituted three-fourths of the cropped plowland indicates also that sugar beets were predominantly planted in rotation with cereals alone.

The situation in the sugar-beet *state* farms was, however, more satisfactory in this respect. Judging by the distribution of the crop area among various crops in the state farms of *Glavsakhar* (Principal Administration of the Sugar Industry), these farms as early as the first half of the 1930's had at least one field under seeded grasses in six- or seven-year rotations, most of which consisted of perennial grasses. Another field was occupied by sugar beets, including beets for production of seed, while only about half of the total crop area was under cereals (*15*, p. 1279). The share of state farms in the total sugar-beet hectarage at this time was, however, 10 per cent at most. Consequently, speaking generally, rational crop rotations could hardly have played an important part in raising yields of sugar beets in the second half of the 1930's.

The increased application of mineral fertilizers on sugar beets was perhaps the principal factor responsible for the improvement of yields in the last prewar years. The disastrous drop in the number of livestock caused by the ruthless mass collectivization in the early 1930's had greatly reduced the supply of available manure, and the application of mineral fertilizers had first to compensate for that deficiency.

According to an official survey taken before collectivization (around 1926–27; *16*, p. 482), about 280 million tons of manure was produced on farms in the European portion of the Soviet Union, 220 million tons of which was used on crops. Most of the manure was used on poor podzol soils in the region north of the chernozem zone. But a substantial portion, nearly one-third of the total, was used on the northern portion of the black-soil zone, the so-called forest-steppe zone, where the principal sugar-beet areas were located. At that time in the forest-steppe zone an average of about 2.7 tons was applied per hectare of the total crop area. But after the mass collectivization the application of manure in fields declined, within a comparable territory of the USSR, to about one-third (75 million tons in 1933 and 87 million tons in 1934; *15*, pp. 337, 1386). Since the crop area was enlarged at the same time, the application of manure per hectare of crop in the sugar-beet zone declined even more. On the basis of official statistics for the principal sugar-beet regions there was applied on the average only about 0.7 ton per hectare of all crops in 1933 and 1934, against 2.7 tons per hectare in 1926–27, the average for the forest-steppe area. The situation improved somewhat in 1935, when in the sugar-beet zone nearly 1.2 tons of manure were applied per hectare of the total crop area, according to the same official statistics. But this was still less than half of the precollectivization rate.

No detailed official statistics on the application of manure in the Soviet Union for later years have been revealed, but some experts who were connected with the Soviet agricultural administration in the Ukraine until the beginning of the war estimate that during the years of the Third Plan (1938–42) some 200 million tons of manure were applied annually on crops in the USSR (*17*, pp. 118–19). Pryanishnikov, in his book published in 1946 (*10*, p. 172), says that 200 million tons of manure were annually applied on crops during the Second Plan, while in 1940 the application of manure had declined, according to the data of the Commissariat of Agriculture, to 125 million tons. He does not explain why the application of manure on crops in 1940 was below the level of 1933–37, while the number of livestock had substantially increased. We therefore assign the figure of 200 million tons to the end of the Second Plan (1937) and the beginning of the Third Plan (1938–39).

Such application of manure is nearly 50 per cent more than the quantity applied on crops in 1935, according to official statistics. On the assumption that the use of manure in the sugar-beet zone increased in about the same proportion, it may be estimated that, during the last years preceding the war, 1.7 to 1.8 tons of manure were applied, on the average, per hectare of cropped plowland, or less than two-thirds of the precollectivization level. The above estimate is in fairly close agreement with the data obtained by the Institute of Sugar Beets on the basis of a sample survey of collective farms within the principal sugar-beet regions for 1937–38 (*16*, p. 541). This survey indicates that the quantity of manure applied per hectare of arable land ranged from 1.2 to 1.9 tons.

It thus appears that although the application of manure in the last prewar years in the sugar-beet regions had much improved when compared with the first half of the 1930's, it was some 30–40 per cent below the precollectivization level. Since a large portion of that manure was produced by livestock owned by members of collective farms individually, it may be assumed that that portion was applied preferably on the plots of land surrounding farmsteads cultivated by kolkhozniki for themselves. It was only in the last prewar years that, under government compulsion, the manure produced by livestock of kolkhozniki was used in the fields of collective farms. In appraising the effect of increased application of *mineral* fertilizers on sugar beets during the second half of the 1930's and the beginning of the 1940's, these facts about the application of manure must be taken into consideration.

It was mentioned earlier that in 1938 mineral fertilizers were applied on the entire sugar-beet area of the Soviet Union, and the rate of application of commercial fertilizers at 7 quintals per hectare was seven times as high as in 1932 and nearly three times as high as in 1934. The sugar beet was the first claimant for mineral fertilizers in the Soviet Union, receiving during the last prewar years more than one-third of the total quantity of mineral fertilizers produced. Only for nitrogen fertilizers did cotton have preference over sugar beets, receiving more than half the total supply. With respect to superphosphate and potash, sugar beets were the largest user.

Applications of mineral fertilizers on sugar-beet crops were thus higher than in the days before collectivization, when only a negligible quantity was used on sugar beets. However, about half of the nutrient elements of these mineral fertilizers was necessary to compensate for decreased use of manure. Furthermore, the rate of application of nitrogen fertilizers was so small that it could not even fully compensate for nitrogen in the lost manure if we assume that, during the six- or seven-year rotation, fields received in 1940 some 6 tons *less* manure per hectare of crop than they received before collectivization. Only potash and particularly phosphate fertilizers were applied in quantities larger than was necessary to compensate for the lost manure. Makhov (*17*) says that with too small rates of application of nitrogen fertilizers a portion of the phosphate fertilizers applied on sugar beets in the Soviet Union was practically wasted. He points out that it would be more rational to apply a part of the phosphate fertilizers on other crops planted in rotation with sugar beet, such as wheat or seeded grasses. But under the Soviet system, when separate crops are under control of special governmental departments, as is true of sugar beets, one may not get a rational distribution even of available fertilizers.

The crucial factor was the absolute shortage of manure. Since there was a deficit balance in the soil between input and output of plant-nutrient elements in the sugar-beet zone, even a substantial application of mineral fertilizers could not result in a satisfactory increase in yields. The rich chernozem soils of the Ukraine cover a part of these deficits from their reserves, especially if rational fallowing of land is practiced. But these deficits under the conditions of Soviet agricultural practice in the sugar-beet zone in the second half of the 1930's were too large to be covered at the expense of the soil resources. The fact that during 1935–39 yields of sugar beets in the Soviet Union, according to official statistics, averaged only about 15 tons, in spite of substantial improvements in techniques and in selection of seed, indicates that the deficit of plant nutrients was too great to maintain the yield at the relatively low level of 20 tons per hectare assumed in the computation of the balance.

However, increased application of mineral fertilizers on sugar beets did permit yields somewhat higher than those obtained before collectivization. Other factors such as selection of better sugar-beet seed and improvement in techniques through mechanization of sugar-beet cultivation also played some role in the achievement of higher yields (see chapter 5).

It is not difficult to accept the Soviet claim that they succeeded in selecting better varieties of sugar beets than those cultivated in Western Europe, and particularly in Germany, from which sugar-beet seed was usually imported into Russia before the revolution. But poor agricultural techniques applied in cultivating sugar beets in the Soviet Union more than offset the gain. Yields remained only half as high as in Western Europe, while the 1935–39 average was still below the level obtained in the estates before the revolution, notwithstanding the much larger use of mineral fertilizers in the later period.

Part of the explanation lies in the great expansion of the area cultivated during the 1930's, which magnified all the problems from harvesting to processing, but particularly that of hauling to factories. The Soviet authorities recognize that even in the best years preceding the war the hauling of beets from fields to factory dumps was the greatest bottleneck in the Soviet sugar industry. The hauling of beets from fields usually lagged behind the harvesting, and a considerable proportion of the crop was stored in fields. The problem of storing the beets under the severe continental climate of the USSR is a difficult one, especially because the sugar campaign in the Soviet Union is much longer than in other countries producing beet sugar, or than it was in Russia before the revolution. The prolonged campaign results from the fact that the Soviet government built only a few new sugar factories before the war and those were built in the new region of sugar-beet cultivation. The doubled area of sugar beets in the old beet regions had to be served by sugar factories inherited from pre-revolutionary times.

In spite of all shortcomings, however, the total production of sugar beets from the greatly expanded hectarage was much increased during the last years preceding World War II, and the Soviet sugar factories could process enough sugar beets to produce an average of 2.5 million metric tons of sugar annually. This permitted the Soviet Union not only to increase per capita consumption of sugar but also to export sugar during the prewar years. Sugar exports from the Soviet Union amounted to 162,800 tons in 1936 and 134,000 tons in 1937 (25).

It is quite possible also that the policy of the Soviet government in the last prewar years was to accumulate some reserves of sugar, as it accumulated grain reserves for use during the war. Technically it was easier to store sugar than grain, and the Soviet sugar industry, concentrated mainly in the Ukraine, was more exposed to the dangers of the war than was the production of grain, which was more evenly spread over the Soviet Union.

EFFECTS OF INVASION

The immediate effect of the German invasion in the summer of 1941 on the Soviet sugar industry was greater than on any other industry in the Soviet Union. The principal sugar region in the Ukraine west of the Dnieper was invaded during the first months of the war, and by the beginning of December 1941 practically all the Ukrainian sugar-beet area and the greater portion of that of the central agricultural region of the RSFSR were in the hands of the enemy. In the beginning of December 1941—that is, still during the first half of the sugar campaign of the Soviet Union, which normally continues until March–April—there remained under control of the Soviet government not more than 220,000 hectares of sugar beets, or somewhat less than one-fifth of the prewar normal hectarage. This comprised about 120,000 hectares in the eastern portion of the central agricultural region of the RSFSR, some 25,000 hectares in the North Caucasus and in the Volga, and some 70,000 hectares in the Asiatic sugar-beet regions and in Transcaucasia. We have seen that the sugar-beet area in the east reached about 90,000 hectares in 1938, but remained practically without change in 1940, and there is no indication that it increased in 1941.[4] Not until 1942 was the cultivation of sugar beets in the eastern regions greatly expanded, to a limited extent offsetting the great losses in the old sugar-beet regions.

The losses of sugar factories were in proportion to the losses of beet acreage or even greater, since the density of factories in the old western regions was greater than in the new regions of sugar-beet cultivation. According to an official statement, there were 199 sugar factories in German-occupied territory (27, p. 25), including 15 to 20 in the territories newly absorbed by the Soviet Union in 1939/40, while in the part of the USSR that was never occupied there were in 1941 fewer than 20 factories. This means that through the 1941/42 sugar campaign about 30 factories remained under Soviet control. Of these, 10 factories were built during 1930–40 in the new sugar territory. Their average production capacity was somewhat larger than that of the old factories in the western regions. Yet the capacity of the sugar-beet factories remaining in the hands of the Soviets at the end of 1941 was not much greater than 15 per cent of the total prewar capacity. Hence, when the Soviet government, during 1942 and 1943, made great efforts to expand the cultivation of sugar beets in the new eastern regions, it was also obliged to build new sugar factories there. We shall see later that this was done to a limited extent, partly by evacuating a few factories from the invaded regions.

Sugar factories were not only lost to the enemy through the occupation, but many were destroyed by the Russians themselves at the time of their retreat, and by the Germans, when they were leaving the occupied regions. According to a Soviet official (27, p. 25), 55 sugar factories were completely destroyed and in two-thirds of the factories the power installation was destroyed. The greatest losses in sugar factories were suffered in the Ukraine east of the Dnieper and in the central agricultural region of the RSFSR. When the Soviet Army liberated the occupied sugar-beet area of the Ukraine and central

---

[4] Kondrashev (26, p. 453) says that the area under irrigated sugar beets in 1941 was 32,000 hectares. In 1940 there were 30,000 hectares of sugar beets in Central Asia and Kazakhstan, practically all irrigated.

Russia during 1943–44, it was therefore necessary to rebuild the great majority of sugar factories—a slow process at best.

It accordingly appears that for the sugar campaign of 1941/42 there remained in the hands of the Soviets very little sugar-beet acreage and few factories. True, some sugar plantations and factories in the Ukraine east of the Dnieper and in Kursk and Orel provinces remained under Soviet control during the early months of the 1941/42 campaign. But under the chaotic conditions then prevailing, little sugar beet could have been processed, and still less sugar could have been shipped out later in the autumn, when the Germans invaded these regions. Practically all sugar that the Soviet government could produce during the 1941/42 campaign was therefore from a sugar-beet acreage equal to about one-fifth of the prewar normal. Even this crop could not have been fully harvested on time and fully transported to factories and processed under conditions created by the invasion. Accordingly, it is hardly possible that the Soviet government could have obtained more than 400,000 metric tons of sugar in the first wartime sugar campaign, as against the prewar average of about 2.5 million metric tons, and the figure may have been substantially lower.

The Germans got still less profit from the sugar plantings of the Ukraine when that region fell into their hands in the summer and autumn of 1941. Although the larger portion of the Ukrainian sugar area was in the hands of Germans before the time of harvest, not much sugar beet was actually harvested and still less was hauled to the factories and processed. A competent German source (*28*) states that of the 850,000 hectares of sugar beets in the invaded territory of the USSR that fell under German control for the 1941/42 sugar campaign (which could have produced some 6 million tons of sugar beet even at the very low level of yield of about 7 tons per hectare estimated for that year) only about 60 per cent was actually harvested (3.5 million tons). Of this only about one-fourth could be delivered to the factories, and much of that was damaged by frost and rotted when the weather became warmer in December. Thus the Germans obtained only about 70,000 tons of sugar in the 1941/42 campaign. The principal reason for such a failure was the extreme shortage of means of transportation for hauling beets to sugar factories.

The Germans were little more successful with production of sugar beets in the occupied USSR in the campaign of 1942/43. According to the same source, in the occupied territory, where normally about 1 million hectares of sugar beets were planted, less than one-third of this was put in and the total quantity of sugar obtained was estimated at about 200,000 tons—this in spite of the fact that for that campaign the Germans could utilize more than 100 sugar factories which they were able to repair. This might suggest the vulnerability of the collective organization of farming under war conditions, when means of production concentrated in big collective farms can be easily destroyed. It must be added, however, that sugar was not the commodity in shortest supply in Germany. With the important sugar industries of Czechoslovakia and Poland under their control, the Germans had sufficient supplies of sugar at their command, and their efforts in the Ukraine were perhaps directed toward obtaining kinds of food which were in shorter supply, such as grain and vegetable oil.

The principal German success thus lay in depriving the Soviet Union of about four-fifths of its sugar industry concentrated mostly in the occupied territory. This stimulated feverish activity by the Soviet government for further expansion of sugar production in the eastern areas, the basis for which had been created during the 1930's.

WARTIME EXPANSION: IRRIGATED BEETS IN THE EAST

Expansion of the sugar-beet area in 1942 and 1943 in the eastern regions, particularly on irrigated lands of Central Asia, was quite impressive. In one or two years the hectarage in the Kirgiz and Kazakh republics was respectively doubled and tripled, and in the Uzbek SSR, where no beet had been cultivated during the prewar decade, one-twelfth of the total cotton area, or about 75,000 hectares, was set aside for beet cultivation (29, p. 10). It was believed that the population, which for generations had had experience with the cultivation of cotton, would know how to cultivate another row crop such as sugar beet. This plan was apparently never totally fulfilled. However, some 40,000 to 50,000 hectares of factory beet were planted in the Uzbek SSR in 1942 and 1943, besides a substantial area of sugar beet for production of seed.

Generally speaking, the production of sugar-beet seed, after the invasion of the Ukraine and the central agricultural region of the RSFSR, was shifted in an organized manner from the old regions of sugar-beet cultivation to Central Asia. During the war this area became the principal producer of seed for practically all the sugar-beet regions still under Soviet control. The fact that the seed problem could be solved, at least partially, on such short notice must be regarded as a substantial achievement. It became possible because of the partial evacuation to Central Asia of the personnel, laboratory equipment, and some breeding materials of most of the sugar-beet experiment stations and of the All-Union Scientific Research Institute for the Sugar Industry. In the first year after evacuation (1942) sugar-beet seed was produced on 6,000 hectares in Central Asia, and the area was much expanded in the following years (*19*, pp. 107–08; *30*, pp. 30–35).

The evacuated personnel also contributed a great deal to the expansion of the cultivation of factory beet on irrigated lands of the three Central Asiatic republics. In 1942 and 1943 plantings reached some 120,000 hectares, or perhaps even 150,000 hectares if sugar beet cultivated for seed is included (*31*, pp. 45–47). This means that the sugar-beet hectarage on irrigated lands of Central Asia increased four to five times. It was achieved by a great increase in the density of sugar beets in the regions of the previous cultivation of beets in the Kirgiz and Kazakh republics, and by an expansion to new regions.

Speaking of the Kirgiz SSR, Sokolovskii (*31*) says that in 1942 and 1943 in some kolkhozes sugar beet occupied 60 to 70 per cent of the total crop area on irrigated lands. Cultivation, originally practiced mainly on low (meadow) lands with subsoil water relatively close, spread to the foothill land, less favorable for beet, and a large proportion of the beet was planted after beet (in 1943, 60 per cent). While even before the war rational crop rotations were not fully mastered by collective farms cultivating sugar beet, during the war they were completely disorganized. The same situation prevailed in the neighboring sugar-beet regions of the Kazakh SSR. There too, in 1943, 50 per cent of the sugar beet was planted after beet, while the acreage of alfalfa in rotations was greatly reduced.

The increase in the density of sugar-beet cultivation resulted in a greatly enlarged load of sugar-beet hectarage per farm worker, while the quantity of work that was previously performed on collective farms by the MTS was much reduced. Furthermore, cultivation was expanded to farms far from railroad stations or from any delivery point, thus increasing the demand upon transportation facilities already strained by the war.

Under such conditions, maintenance of the relatively high prewar level of sugar-beet yields on the irrigated lands of Central Asia was impossible during the war. This was particularly so because shipments of mineral fertilizers, mostly from the European parts of the USSR, were inevitably greatly reduced by the war, and the supplies of fertilizers from local sources could not replace them.

In the best sugar-beet area of Central Asia (Kant district of the Kirgiz SSR) the yield of sugar beet in 1942 was reportedly 168 quintals per hectare as against 349 quintals in 1938 and 471 quintals in 1940 (*33*, p. 10). The Soviet authority reporting this yield took pains to explain that the decline of the yield was not caused by unfavorable weather conditions in this particular year, but by unsatisfactory farm techniques and the reduction in application of fertilizers. The data indicate that the number of cultivations of sugar beet was reduced on the average from ten in 1940 to below three in 1942, the number of waterings from ten to five, and the quantity of applied mineral fertilizers from 14.5 to 5.2 quintals per hectare. If such a decline of yields took place in the most advanced sugar-beet region of Central Asia, and in the first year following the beginning of the war, the results in other regions were presumably still less satisfactory. In neighboring regions of the Kazakh SSR (Dzhambul oblast), for instance, the average yield of beets for 1943/44 was reported as low as 67.5 quintals per hectare (*34*, p. 2).

Much poorer results were obtained from the very ambitious plan of introducing sugar-beet cultivation on some 75,000 hectares in the Uzbek republic, starting from nothing. For one thing, sugar-beet cultivation in the Uzbek SSR was at first dispersed among too many collective farms, which made the guidance of inexperienced growers particularly difficult. The chief of the sugar-beet procurement of Glavsakhar revealed that sugar-beet cultivation in the Uzbek republic was completely reorganized in 1944. It was then concentrated on fewer than 400 collective farms located within short distances from the newly erected sugar-beet factories, while in 1942 and 1943 more than 1,000 collective farms, many of which were far from the delivery points, had cultivated sugar beet. Consequently, in these earlier years the results were quite unsatisfactory—very low yields and still smaller deliveries.[5] This was caused partly by poor selection of lands for sugar beet. Apparently, selection was made first on the assumption that sugar beet can be cultivated in saline soil, while later experience has shown that sugar beet is to some extent resistant only to a surface salination of soil, while deeply salined soils are even more damaging to sugar beets than to cotton (*33*, *passim*, esp. pp. 71–73). Consequently, sugar-beet crops were very uneven in the fields of the Uzbek SSR, and in some places they completely failed. In no year was the

[5] Fridman (*35*, p. 7) says that as a result of the reorganization the delivery of sugar beet in the Uzbek SSR in 1944 was three times larger than in the two preceding years. This points to a very small harvest in those earlier years.

planned area of 75,000 hectares reached,[6] a fact which points to a large abandonment of crops and, of course, to low yields. Yields of sugar beet per hectare in the Uzbek republic in the early war years could in no case have been on the level of those in the Kirgiz SSR, where the growers had had experience in cultivation of beet for a decade. And even there the yield in 1942 had declined to half or a third of the prewar level, and was about on the level of unirrigated sugar beet in the prewar years. If cotton yields per hectare had declined in Central Asia during the war years (1942, 1943) to the level of 1930 (that is, were reduced to half) mainly because cotton was not receiving enough mineral fertilizers, the situation with irrigated sugar beet, a new crop in the area, could certainly not have been better.

The missing mineral fertilizers could not be replaced in Central Asia by manure, since there was generally very little of it in this area and its preservation was very inefficient. Green-manure practices also were only in the experimental stage (*29, passim*). Consequently, yields of irrigated sugar beet on the greatly expanded area could not even approach the high yields prevailing before the war on a relatively small area of better lands.

Furthermore, much of the sugar beet harvested was lost or greatly damaged by long storage in the fields during the war years. Given the continental climate of Central Asia with its extreme seasonal and even diurnal temperature variations, the problem of storing beets in fall and winter is extremely difficult, while the sugar production campaign normally continues six or seven months into April. An authority who studied harvesting, hauling, and storing of sugar beets in Central Asia during the war years says that the hauling of beets from the field frequently continued until April, and the storing of beets in the field usually did not follow the prescribed technique. Consequently, factories received little sugar beet in good condition during the second half of the campaign, and the proportion of beet not fulfilling required conditions sometimes exceeded 50 per cent (*22*, pp. 163–64). It appears from this that while the expansion of the area of irrigated sugar beet in the early war years was quite impressive, the increase of production was much less significant.

Expansion of irrigated sugar beet in Transcaucasia was much more modest—from 6,000–7,000 hectares in prewar years to about 10,000–11,000 hectares in 1943 (*7*, pp. 44–49).

---

[6] Lupinevich and Korobov (*7*, pp. 44–49) mention the sugar-beet area in Uzbek SSR in 1942 at 42,000 hectares and in 1943 at 53,000 hectares.

WARTIME EXPANSION: UNIRRIGATED BEETS

Expansion of sugar-beet area during the early war years on non-irrigated land was less spectacular. In the never-occupied eastern margin of the old sugar-beet area in the central agricultural region of the RSFSR (eastern portion of Voronezh oblast, as well as Tambov, Penza, and Ryazan oblasts), where some ten sugar factories remained under Soviet control, great efforts were made to expand sugar-beet plantings in order to utilize the remaining capacity of the sugar factories. A similar situation existed in neighboring oblasts along the western bank of the Volga (Saratov, Stalingrad, and Ulyanovsk), which could ship their beets by railroads to factories relatively short distances away, in the above-mentioned provinces. However, climatic conditions in the Volga provinces, particularly insufficient moisture, were less favorable for beet cultivation.

Expansion of sugar-beet planting in other parts of the unoccupied European territory of the USSR was limited not only by unfavorable climatic conditions, but also by remoteness from sugar factories. There were no factories north of Tula and only one east of the Volga (in Kuibyshev oblast). This factory, which was under construction before the war and apparently completed during the war, became the basis for expansion of sugar in the Transvolga area up to the foothills of the Urals. But that area is too dry for sugar beet. Normal cultivation of this crop in the Transvolga steppe had been considered feasible only with irrigation, and apparently the unfortunate wartime experience only confirmed the conviction (7, pp. 44–49). The conditions in the foothills of the Urals, particularly in the Bashkir ASSR, were regarded as more favorable, and in 1942 and 1943 about 10,000 hectares were planted to beets. However, to be processed for sugar, the beets would have had to be transported the fairly long distance to the only sugar factory east of the Volga, since the sugar factory under construction in the Bashkir republic was not completed before 1945. Furthermore, yields of beets even here were reported as extremely low in some years, notably 1943 (7, p. 48).

The growing of sugar beet was undertaken during the war also in the northern provinces around Moscow and farther north, as well as in northeastern provinces west of the Urals, and claims were made that yields and sugar content of beets in these areas were comparable with those in the old sugar-beet regions (*36*, pp. 33–36). But these beets were not processed for sugar in factories but for syrup in small, primitive plants on collective farms. In 1944 there were between 1,000 and 2,000 such syrup plants, and contracts were made with collective farms for the cultivation of sugar beet on some 15,000 hectares for this purpose. Such a production was less developed in 1942 and 1943, but several thousand hectares of beets for production of syrup were cultivated in these earlier years.

Expansion of cultivation of nonirrigated sugar beets in Siberia was apparently even less successful than in the European part of the USSR. However, in Western Siberia, mainly Altai, the sugar-beet cultivation rose from 27,000 hectares in the last prewar years to nearly 40,000 hectares in 1943 (7), and two additional sugar factories were completed, one in Western Siberia (Omsk) and another farther to the east.

Vesherovich, the chief engineer of the Altai sugar trust, says, however, that during the war two Altai sugar factories normally had not enough beets to work at full capacity (37), because they no longer received beets from Kazakhstan, where new sugar factories were opened. The fact that the 30,000 to 40,000 hectares in Altai could not supply enough beets to keep busy two sugar factories with a combined daily processing capacity of 2,000 tons of beets indicates that wartime yields in Altai must have been below the prewar low average of about 8 tons per hectare. There are indications that a considerable proportion of the harvested beets could not be hauled to factories in this area.

Practically no expansion of the sugar-beet area took place in the Far East, where it remained at the prewar level of 4,000 to 5,000 hectares. That the cultivation of sugar beets in this area was not sufficiently mastered by collective farms, in spite of a better supply of mechanical equipment than in the old sugar-beet regions, and that yields of beet continued to be low and unstable, is confirmed by such an authority as the Deputy Minister of Technical Crops of the RSFSR, Nazartsev (38, p. 3).

PRODUCTION IN 1942 AND 1943

The general expansion of the areas of both irrigated and nonirrigated beets in the new regions of production was, however, so large in 1942 that it brought the total area of sugar beets under Soviet control to over 300,000 hectares, as against the estimated 220,000 hectares in 1941. This increase took place in the face of new losses of about 50,000 hectares of plantings in the further German invasion in Voronezh province and North Caucasia.

In 1943 the total area of sugar beets under Soviet control made a further substantial increase. No definite figure on the area in this year was officially revealed, but on the basis of statements by some officials it may be roughly estimated to have reached nearly 450,000 hectares.[7] This time the principal expansion took place not in the eastern areas, where it remained about on the 1942 level, but in some portions of the old sugar-beet regions, liberation of which began in the spring of 1943. By April 1, the western portion of Voronezh oblast and about 90 per cent of Kursk oblast, both important sugar-beet producers, had been liberated, as well as practically the entire North Caucasus, where before the invasion sugar beets also were grown on a moderate scale. At that date only a small portion of the eastern Ukraine was liberated, where normally no sugar beets were cultivated. Liberation of the important sugar-beet area of the Ukraine east of the Dnieper took place only during the latter months of 1943, and, consequently, could hardly be used for cultivation of beets for the 1943/44 campaign. Of course, some land in this region was planted to beets during the period of German control, and later fell into Soviet hands. However, cultivation of sugar beet east of the Dnieper during the occupation had been very limited. The Ukraine west of the Dnieper, where more beets were produced while in German hands, was not liberated until the first half of 1944. (See accompanying map.)

Although in 1942 and 1943 about 40 and 30 per cent respectively of the total sugar-beet area was irrigated, as against only 3 per cent in 1940, this does not indicate that yields of beet per hectare in these years were above the prewar level. We have seen that in 1942 in the best region of irrigated sugar beet the yield declined to the level of the prewar average for nonirrigated beets, while in most of the other irrigated regions yields were much lower. Furthermore, expansion of nonirrigated beets took place in the low-yield eastern areas, or in quite new regions with no experience with sugar beet. Consequently, yields of beets must have declined below the prewar level in the very first year of the war, and continued declining for a few years, since the restoration of sugar-beet cultivation in the liberated area could proceed only under very difficult conditions.

Under the circumstances total production of sugar beets and particularly deliveries to the sugar factories probably were even smaller in 1942 than in 1941, in spite of the fact that the sugar-beet area under

---

[7] I. Benediktov (*39*, pp. 32–39), Commissar of Agriculture of the USSR, stated that the area under sugar beets in the regions behind the front increased in 1943 by 138 per cent over 1940. On the other hand, it was said (*40*) that in 1944 there were sown 270,000 hectares more sugar beets than in 1943, while the 1944 area may be estimated on the basis of other information at about 700,000 hectares.

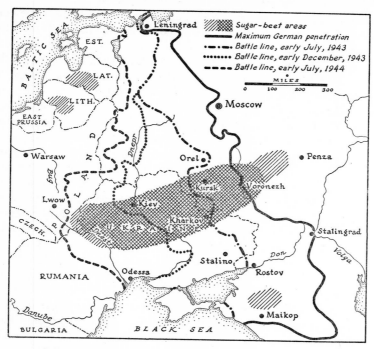

STAGES IN THE LIBERATION OF SOVIET SUGAR-BEET AREAS
FROM GERMAN OCCUPATION IN WORLD WAR II

Soviet control increased by some 40 per cent. Direct statements by
the Soviet authorities indicate that further expansion of the sugar-
beet area in 1943 by about the same percentage as in 1942 had not
increased deliveries of sugar beet to the sugar factories in that year
either. Indeed, Fridman, Chief of the Sugar-beet Division of Glavsakhar, stated (35, p. 7) that *only in 1944* did the procurement of
sugar beets for factories begin to increase after the great decline
during 1941–43. Zotov, Commissar of Food Industry, said that during the 1943/44 campaign only 28 sugar factories were in operation.
Since nearly 20 sugar factories were at that time in the new eastern
regions—ten constructed before the war and perhaps that many after
the beginning of the war (4)—fewer than ten factories were in operation in the old regions of the sugar industry. In the best prewar years
the average production of sugar per factory per campaign was about
13,000 tons. The production per factory in the 1943/44 campaign,
when two-thirds of the factories were working under less favorable

conditions in the new eastern areas, and one-third in the chaotic conditions of the newly liberated areas, could hardly have reached the prewar performance, even though the processing capacity of some eastern factories was above the average for all sugar factories of the USSR. These reports indicate that total sugar production during the 1943/44 campaign could not have reached 400,000 tons and probably was substantially below this level. Later, at the Nineteenth Congress of the Communist Party in 1952, Mikoyan, then in control of the food industries of the USSR, indicated that the decline of sugar production during 1942/43 was even more disastrous than was implied by Soviet efforts to expand sugar-beet cultivation in the new areas. Indeed, according to his statement, output of sugar in 1943 fell below 150,000 tons.[8] If some outside observers appraised the supply of domestic sugar in the Soviet Union during these years at a substantially higher level, we are inclined to believe that these additional quantities were obtained from stocks accumulated before the war, and not from larger wartime production.

RESTORATION IN THE LIBERATED AREAS

The restoration of sugar-beet cultivation and the sugar industry in the liberated areas may be divided into two periods: (1) from 1943 to 1946, when it proceeded quite unsatisfactorily, despite the claimed rapid expansion of the sugar-beet area, and (2) after 1947, when it was more successful, although still below the goal established by the Fourth Five-Year Plan. In the first period the greatest difficulties were experienced in restoring the production of sugar beets, the restoration of the processing capacity of factories proceeding ahead of the supplies of beets (41); in the second period the supplies of beets in the fields in some years apparently exceeded both the facilities for transporting them and the processing capacity of the factories.

The restoration of the liberated sugar-producing areas began as early as 1943, when, early in the spring, the central agricultural region of the RSFSR was free from occupation. However, the condition of the liberated regions, and the situation with respect to the total supplies of agricultural machinery, draft power, and fertilizers, as well as labor, were so deplorable that it is more accurate to say that the real work of restoration did not begin before 1944. Practically the

---

[8] In his discussion of Saburov's report on the Directives for the Fifth Five-Year Plan, Mikoyan said that "the output of sugar has declined in 1943 nearly twenty times as compared with the prewar level." Soviet sugar production for 1935/36–1940/41 averaged about 2.5 million tons (cf. *24*, pp. 3–4).

entire old sugar-beet-producing area of the USSR, including the most important Ukrainian region west of the Dnieper, was under Soviet control in the spring of 1944.

With the entire old sugar-producing area under its control, but with an extreme shortage of all factors of production, the Soviet government had to reverse its policy. Forced expansion of the sugar industry in the less favorable eastern areas was immediately abandoned, to a certain extent even reversed, and the total effort was concentrated on restoring the old producing areas, particularly that of the Ukraine. After 1943 the sugar-beet hectarage in the eastern areas began to decline even in the irrigated regions, since it was overexpanded there in proportion to the available productive forces and the land resources. On the other hand, the sugar-beet area was rapidly being restored in the old regions, particularly the Ukraine. When measured by expanding hectarage of sugar beet, the restoration appears rapid even before 1947. But when the total sugar-beet production, deliveries to factories, and processing into sugar are considered, the results were far from satisfactory.

Development of the sugar-beet hectarage in the USSR during the period of restoration, based on various official and semiofficial statements, is shown by the following tabulation (in 1,000 hectares):

Region	1943 around	1944	1945	1946	1947	1948	1950 planned
USSR, total	450	702	821	938	1,086	1,226[a]	1,354[b]
Ukrainian SSR	?	280	420	527	639	824[a]	838
Irrigated sugar beet .	130	...	80	...	...	...	90

[a] The area sown to sugar beet in 1940, officially claimed to have been restored by 1948.

[b] The area actually sown in 1950 was about 1,290,000 hectares, since it is officially stated that the area of sugar beet increased 57 per cent during the Fourth Five-Year Plan period (*74*, p. 263).

The tabulation shows that practically the entire expansion of the sugar-beet hectarage from 1944 to 1948 took place in the Ukraine. The cultivation of sugar beets elsewhere in some of these years had even been limited, since the reduction of the cultivation in the new sugar-beet regions in the east and north was greater than the restoration in the old sugar-beet region of the central agricultural area of the RSFSR. The share of the Ukraine in the total sugar-beet hectarage of the USSR increased from about two-fifths in 1944 to two-thirds in 1948, when, according to the Soviet press, the prewar sugar-beet hectarage was restored both in the Ukraine and in the USSR as a whole (*42; 43*).

It is significant that the cultivation of sugar beets on irrigated land rapidly declined; in 1945 it was only 60 per cent of its top level in 1943. It made up in 1945 less than 10 per cent of the total sugar-beet hectarage of the USSR, as against 30 per cent in 1943. Furthermore, no significant expansion of sugar beets on irrigated lands was planned for 1950 by the Fourth Five-Year Plan. This is an additional indication that the results of rapid expansion of sugar beets there early in the war years were disappointing. The possibility is not excluded, however, that with completion of several irrigation projects in the Kazakh and Kirgiz republics, further expansion of cultivation of sugar beets on irrigated lands will take place. There was discussion in the Soviet professional press of the possibility of supplying the total sugar requirements of Central Asia and Western Siberia from this source (*44*). However, the restoration of the sugar-beet area, as it proceeded during 1945–50 according to the Fourth Plan, represented a return to the late prewar geographical pattern rather than a further shift to the east. In this respect the plans with regard to the sugar industry differed from those for several others, particularly for heavy industries such as coal and iron.

The restoration of the prewar area of sugar beet in three years may be regarded as gratifying, particularly considering that even according to optimistic statements in the Soviet press there were in the Ukraine in 1944 only about 40 per cent of the prewar number of tractors and this percentage increased only to 60 in 1946, as computed by the writer on the basis of Klimenko's (*45*, p. 58) information. The situation with respect to horses and machinery was even worse.

However, the quality of work in the sugar fields under such conditions was far from satisfactory. Plowing was not deep enough and until 1949 the timing of operations on crops was poor. According to authoritative testimony of Vitovsky, Chief Agronomist in the Principal Administration on Sugar Beets in the Ministry of Technical Crops, only 40 per cent of the land planted to sugar beets for the 1946 crop had been plowed in the fall. Furthermore, spring plowing and consequently the plantings were done late, and the fields were infested by weeds (*46*). With such unsatisfactory agricultural techniques yields were necessarily low, especially because the application of manure and mineral fertilizers declined far below the prewar level, which had been far from satisfactory.

According to an optimistic official statement, in the entire USSR only 100 million tons of manure were applied to the 1946 crops (*47*). This is only one-half the low prewar application of manure. However, by far the largest portion of this manure was applied outside the sugar-beet area. More important is another official statement that in the Ukraine, in the same year, 11 million tons were applied, or less than half a ton per hectare of crops. And this was presented as a great achievement, since in 1945 only 4 million tons of manure were brought into the field (*48*).

The situation with respect to application of mineral fertilizers was certainly no better, since the total Soviet production of mineral fertilizers in 1946 was only about one million tons (*49*), or about one-third of the prewar production; in 1945 it was even lower, about 700,000 tons (*50*).

Following the report of a member of the Politbureau, Andreev, on the disastrous situation of Soviet agriculture in 1946, a special decision of the Communist Party on February 28, 1947 assured 397,000 tons of mineral fertilizers for the 1947 sugar-beet crop. This was only 40 per cent of the prewar supply of fertilizers for sugar beets, whereas the acreage was restored to nearly 90 per cent of prewar. In 1946, applications of mineral fertilizers by some collective farms in quantities of 1.5 to 2 quintals per hectare of sugar beets were mentioned among the best achievements (*51*), while in the last prewar years the average application of mineral fertilizers was 7 to 8 quintals per hectare of beets.

The position with respect to mineral fertilizers was much improved in 1949, however, when production in the USSR reached the prewar level (*49*, p. 14). By Soviet claims, in 1949 Soviet agriculture received 33 per cent more mineral fertilizers than in 1940, and prewar applications were allegedly exceeded even in 1948 (*52*, pp. 144–45). Since total production of mineral fertilizers in the USSR in 1948 was below the prewar level, these claims are difficult to accept.

POSTWAR YIELDS AND OUTPUT

It is no wonder that under such conditions sugar-beet yields were very low during 1944–46. The disastrous situation in the last year was caused partly by the drought of 1946, which was particularly severe in the principal sugar-beet areas. The Soviet government this time did not try to camouflage the disaster. On the contrary, there are reasons to believe that it overemphasized the effect of the 1946 drought, since it wished to have UNRRA's aid to the Ukraine and Belorussia extended for another year.

Since total production of sugar in 1946 amounted to about 500,-000 tons from a total sown area of beets approaching 940,000 hectares, yields of sugar beets as low as 5 to 6 tons per hectare are indicated.[9] Nor was the situation much better in the two preceding years. Andreev's report of February 28, 1947 (55) implied that the average "biological" yield of beets in 1945 was about 10 tons per hectare, and that the total production of sugar beet amounted to only about one-third of what was planned for 1950, or 8.7 million tons. As the Soviet press openly recognizes that the difference between the biological yield of beets and actual delivery during these years represented 2–3 tons per hectare (53), it appears then that even according to Andreev's statement no more than 6 million tons of sugar beets was actually harvested in 1945. The picture appears even darker, judging from statements of other Soviet high officials. Benediktov, Soviet Minister of Agriculture, discussing the future development of Soviet agriculture according to the Fourth Five-Year Plan (56, p. 3), said that "gross production of sugar beets planned at 26 million tons will exceed the prewar level by 22 per cent and the gross production in 1945 by 300 per cent." It appears from this that gross (biological) production of sugar beets in 1945, according to Benediktov, was 6.5 million tons, and consequently the biological yield of beets per sown hectare was below 8 tons. This is in agreement with the statement by Fridman, Chief of the Sugar-beet Division of Glavsakhar, that the planned 1950 procurement of sugar beets must increase more than five times over 1945 (35, p. 7). On the basis of this statement the procurement of sugar beets in 1945 must have been about 4 million tons, since only 20 million tons of beets would be needed for the production of 2.4 million tons of sugar, planned for 1950, even with the low extraction rate of 12 per cent.

The fact that deliveries of sugar beets to factories in 1945, a year when weather conditions were satisfactory, were as low as 5 tons per hectare may appear unbelievable. It is confirmed, however, by another statement in the official publication of Glavsakhar. Although the gross production of sugar beet in 1946 was smaller than in 1945 owing to unfavorable meteorological conditions, it was hoped that better conditions of hauling from the fields might permit sugar production to equal that in 1945 (57). This means that the Soviet sugar production in 1945 could have been only a little larger than 500,000 tons.

[9] The leading article (53), mentioning that one collective farm in Tambov oblast obtained 64 quintals of beets per hectare, qualifies this yield as "not bad." The yield of beets in the high-yielding area of Kamenets-Podolsk oblast in the west Ukraine averaged 48 quintals per hectare (54).

Since the same Soviet official states elsewhere that the 1945 procurements of sugar beets were several times larger than during the years 1941–43 (*58*, p. 7), our previous estimates of Soviet sugar production during these earlier years may appear too optimistic. They are nevertheless substantially lower than contemporary appraisals of Soviet sugar production by outside observers.

The situation in 1947 was much improved. More favorable weather in this particular year was the principal factor, but there was also a further expansion of the sugar-beet area by about 150,000 hectares and some improvement in agricultural practices. However, we know that even the planned supplies of mineral fertilizers for sugar beets could assure a rate of application per hectare of crop equal to only half the prewar level, and the supplies of tractors and machinery could have been only slightly improved in 1947 since the restoration of their production proceeded slowly in the first two years of the Fourth Plan period and the plans for their production had not been fulfilled.

Nevertheless, a relatively large crop of sugar beet was grown in 1947. It was officially claimed that the gross production in 1947 was 190 per cent larger than in 1946 (*42*; *52*, pp. 130–31); that is, the yield per hectare was two and a half times as large as the very low yield of 1946. Although this did not bring the yield to the prewar level, there is no question that the 1947 sugar-beet crop was fairly good.

Unfortunately, a large portion of the crop was lost in the fields, since there was not enough transport to haul it to the factories, and a considerable portion remained in the fields, poorly stored, until the spring. According to Panasyuk, Deputy Minister of Agriculture of the USSR, more than 4 million tons of harvested beets were lying in the fields waiting for hauling in the middle of November, a fairly late date under the climatic conditions of the Ukraine (*59*). Late in November about one-third of the Ukrainian crop and from two-fifths to a half of the crop in the principal sugar-beet area of the RSFSR still remained in the fields (*60*). Consequently, an unusually large proportion—even for Soviet conditions—of the good 1947 crop was completely lost or greatly damaged. The percentage of sugar extraction from the processed beets must have been lower than usual.

In 1948 the Soviet government continued its program of expanding the sugar-beet area, and it was officially claimed that the prewar area was reached in that year (*61*). If this was not an overstatement, then apparently the yield of beets per hectare did not increase in that year and may even have declined as compared with 1947. It was claimed by the Soviet press that the production of sugar from the 1948 crop increased over the preceding crop year by 17 per cent for the USSR as a whole and by 30 per cent for the Ukraine (*62*). Such an increase in total sugar production points to a reduction in yields of beet per hectare in 1948, if one takes into consideration that in order to reach the prewar hectarage the sugar-beet area in the USSR as a whole had to increase from the 1947 level by 13 per cent and in the Ukraine by nearly 30 per cent. Furthermore, there are indications that the extraction rate of sugar from beets also rose moderately from about 12 per cent in 1947 to a more normal level, approaching 13 per cent, in 1948 (*63*, p. 18).

Dissatisfaction with the level of sugar-beet yields from the 1948 crop is clearly evident from several statements of officials (*64; 43; 65*, pp. 5–11) connected with the later government-sponsored campaign, calling upon thousands of kolkhozy to accept the obligation to achieve, in the next crop, yields of a certain level—20 tons of beets per hectare on the average for the principal sugar-beet-producing oblasts of the Ukraine. This campaign for raising yields was apparently associated with the relaxation of extreme pressure for further expansion of the sugar-beet area. A similar policy prevailed during the late 1930's, after the setback following the rapid expansion of crop areas simultaneously with the mass collectivization during the early 1930's.

The stabilization of the sugar-beet area at the time when the supplies of tractors, agricultural machinery, trucks, and mineral fertilizers began to increase rapidly during the later years of the Fourth Five-Year Plan period could be expected to result in a better and more timely performance of field work. In an authoritative report on the development of the Soviet national economy it is claimed that as early as 1949 the Machine Tractor Stations performed 19 per cent more work on collective farms than they did in the prewar year 1940, and that their work increased further by 18 per cent during the first three quarters of 1950 (*66*, p. 9). The improvement in supplies of tractors and machinery resulted also in better cultivation of sugar beet. It was claimed in the Soviet press that the entire sugar-beet crop of 1949 was planted on ground deeply plowed in the fall and that the same was true of the 1950 crop (*67; 68*).

Beginning with 1949, sugar beets could receive the same quantity of mineral fertilizers as in the last prewar years and by 1950 their application could even be above the prewar level. The claim that the receipts of mineral fertilizers by the entire Soviet agriculture in 1950 were nearly double those in 1940 (*66*) must, however, be discounted. An analysis of the data on agricultural practices in the state sugar-beet farms, relating to 1949 or 1950 (*69*, pp. 32–36), indicates that in the selected farms, which of course could not be below average, the application of mineral fertilizers per hectare of sugar beet was about on the prewar level of 7 to 8 quintals. State farms are normally better supplied with fertilizers than collective farms.

There is no question that increased supplies of tractors and trucks greatly contributed to the better and more timely harvesting of sugar beets. Mechanization of some steps of harvesting was also a factor. Whereas in 1946 the Soviet government could not even plan to harvest more than 30–40 per cent of the total beet area with the help of beet lifters (*70*), drawn mostly by horses, in 1950 it was expected that 80 per cent of the sugar beets would be harvested with sugar-beet lifters and combines drawn by tractors.

The 1949 sugar-beet crop was probably the first one since the war that was harvested and hauled to factories without much delay and excessive losses. The 1948 crop had been harvested and delivered better than the previous crops, but still the Soviet press complained that the contracts for delivery of beets to factories were fulfilled that year only to 62 per cent of the goal. Concerning the 1949 sugar-beet crop in the Ukraine, the principal officials of the republics reported to Stalin the completion of harvesting and transporting of beets before the fixed term (*68*). But they were obliged to report at the same time that *not all obligations* accepted by collective farms the preceding spring were fulfilled, including that to produce in the principal sugar-beet-growing provinces on the average not less than 20 tons of beets per hectare. In the same report it was stated that in 1949 there were delivered to the beet-receiving points in the Ukraine 42 per cent more sugar beets than in the previous year. This should not be understood to mean that the crop was that much larger. It means that deliveries were proceeding much better in 1949 than in the previous years, when they were completed, as usual, only in the winter.

On the basis of the official statements, the gross production of sugar beet in 1949 may be put at about 22.3 million tons and in 1950 at about 23.5 million tons.[10] This means that the goal of the Fourth

---

[10] N. K. Bulganin in his speech on November 6, 1950 (*71*) said that the 1950 sugar-beet crop promised to be 2.5 million tons larger than that of 1940, which was officially reported at 21 million tons (*72*, p. 66). Total production of sugar beet in 1950 exceeded the 1949 production by 1.2 million tons (*73*, p. 1). The estimate of the 1950 sugar-beet production at 23.5 million tons is substantiated also by the state-

Five-Year Plan to raise the gross production of sugar beets up to 26 million tons in 1950 was missed by about 10 per cent.

On the basis of the available information it is difficult to determine whether the low yield per hectare or the small hectarage was responsible for nonfulfillment of the plan. It was never announced by the Soviet government that the planned hectarage for sugar beet had been reached. It was only claimed that the prewar area of sugar beets had been restored in 1948, and then it was reported that the area was considerably increased in 1950. It may be inferred from this that the 1950 sugar-beet area exceeded the prewar area but did not reach the goal of 1,354,000 hectares fixed by the Fourth Five-Year Plan. From the official report on the fulfillment of the Fourth Plan, it may be inferred that the 1950 sugar-beet area was still slightly below 1.3 million hectares (see footnote to tabulation on p. 305).

Furthermore, it must not be forgotten that the Soviet goals for yields are in terms of "biological" yield and that actual deliveries of sugar beets differ from biological yields, according to the Soviet press, by 2–3 tons per hectare. Consequently, with the gross production of sugar beets in 1950 equal to 23.5 million tons, the delivery of beets to factories would be around 20 million tons on the basis of Soviet estimates. Actually it could be substantially smaller. If beet deliveries to the sugar factories did not fall much below 20 million tons, the Soviet claim that the 1950 planned production of sugar of 2.4 million was overfulfilled (*76*, p. 1)[11] may be correct, since there are indications that the extraction rate of sugar from beets in the USSR in recent years exceeded 13 per cent. Less than 18.5 million tons of beets would be needed to reach the goal at that extraction rate. Consequently a crop with the harvested yield of about 15 tons per hectare on an area somewhat exceeding the prewar hectarage would suffice to reach the planned goal of sugar production in 1950.

To achieve this it would be necessary also to have the processing capacity of sugar factories restored to the prewar level. As was mentioned earlier, the processing capacity of the available sugar factories was more than sufficient to process all beets that could be delivered

---

ment in the official report on the results of fulfillment of the Fourth Five-Year Plan (*74*) indicating that the total crop of sugar beets increased during the Fourth Plan period 2.7 times, while the 1945 crop was implied in Andreev's speech of February 27, 1947, as 8.7 million tons.

[11] An official communication on the results of the fulfillment of the Fourth Plan (*77*, p. 7) says that the 1950 sugar production exceeded the prewar (1940) production by 17 per cent. This means that it was equal to 2.5 million tons.

to factories in the early years of restoration of the industry. The small deliveries of beets to the sugar factories during 1944–46 indicate that less than half the prewar capacity would do the job, while according to Soviet information the productive capacity of the sugar factories that remained intact, or that had been restored before 1946, exceeded 50 per cent of the prewar capacity. This follows from the statements that during the Fourth Five-Year Plan period the productive capacity of the sugar industry increased nearly twofold (76, p. 1), and that in 1950 the productive capacity of the sugar factories was restored to the prewar level even in the regions that suffered most from war devastation.

POSTWAR PROCESSING CAPACITY

Data on the number of factories in operation in postwar sugar campaigns and the productive capacity of the sugar factories ready for work (in per cent of prewar capacity), based on the Soviet technical publications, are given below (75):

Sugar campaign	Number of factories in operation	Approximate processing capacity (*per cent of prewar*)
1945/46	168	over 50
1946/47	145	?
1947/48	167	72.5
1948/49	178	about 80.0
1949/50	196	about 95.0
1950/51	over 200	100

The large number of factories reported as being in operation in the 1945/46 campaign point to a low rate of operation for many of them. Presumably several factories built in the east did not have enough beets to supply their capacity, since the sugar area in the east declined considerably from its top level in 1943 and the yields could hardly have improved much. The decline of the number of factories in operation in the 1946/47 campaign reflects the poor crop of 1946. The fact that 94 of the 102 sugar factories that were in operation in the Ukraine completed their work before the middle of January (78) indicates that sugar-beet supplies were short in the Ukraine.

The fact that in the 1947/48 campaign there were in operation only as many factories as in the 1945/46 campaign shows that in preceding years many factories put in operation really were not fully restored. But the 167 factories that were in operation in 1947/48, that is, 80 per cent of the prewar number, with a productive capacity of 72.5 per cent of the prewar, could handle sharply increased supplies of beets from the good 1947 crop if they were well distributed with regard to supplies of beets. There would have been, however, a shortage of processing capacity in the Ukraine east of the Dnieper and in the central agricultural area of the RSFSR, since in these regions the productive capacity of sugar factories, according to the technical trade press, was only 45 to 60 per cent of the prewar level. The 1947/48 campaign was apparently a difficult one for many factories also because of irregular deliveries of beet to factories. One Soviet expert, speaking of that campaign, said that sugar beets hauled from fields during December–April lost nearly 50 per cent of their sugar content and that consumption of fuel in sugar factories was much above the established norms (*79*, pp. 4–6).

The restoration of the prewar sugar-beet area for the 1948/49 campaign could easily have created serious difficulties for the processing industry, then only about 80 per cent reconstructed, had not relatively poor yields saved the situation.

The 1949/50 campaign was apparently much more satisfactory. In preparation for this campaign it was planned to bring the processing capacity of the sugar industry up to the prewar level, but according to the technical trade press this goal was not reached in important regions of the eastern Ukraine and the central agricultural area of the RSFSR (*80*, p. 1) that suffered most from the war. However, the number of factories in operation increased about 10 per cent from the previous year and the processing capacity of factories was substantially raised. Particular attention was given to enlargement of the refining capacity of sugar factories, which was lagging behind other phases of processing. As stated above, deliveries of sugar beet to factories were much more satisfactory than in the previous years. Technical performance by the sugar industry accordingly improved. The trade press claims that the 1949 plan for sugar production was fulfilled and that in the second half of 1949—that is, in the first half of the 1949/50 campaign—the production of sugar was not less than in the corresponding period of the better prewar years (*81*, pp. 1–3).

The 1950/51 campaign opened with the processing capacity of sugar factories fully restored to the prewar level. The trade press claimed that technical equipment of factories was much modernized and improved, not only in the fully reconstructed factories, but also in those that escaped war destruction. Special attention was directed to mechanization of processes requiring heavy work, such as the unloading of beets delivered to factories, the hauling from dumps to factories, etc., and to establishing automatic controls over various processes of manufacture.

There is no question that the industry, two-thirds of which was completely rebuilt, must be much more modern than it was before the war. For reconstruction the Soviet government had at its disposal not only its own machine-building industry but also those of Germany and Czechoslovakia, the two countries most specialized in the construction of beet-sugar factories in Europe. The experience of other countries, including America, was not disregarded. In the very beginning of the reconstruction, early in 1946, there was organized a special conference of technicians of the Soviet sugar industry, the principal purpose of which was to select the most modern and efficient equipment to be used in the reconstruction of the industry. The work of this conference was based on special reports of Soviet technicians who studied, especially for this purpose, the German, Czech, and American sugar industries (*82*, pp. 3–21).

It must be said, however, that there was no radical change in the structure of the Soviet sugar industry, as one might have been led to suppose from the speech of Khrushchev, then Secretary of the Communist Party of the Ukraine, at the conference of the workers of the sugar industry in the summer of 1944 (*83*, pp. 19–24). At that time a plan of two-step processing of sugar was under serious consideration. It was planned to produce, first, sugar briquets in small plants located near the sugar-beet fields and to process these later into final products in larger sugar factories. Economy in transport of beets was expected with such a system. However, only a few experimental plants of this type were built (*84*, pp. 12–14), and the reconstruction of the sugar industry proceeded on the basis of old patterns and generally in the old geographical locations. It is of interest to note that of ten new sugar factories included in the Fourth Five-Year Plan, six were planned for the old sugar-beet regions (four of which were in the Ukraine), and only four in the new regions of both the east and the west (*27*, p. 64). Apparently sugar factories built in the new eastern regions before and during the war can process more sugar beets than they can procure. At least, while it was claimed that the goal of 2.4 million tons of sugar production in 1950 for the USSR as a whole had been overfulfilled, it was announced that the sugar industry outside of the Ukraine had not fulfilled its plan, because of insufficient procurement of beets (*73*, p. 1).

CONCLUDING OBSERVATIONS

Five years after the end of the war the Soviet sugar industry was nearly restored, although the 1950 production of sugar beet was still some 10 per cent short of the goal. This may be regarded as a notable achievement, since about 90 per cent of the prewar sugar-beet hectarage and an even larger percentage of the sugar factories were in the area of the German invasion of 1941–42, which brought destruction to both agricultural and manufacturing sectors of the industry.

However, it is not possible to say that beyond some modernization of sugar factories the industry has been restored on a more solid foundation than in prewar years. Its agricultural basis continues unsatisfactory, sugar-beet yields are low, and the organization of production is uneconomical. The basic reason is that the Soviet agricultural organization, based on *collective* and *state* farms, has not yet been able satisfactorily to master the organic processes of agriculture. It has been least successful in the restoration of animal husbandry and the introduction of rational crop rotations.

The unsatisfactory condition of the Soviet sugar industry's agricultural basis is further demonstrated by the development of sugar-beet production during the Fifth Five-Year Plan period, 1951–55. Indeed, after a rapid increase of production during the Fourth Plan a period of stagnation followed during the next four years. Khrushchev, in his report to the Twentieth Congress of the Communist Party, supplies the following information[12] on the gross production of sugar beets in the Soviet Union as a percentage of production in 1950 (*14*):

1950	1951	1952	1953	1954	1955
100	114	107	111	95	147

During the first four years of the new plan-period, production evidently fluctuated around a level only about 7 per cent higher than in 1950, and even fell below that year's production in 1954 because of drought in the Ukraine. The ambitious plan to raise sugar-beet production 65 to 70 per cent by 1955 completely failed, although Khrushchev claims a great and sudden increase in the beet production of this particular year. That sudden increase, even if substantiated, must be

---

[12] These figures do not always agree with previous statements by Khrushchev. For instance, in his report to the Plenum of the Central Committee of the CPSU on September 3, 1953 he indicated the 1952 gross production of sugar beets at 22 million tons, or about 6 per cent below the 1950 production calculated on the basis of the previous official statements at 23.5 million tons (see p. 311). The later report implies instead a 7 per cent increase.

regarded as exceptional, explained this time by unusually favorable weather conditions in the Ukraine.[13] Indeed, Kirichenko, in his report to the Nineteenth Congress of the Ukrainian Communist Party on January 18, 1956 (*18*), said that yields of sugar beets in the Ukraine in 1955, averaging about 20 tons per hectare, were at a record high. This record level of yield of sugar beets in 1955 was, however, 6 tons (or more than 20 per cent) below the target of 25.5–26.5 tons per hectare planned for the Ukraine for that year (*21*). It appears, thus, that the raising of yields of beets per hectare was lagging behind the planned target even more than total production.

The lag in fulfillment of the new plan for sugar-beet production eased the problem of additional processing capacity, which, as we know, was about restored to the prewar level by 1950. Some further increase in the total processing capacity of the Soviet sugar industry is indicated by the fact that the sugar-processing campaigns in the first three years of the new plan-period were substantially shorter than before the war: 118 days in 1951/52, 113 days in 1952/53, and 108 days in 1953/54 according to Soviet sources (*85*, pp. 285–86), as against over 150 days prewar. There is also direct information from Soviet sources that during the 1953/54 sugar campaign 217 sugar factories were in operation in the Soviet Union (*32*, p. 6), a few more than in 1950.

However, when the problem of increasing the supply of consumers' goods arose in 1953 under Malenkov's government, and it was expected that 1954/55 sugar production would be raised to 4.3 million tons—a target far from attained partly because of the drought in the Ukraine—plans were rushed to build 25 new sugar factories and, for the time being, to extend the processing campaign by 20 to 25 days (*85*)—a return to the prewar practice. This extension of the campaign became necessary to cope with the large 1955 beet crop, but led to increased losses of sugar through prolonged storage of beets (*87*, p. 62).

The fact that 15 of the 25 sugar factories planned for construction were to be located in the western Ukraine and Moldavia indicates that the prewar plan for the expansion of sugar beets to occur mainly in the eastern parts of the USSR were still checked, at least tempo-

---

[13] Soviet sugar production in the calendar year 1955 was recently officially reported at 3.4 million tons, equal to the output in 1953 and exceeding that of 1950 by 37 per cent (*92*, p. 91). This indicates that the Khrushchev index of 147 for sugar-beet production in 1955 may be an overstatement, although it is reproduced in the new official source (*92*, p. 101).

rarily. This is also indicated by the direct statement of Khrushchev in his report to the Plenum of the Central Committee of the CPSU in February 1954, in which the western Ukraine is placed first, Belorussia and the Baltic States second, and the eastern regions of the USSR last as appropriate regions for proposed expansion of at least 300,000 hectares in the sugar beet area during the next 2 to 3 years.

The restoration of Soviet sugar production to the prewar level and above does not mean that the per capita consumption of sugar will be raised in proportion. The Soviet Union has incorporated in its boundaries new territories with a prewar population of 20–25 million. At the present time the population of the USSR as a whole must be substantially above the prewar level in spite of the heavy losses during the war.[14] On the other hand, it must be mentioned that the Soviet Union at present controls the sugar industries of such important sugar producers and exporters as Czechoslovakia, Poland, and the Soviet zone of Germany. Substantial or even preponderant portions of Polish and Czech exports of sugar have been directed to the USSR during the postwar years, and still larger quantities of sugar have been obtained by the Soviet from occupied Germany as reparation payments.

According to F. O. Licht's *Sugar Information Service,* imports of sugar into the USSR are estimated for 1947/48–1949/50 in the following quantities, in thousand metric tons (*88*, II, pp. 24–26; V, pp. 1, 4):

	1947/48[a]	1948/49[a]	1949/50[a]
Total imports	200	500	400
From Czechoslovakia	33.3	103.8	65.1[b]
From Poland	88.1	126.3[c]	?

[a] September–August years except as noted.
[b] September–January.
[c] September–May.

Since no statistics of Soviet foreign trade are officially revealed, the above data on total sugar imports must be regarded as approximations by a competent sugar service. To a certain extent they are confirmed by the fact that the 1949 imports of sugar into the USSR were estimated by the secretariat of the International Sugar Council at about the same level—400,000 metric tons for the calendar year (*90*, p. 35).

[14] The Soviet population as of April 1956 is officially estimated at 200.2 million, compared with 170.6 million according to the census of January 17, 1939 in the prewar boundaries (*92*, p. 17).

From the tabulation above it may be inferred that in spite of substantial imports of sugar from Czechoslovakia and Poland, the principal source of Soviet sugar imports since the war has been the Soviet zone of Germany. According to Licht's estimate (*91*, p. 77) the Soviet Union continued to be a net importer of sugar during the following years, and for the period 1950/51 to 1954/55 its net imports averaged about 300,000 tons annually. This estimate appears reasonable in spite of official statistics of the International Sugar Council which show the Soviet Union in 1954 as a net exporter to the extent of 53,000 tons (*86*, p. 37). Imports from Czechoslovakia and Poland into the USSR are not reported to the Council, and these were probably large, particularly the imports from Poland. Heavy net imports of sugar into the Soviet Union, exceeding 420,000 tons, are in fact reported also by the International Sugar Council for 1955, when the USSR became a heavy purchaser of sugar from Cuba and other Western sources. With possible imports from Czechoslovakia and Poland, actual net imports in 1955 may even have exceeded the Council's figure.

The fact that large portions of the Czech and Polish export surplus in the postwar years were absorbed by the Soviet market, and that Soviet takings of sugar from Germany even exceeded that country's actual surplus, had a number of effects on the world sugar situation. Czechoslovakia and Poland were significant exporters to the world market before the war, and about half a million tons consumed in western Germany originated within the Soviet-occupied East zone. During the years of severest shortage in Western Europe, supplies from these customary sources were no longer forthcoming, and availability for the future remains seriously in doubt.

Even while the Soviet Union was importing large quantities from its European satellites and its own sugar industry was far from restored, it was exporting substantial quantities of sugar, presumably to its traditional markets in the neighboring Asiatic areas. These exports were estimated for 1948/49–1949/50 at 100–150 thousand metric tons (*63*, pp. 24–26). The Soviet intentions to restore exports of sugar on Asiatic frontiers, not only to its satellites but to other countries as well, may be inferred from the fact that in November 1950 the Soviet Union concluded a trade pact with Iran, according to which it agreed to send 75,000 tons of sugar to that country during the year beginning November 10, 1950. There were also exports of that magnitude in 1954 and 1955 (*86*).

Under the International Sugar Agreement of 1937, the combined export quota of Poland, Czechoslovakia, and the USSR (excluding shipments to Mongolia, Sinkiang, and Tannu Tuva) amounted to more than 600,000 tons, and prewar Germany added 120,000 tons to the total. Not only do these historical quotas strengthen the Soviet hand in postwar international sugar conferences, but Communist control of the Chinese mainland places an important sugar-deficit area within the Soviet orbit. The way is paved for a substantial intra-Soviet sugar trade. The commerce might be based either on direct shipments to China from Poland and Czechoslovakia (e.g., as a return cargo on vessels carrying soybeans west of Suez), or on USSR absorption of sugar from its western satellites and simultaneous exports on its eastern frontiers. So interpreted, Soviet plans to expand the cultivation of sugar beets on the irrigated lands of Central Asia may be calculated to serve an Asiatic export market.

CITATIONS

1 K. T. Voblyi, *Essay on the History of the Beet-Sugar Industry of the USSR* (Moscow, 1928), I.

2 M. Dolgopolov, "Expansion of Technical Crops in the USSR," *Socialist Agricultural Economy* (Moscow), April 1940.

3 Central Office of National-Economic Accounting, *National Economy USSR: Statistical Handbook 1932* (Moscow, 1932).

4 G. S. Benin, "Some Problems of Restoration and Development of the Sugar Industry," *Food Industry* (Moscow), 1944, Nos. 7–8.

5 G. T. Selyaninov, "Specialization of Agricultural Regions by Climatic Characteristics" in *Plant Breeding of the USSR* [*Rastenievodstvo SSSR*], (All-Union Institute of Plant Breeding of the People's Commissariat of Agriculture of the USSR, Moscow and Leningrad, 1933), I, particularly an attached map.

6 *Transactions of the First All-Union Conference on the Distribution of Productive Forces of the USSR* [*Trudy Pervoi Vsesoyuznoi Konferentsii po razmesheheniyu proizvoditelnykh sil SSSR*] (Moscow, 1933), IV, *Ural-Kuznetsk Combine*.

7 I. Lupinevich and Z. Korobov, "The Problem of Location of Sugar-Beet Crops," *Socialist Agricultural Economy* (Moscow), October 1944, hereafter cited as *Soc. Agr. Econ.*

8 A. Silin, "Results and Prospects of Sugar-Beet Cultivation in Altai Krai," *Soc. Agr. Econ.*, October 1940.

9 *Review of 23* by P. Pleshkov in *Soc. Agr. Econ.*, February 1948.

10 D. N. Pryanishnikov, *Nitrogen in the Life of Plants and in the Agriculture of the USSR* (Moscow, 1946).

11 I. V. Zilberman and P. Petersen, "On the Geographical Distribution of Sugar Beet Crop," *Soc. Agr. Econ.*, November-December 1945.

12 D. N. Pryanishnikov and I. V. Yakushkin, *Field Crops* (Moscow, 1936).

13 E. Yaroshchuk and V. Yakovlev, "Means of Improving Sugar Beet Rotations," *Soc. Agr. Econ.*, January 1938.

14 Pravda, Feb. 15, 1956.

15 USSR People's Commissariat of Agriculture and People's Commissariat of State Farms, *Agriculture of the USSR Yearbook 1935* (Moscow, 1935).

16 D. N. Pryanishnikov, *Agrochemistry* (3d ed., Moscow, 1940).

17 Gregory Makhov, "Memorandum on Fertilizers and Crop Rotations in the USSR" (Manuscript, 1950).

18 Pravda Ukrainy (Kiev), Jan. 19, 1956.

19 N. I. Orlovskii, "Development of the Selection of Sugar Beet Seed in the USSR," *Agrobiology* (Moscow), November-December 1947, No. 6.

20 Naum Jasny, *The Socialist Agriculture of the USSR: Plans and Performance* (Food Research Institute, Grain Econ. Ser. 5, Stanford, Calif., 1949).

21 Socialist Agriculture (Moscow), Oct. 10, 1952.

22 B. A. Rubin, *Storage of Sugar Beet* (Moscow, 1946).

23 S. G. Kolesnev, *Organization of Socialist Agricultural Enterprise* (Moscow, 1947).

24 Socialist Agriculture, Oct. 15, 1952.

25 USSR, *Statistics of Foreign Trade* (Moscow), No. 12, for 1936 and 1937.

26 S. K. Kondrashev, *Irrigated Agriculture* (Moscow, 1948).

27 V. P. Zotov, Minister of Food Industry of the USSR, in his pamphlet, *Development of Food Industry in the New Five-Year Plan* (Moscow, 1947).

28 F. O. Licht's *Sugar Information Service*, in a circular letter of April 1946 (in German).

29 D. N. Pryanishnikov, "Concerning Crop Rotations and the System of Fertilization in the Cotton Regions of Central Asia," in *Report of the All-Union Institute of Fertilizers, Agrotechnique and Soil Science in the name of K. K. Gedroits*, for 1941–42, p. 10 [*Nauchnyi Otchet Vsesoiuznogo Instituta Udobrenii, Agrotechniki i Agropochvovedeniia imeni K. K. Gedroits za 1941–42 g.*, Ogiz-Selkhozgiz, 1944].

30 P. Petersen, article in *Soc. Agr. Econ.*, September 1943.

31 P. Sokolovskii, "Concerning Sugar Beet Cultivation in Irrigated Regions," *Soc. Agr. Econ.*, April 1944.

32 F. O. Licht's *Sugar Information Service*, July 2, 1954.

33 A. N. Rozanov and A. A. Lazarev, *Soil Conditions in the Regions of Sugar Beet Cultivation in Central Asia* (Moscow, 1945).

34 Socialist Agriculture, May 29, 1945.

35 S. E. Fridman, article in *Sugar Industry* (Moscow), June 1946.

36 I. Sipiagin, "Successes of Sugar Beet Cultivation during the War in the Nonblack Soil Zone," *Soc. Agr. Econ.*, January-February 1945.

37 Vesherovich, "Prospects for Development of Sugar Industry in Altai and in Other Parts of Western Siberia," *Sugar Industry*, January 1948.

38 N. Nazartsev, "Cultivation of Sugar Beets in the Far East," *Socialist Agriculture*, Oct. 24, 1946.

39 I. Benediktov, article in *Bolshevik* (Moscow), March 1944.

40 Izvestiya, June 7, 1944.

41 Sugar Industry, April 1948.

42 Leader in *Socialist Agriculture,* Mar. 24, 1949.

43 G. P. Butenko, Minister of Agriculture of the Ukrainian SSR, in *Socialist Agriculture,* Jan. 5, 1949.

44 I. V. Zilberman, articles in *Sugar Industry,* January and July 1947.

45 N. Klimenko, "Agriculture of the Soviet Ukraine on the Way to Prosperity," *Soc. Agr. Econ.,* November 1947.

46 I. Vitovsky, article in *Socialist Agriculture,* Feb. 1, 1947.

47 Leader in *Socialist Agriculture,* Jan. 29, 1947.

48 G. P. Butenko in *Socialist Agriculture,* Apr. 20, 1946.

49 Computed on the basis of a statement in *Chemical Industry,* 1947, No. 4, as quoted in the United Nations, Dept. of Econ. Affairs, *Economic Survey of Europe 1949* (Geneva, 1950), in a footnote to the table on p. 14, given in Appendix B, pp. 231–32.

50 Leader in *Socialist Agriculture,* Feb. 25, 1947.

51 Socialist Agriculture, Apr. 21 and May 8, 1946.

52 S. Gurevich and S. Partigul, *The New Economic Upswing of the USSR in the Postwar Five-Year Plan Period* (Moscow, 1950).

53 Leader in *Socialist Agriculture,* Sept. 13, 1946.

54 Socialist Agriculture, Dec. 1, 1946.

55 Izvestiya, Mar. 7, 1947.

56 Socialist Agriculture, June 20, 1946.

57 Sugar Industry, January 1947.

58 Sugar Industry, June 1946.

59 Socialist Agriculture, Nov. 15, 1947.

60 Socialist Agriculture, Nov. 25, 1947.

61 Socialist Agriculture, July 3, 1948, Jan. 5 and Mar. 24, 1949.

62 F. O. Licht's *Sugar Information Service,* Aug. 2, 1949, p. 4, with reference to *Izvestiya* and Gurevich and Partigul (see 52, above, pp. 135–36).

63 F. O. Licht's *International Sugar Statistical Year and Address Book 1950* (Ratzeburg), Vol. I, Part II.

64 Leader in *Pravda,* Feb. 12, 1949.

65 N. V. Vinogradov and others, article in *Sugar Industry,* August 1949.

66 Editorial in *Planned Economy* (Moscow), November-December 1950, No. 6.

67 Socialist Agriculture, Jan. 5, 1949.

68 Pravda, Nov. 17, 1949.

69 M. K. Melnik, "Possibility of Increasing Productivity and Marketability of Sugar Beet State Farms," *Sugar Industry,* February 1951.

70 Socialist Agriculture, Sept. 10 and Oct. 12, 1946.

71 USSR Information Bulletin (Embassy of the Union of Soviet Socialist Republics, Washington), Nov. 24, 1950.

72 S. Demidov, "Agriculture in the New Five-Year Plan," in the symposium *Concerning the Five-Year Plan of Restoration and Development of the National Economy USSR in 1946–50* (Moscow, 1946).

73 Sugar Industry, March 1951.

74 USSR Information Bulletin, May 1, 1951.

75 Sugar Industry, various issues 1948–51.

76 *Sugar Industry,* January 1951.

77 *Planned Economy,* March-April 1951, No. 2.

78 *Izvestiya,* Jan. 17, 1947.

79 M. K. Melnik, article in *Sugar Industry,* October 1948.

80 Leading article in *Sugar Industry,* June 1949.

81 *Sugar Industry,* January 1950.

82 *Sugar Industry,* July-August 1946.

83 *Soc. Agr. Econ.,* August-September 1944.

84 I. P. Lepeshkin, "Advanced Technique in Action," *Sugar Industry,* October 1948.

85 *International Sugar Journal* (London), October 1954, referring to *Sugar Industry,* January 1954.

86 *Statistical Bulletin of the International Sugar Council* (London), April 1956.

87 C. Czarnikow, *Sugar Report No. 247,* Apr. 12, 1956.

88 F. O. Licht's *International Sugar Statistical Year and Address Book 1950,* Vol. I, *World Sugar Statistics* (Ratzeburg, mimeo.), Parts II and V.

89 V. P. Timoshenko, *The Soviet Sugar Industry and Its Postwar Restoration* (Food Research Institute War-Peace Pamphlet No. 13, Stanford, Calif., August 1951).

90 United Nations, *Review of International Commodity Problems 1950* (New York), January 1951, Appendix A.

91 F. O. Licht's *Sugar Information Service: East European Special Sugar Edition,* July 1955.

92 Central Statistical Administration at the Council of Ministers of the USSR, *National Economy of USSR. Statistical Handbook* (Moscow, 1956).

CHAPTER 11

THE FREE MARKET AND THE INTERNATIONAL SUGAR AGREEMENT OF 1953

Postwar discussions of principles and protracted negotiations on text produced three international commodity agreements in 1953. While tin could count at best on the tacit support, and wheat on the tacit opposition, of the respective major importer, the new International Sugar Agreement (ISA) received considerable acclaim on both sides of the Atlantic (1, p. 56; 2, p. 436; 3, p. 3). Participation by the Soviet bloc was regarded by some as that long-awaited "hopeful augury of international cooperation" in world trade (4, p. 1). Those who feel uncomfortable about a "restrictive quota agreement" that relies on production control and export quotas, or who see here "a gigantic international cartel, an adventure in international socialism" reflect only a small minority of published opinion (5, p. 53; 6, p. 42).

Operationally the Agreement was indeed of the standard export-quota form modified only by consumer representation and a target "zone of stabilized prices."[1] Price in the world "free market," i.e., that part of international trade enjoying no special protection or privilege in the country of destination, was to be held between 3.25 and 4.35 cents per pound, f.a.s. Cuba, by appropriate adjustment in the level and distribution of quotas. Price stability so defined was the primary objective, but there were two minor themes. One was the notion that prices, quotas, and producer incomes under the Agreement would conform to some principle of equity. The other, in accordance with accepted principles governing commodity-control agreements (10; 11), was represented by a few clauses directed at making those longer-run adjustments that contemporary surpluses required.

While an international arrangement for sugar is easier to justify in the abstract and holds far fewer dangers than is common with commodity agreements, the expressed hopes of the agreement's sponsors and the fears of its detractors were equally likely to be disappointed.

---

[1] For full text of 1953 Agreement, see 8, pp. 23–28; for prewar text, 9, pp. 26–45. A portion of the discussion in this chapter originally appeared in 20, pp. 837–53.

Importing countries accepted few obligations; free-market exporters enjoyed, for the most part, only token privileges. Export capacity (especially in Cuba) far in excess of prospective requirements of the free market—a legacy of the Second World War and national sugar policies—was severely to test the International Sugar Council's ability to hold prices above the agreed minimum for some time. No clauses effectively dealt with the basic problems, whether by prying open protected markets, reducing margins of protection, or even preventing expansion of subsidized output and further contraction of the free market. Indeed, there was less ground for concern over "the dangers of artificially maintained prices" or the risk of "introducing undue rigidities in the pattern of production and trade" than over the extreme looseness, if not the futility, of the entire arrangement (*12*, pp. 1–2; cf. *7*, pp. 207–11).

THE "FREE" SUGAR MARKET

Crucial to the necessity for and operation of any international sugar agreement is the volume of the commodity that is admitted into export markets on a nonpreferential basis. The gradual erosion of the "free" market, characterized chiefly by the expansion of preferential cane-sugar suppliers, has been the outstanding feature of the world sugar economy in the twentieth century (*13*, chap. IV). While the United States provides a case of great expansion in consumption with insignificant resort to free market sources, in the British Commonwealth (and especially in the United Kingdom) there has been a persistent withdrawal. England had been the hub of the free market in the late nineteenth century, for her rise to modern levels of per capita sugar consumption had come largely before 1900 and without benefit of any domestic sugar production whatsoever. The mere threat that she would impose a countervailing duty against subsidized exporters was the main sanction behind the Brussels Sugar Convention of 1902. Except for the consequent removal of unfair competition from subsidized beet sugar, her own cane colonies received no special consideration. Sugar duties reintroduced in 1901 were light, nondiscriminatory, and for revenue purposes only. Well over half the sugar imported by England on the eve of World War I was beet sugar, most of it from Germany and Austria-Hungary, while Cuba alone provided more cane sugar than all British territories combined. In Asia, similarly, British India was a major outlet for free-market sugar, especially from Java. While liberal trade policy still flourished, the major British protectionist was Canada, which encouraged shipments from the British West Indian possessions by introducing imperial preference in 1898.

The trade disruption associated with World War I paved the way for a series of retreats from Britain's open-market policy, already described in chapter 8. By 1937, England was importing more sugar from the Commonwealth than from elsewhere. When expansion proved less easy for the colonial territories than for South Africa and Australia, both of whom shifted from a net-import to a net-export basis, the standard colonial rate was reduced below that applicable to Dominion sugar, and a still further concession was granted on a specified volume of colonial sugar. Comparable developments occurred elsewhere. In Asia, Japan developed Taiwan as its major source of Empire supplies, and India resorted to a prohibitive tariff in 1931. Expansion of output in United States overseas territories was in part merely at the expense of Cuba, itself a preferential supplier, but Cuban sugar displaced from the American market became in turn a depressant in remaining markets overseas. In aggregate terms, world exports increased some 5 million metric tons between 1909–13 and 1924–28, but almost three-quarters of this was for preferential destinations (*13*, p. 132). In the succeeding decade, the preferential sector (particularly the British) continued to grow, but total world trade in sugar declined some 10 per cent (*13*, pp. 164–65) and the free-market portion correspondingly more. It was against this background that the International Sugar Agreement of 1937 was negotiated.

After the dislocations of World War II had largely been overcome, the aggregate size of the free market for sugar stood somewhat larger than it had been immediately before the war. Actual requirements of the free market in the first quota year of the ISA of 1937 and estimated requirements for the first quota year under the postwar Agreement are as follows, in thousand metric tons, raw value (*14*, pp. 77–80; *15*):

Area	1937/38	1954
British Commonwealth (net) ex-India and Pakistan.	1,277	212
French Union	270	0
Other Europe	795	890
America	286	364
Africa	65	110
Asia and Oceania	676	2,293
Subtotal	3,369	3,869
Statistical adjustment	— 69	— 5
Total	3,300	3,864

The improvement was in many respects more superficial than basic. Consumption in Europe and America continued to expand more rapidly than imports. Indeed, such smaller nations as Finland, Greece, Norway, and Switzerland represented well over 50 per cent of the European "free" market. Free-market exporters contributed less than one per cent of a United States consumption amounting to some 8 million tons. The Asiatic total was inflated by virtue of the fact that Japan had lost her sugar empire and now was dependent almost exclusively on "free" imports. But Japan's holdings of foreign exchange were not large, and the place of sugar imports as a claimant upon them was highly marginal. India's and Pakistan's imports, which had risen in 1954 to some 700,000 tons from negligible quantities before the war, were also an unstable factor, by virtue both of dubious external purchasing power and of rising domestic output. Against all questionable gains stood the relentless decline in the net takings of other portions of the British Empire.

SIGNIFICANCE OF THE COMMONWEALTH SUGAR AGREEMENT

Although the United Kingdom came to depend less on Empire sources in the course of World War II (Chart 4 B), wartime arrangements in fact strengthened the ties between London and the colonial and Dominion sugar suppliers overseas. While bulk purchasing by the Ministry of Food during the war was an instrument for assuring minimum Commonwealth production consistent with severe consumer rationing and British overseas commitments (see chapter 8), it was shortly converted into a device for promoting a considerable expansion of output. "World" sugar prices stood considerably above prewar parity even after the postwar years of intense shortage had come to an end. Moreover, most available sugar sold only for dollars. From the English point of view, price and currency considerations warranted a greater dependence on sterling sources, while sugar purchases were a form of economic aid to British colonial territories. Moreover, since producers would otherwise have been unwilling to undertake the fixed investment in field and factory, the term of the bulk purchase arrangements was necessarily lengthened. For the period January 1, 1948 to December 31, 1952, the Ministry of Food guaranteed Commonwealth producers a market for their entire output, at prices to be negotiated annually. Derationing of sugar in the United Kingdom had itself to be delayed until the response to this policy had satisfactorily been demonstrated. The incentive to expand was particularly strong, inasmuch as export performance during 1948–52 was to determine the minimum figure that individual regions would enjoy in any future Commonwealth arrangement (*16*, p. 6).

The culmination of this process was the Commonwealth Sugar Agreement (CSA) entered into by the United Kingdom and the exporting colonies and Dominions on December 21, 1951 (*17*, pp. 95–99; *18*, pp. 3–8). Under its terms, the exporters were assured of an outlet in the United Kingdom for more than 1,500,000 long tons of sugar, their so-called "negotiated-price quota." In addition to this allotment, which was to be purchased at an annually negotiated price, a further sale of some 800,000 tons was guaranteed by the United Kingdom, but only at the world price plus imperial preference. The entire arrangement was to last at least through 1959, and could be renewed each year in order to continue commitments for a full eight years in advance. At expected levels of unrationed consumption in the United Kingdom, the negotiated-price quota together with domestic beet production of 700,000–800,000 tons raw value left only some 250,000 tons to come from free-market exporters. Indeed, the United Kingdom agreed to raise the negotiated-price quotas proportionally to any excess in consumption above 2,550,000 long tons. The contrast between the prewar and postwar sugar position in the United Kingdom is shown in the following tabulation, in thousand long tons, raw value (*40*):[2]

Source	Average 1937–39	1950	1951	1952	Average 1950–52	Negotiated-price quota, CSA[a]	1953
Domestic production							
beet sugar	428	753	668	639	687	—	800
Imports, total	2,262	2,100	2,281	2,045	2,142	—	3,032
Less: refined exports ..	374	836	806	755	799	—	736
Imports, net	1,888	1,264	1,475	1,290	1,343	—	2,296
Imports, from Commonwealth	1,230	812	862	888	854	1,568	1,439
Australia	370	247	175	118	180	300	496
Br. W. Indies	235	302	308	378	332	} 640 {	492
Br. Guiana	73	67	83	132	94		129
Mauritius	280	164	260	251	225	335	250
Union S. Africa	188	15	20	—	12	150	20
Fiji	70	17	16	9	14	120	51
Others	14	—	—	—	—	23	1

[a] Excluding 75,000 tons destined for New Zealand.

[2] The heavy volume of imports in anticipation of derationing, effective September 1953, conceals the full extent to which foreign sugars have been displaced from the United Kingdom market. Net imports were one-third lower in 1955 than they had been in 1953. The absolute decline from Cuba alone amounted to about one million tons, although Commonwealth exporters actually showed an increase of some 250,000 tons (*29*, p. 9).

So far as relative advantage to individual producing sectors was concerned, the Agreement represented a nice compromise. In general, over-all quotas were about 50 per cent above prewar exports. So close an approach to uniform increases was of itself more favorable to the two Dominions, with their large and growing domestic markets, than to the export-dependent colonies. If South Africa's quota was slightly below prewar, this was offset by the provision that permitted exports within its geographic region (such as to South-West Africa and the Rhodesias) in excess of quota. By comparison with postwar performance, the tabulation indicates that the Union was dealt with extremely generously. As between Australia and the British West Indies, the latter experienced a more steady postwar expansion but seemed to have less capacity for long-term growth. The main protest against the CSA figures in fact came from the West Indies Sugar Association, partly because a substantial increase in Jamaican production (paralleling decline in the local banana industry) made prewar relationships somewhat inappropriate. Yet the British West Indies were assured of the contract price on two-thirds of their exports, as against only one-half for Australia. As for the United Kingdom's own beet sugar, no further expansion was anticipated and pressure to build new factories was resisted (*19*, p. 8).

The price features of the CSA are of special interest, for their bearing on internal aspects of British sugar and on the world market alike. The principle to which the guaranteed price was to conform was that it be a "reasonably remunerative price to efficient producers." This was not entirely jargonistic. A uniform price applied to all Commonwealth producers. This was unlike the practice in other bulk-purchase contracts, and contrary to the prewar policy that allowed a higher preferential on sugar from colonial than from Dominion sources. A uniform price was specifically intended as an inducement to and reward for efficiency, notwithstanding differences in the current cost position of various regions (*21*, pp. 26–27). Moreover, annual adjustments in price were based on a formula that reflected cost changes in the exporting regions as a group, and there was a provision for rewarding areas that best responded to the contract prices, if quotas were subsequently raised. Leaving part of the export sugar to receive only the free-market price was in itself a means of bringing some independent indicator of efficiency to bear, although such policies as that reflected by the CSA were making the world price perform that role less satisfactorily. Finally, the negotiated price was closely tied to the internal price structure within the United Kingdom. The Ministry of Food, the exclusive importer, resold raw sugar at prices meant to equate the cost of Commonwealth and foreign raw, and that rate was the basis for pricing domestic beet sugar as well.

Important changes in sugar price relationships lie behind the financial terms actually arrived at, as the following tabulation indicates, per long ton (*21*, pp. 22, 26–27, 44; *41*, p. 883; *42*, p. 991):

Year	Cuban raws c.i.f. U.K.			Commonwealth bulk purchase			Australia home consumption			Australia export price			U.K. sugar beets		
	£.	s.	d.	£.	s.	d.	£.	s.	d.	£.	s.	d.	£.	s.	d.
1938	5	7	11	—	—	—	24	0	0	8	4	3	2	6	1
1939	7	4	2	—	—	—	23	12	6	10	7	6	2	9	6
1947	31	11	8	24	5	0	24	0	0	29	12	6	5	11	2
1948	30	5	0	27	5	0	23	1	0	28	2	0	5	8	8
1949	29	6	8	27	5	0	24	6	0	29	7	6	5	4	11
1950	40	16	8	30	10	0	24	11	0	32	16	6	5	12	6
1951	49	11	8	32	17	6	33	14	0	36	15	6	5	14	6

Cuban raws are 96°, ex-duty; Commonwealth bulk purchase is for 96° c.i.f. United Kingdom at prewar freight rates; and Australian rates are returns to producers, 94 net titre. Beet prices are weighted average of growers' prices for beets delivered under contract to factories.

The great increase in the sterling price of free-market sugar is particularly notable. Even before the Korean inflation made its special contribution, Cuban sugar in London held 500–600 per cent above its 1938 price. By virtue of the prewar tariff preference, Empire sugars received a premium of about 50 per cent over free-market rates in the late 1930's. In the late 1940's, they climbed merely to parity with foreign sugar, and then only belatedly. Though the proportional increase was necessarily less than that enjoyed by Cuba, it amounted to about 200 per cent, and in fact was more generous than the percentage increase allowed to domestic beet growers. Moreover, the external price looked extremely attractive to Australian producers. The controlled price on their home sales remained almost rigid, and as late as 1948 actually stood below the depression level of 1933 (*22*, p. 70). Whereas exports were dumped at one-third to one-half the domestic price before the war, they actually represented the more attractive outlet in the postwar years tabulated above. Though the discount below world-market prices widened substantially in 1951, Commonwealth exporters were doing reasonably well on sales to the United Kingdom.

A new ISA agreement was nevertheless of considerable importance to Commonwealth exporters. Revenue from the final third of their CSA quotas depended on the world price, plus the small imperial preference to which they were entitled in certain Commonwealth countries. Assured of a destination for their own exports, they had a single-minded interest in seeing the world price at the highest possible level, regardless of the burden placed on free-market exporters in supporting that figure. The United Kingdom, though a net sugar importer, had also an important interest in a high free-market price. A good part of the preferential sugar would in fact be taken up by Canada, not itself a party to the CSA. The more satisfactory the return earned on this portion of their sales, the less hard would Commonwealth exporters bargain at annual price negotiations with the United Kingdom. Not only would Canada help bear the burden of Commonwealth support, but its contribution would be made in hard currency. Indeed, there were occasional protests from the British West Indies that restrictions on their imports from Canada in the interests of the sterling area were jeopardizing their position in that market. In sharp contrast with the situation in 1898, Canada was now the Commonwealth's chief advocate of a liberal trade policy. In the contemporary environment, a tariff structure favoring Empire sources seemed almost no protectionism at all.

The United Kingdom had an interest in the free-market price for other reasons as well. The business of re-exporting sugar refined from imported raw reached a volume of some 700,000 tons as part of the postwar export drive and even at the cost of starving the domestic market. Sugar sold abroad had to compete on a free-market basis. Moreover, there was the broader though related policy of promoting London's entrepôt trade and terminal commodity markets. Sugar could not be expected to fulfill its anticipated role so long as imports were exclusively on government account. Partly because of the special commitments assumed under the CSA and partly because of the risks involved in liberalizing the trade in a dollar commodity, shifts in policy came slowly. It was not until October 1952 that British refiners were permitted to purchase any raw sugar on private account, and then only for re-export to designated nonsterling areas. The Ministry of Food did not relinquish its responsibility for exports to certain sterling territories until January 1, 1954 (*23*, p. 5). By July 1954, liberalization had reached the stage where raw (but not refined) sugar originating in the dollar area could be sold to nonresidents for sterling or the appropriate local currency (*24*, p. 34). Accordingly, sterling became in effect convertible in terms of raw sugar, with the Bank of England's supervisory powers over the trading community replacing more direct forms of commercial restrictions. Finally, in 1955 a plan was announced for honoring the CSA negotiated-price quotas and yet bringing to an end the practice of government purchase of all sugar imported for domestic consumption. A Sugar Board was to purchase the relevant Commonwealth sugars in export locations, absorb the losses (and occasional profits) involved in reselling them immediately at going commercial prices, and all shipment was then to proceed on a private basis. All sugar would accordingly arrive in the United Kingdom on world price terms (*25*).

The quota provisions and not merely the price features of the CSA had an important bearing on a new ISA. The full CSA figures, at 2,375,000 long tons, represented an increase of 50 per cent over the level of Commonwealth exports established under the ISA of 1937 (*9*, p. 31). With negotiated-price quotas fulfilled and domestic beets supplying 700,000–800,000 tons of sugar, raw value, an unrationed market in the United Kingdom could be expected to get along with only about 250,000 tons of foreign sugar. That figure was 400,000 tons below the net requirements of a smaller prewar consumption. Moreover, the CSA quotas were to be considered an irreducible minimum (Art. 10) in any new ISA, and indeed there was an implied threat that Commonwealth export might go still higher if negotiation of a new international instrument was delayed. Were it not for the growing requirements of some portions of the Commonwealth overseas, the further erosion of the free market here at work would have been more serious. The prewar and postwar positions of relevant portions of the British Commonwealth, in thousand long tons raw value, were as follows (*40*, p. 23):

Trading area	Average 1937–39	Average 1950–52	Full quota CSA	1953
Deficit areas—net imports	2,876	2,723	—	3,726
Including:				
United Kingdom	1,888	1,343	—	2,296
Canada	437	531	—	518
Malaya	123	180	—	159
Ceylon	84	139	—	155
Surplus areas—net exports	1,675	1,718	2,375	2,333
Including:				
Australia	426	333	600	717
British Guiana	181	196 ⎫ 900	⎰ 212
British W. Indies	414	576 ⎬	⎱ 646
Jamaica	102	213	...	278
Trinidad	125	128	...	130
Barbados	126	159	...	151
Windward and Leeward Isles	61	76	...	87
Mauritius	295	438	470	474
Union South Africa	214	43	200	93
Fiji	127	107	170	178
Commonwealth net imports	1,201	1,005		1,393

INTERNATIONAL AGREEMENT OF 1953: BENEFITS AND BURDENS

Tests of equity perhaps ought not to be applied to an agreement that passed through several drafts over a period of years and was hammered into its final form only after six weeks of hard bargaining. But two aspects of the question can hardly be ignored. One is the balance established between consumer and producer interests; the other, the distribution of benefits and burdens among exporting countries.

Equality of voting rights of importing and exporting countries was officially respected in the agreement:

Importing countries		Exporting countries	
United Kingdom	245	Cuba	245
United States	245	USSR	100
Japan	100	Dominican Republic	65
Canada	80	China (Formosa)	65
Germany (Fed. Rep.)	60	Brazil	50
Switzerland*	45	Australia	45
Ceylon*	30	Czechoslovakia	45
Norway*	30	Indonesia*	40
Portugal	30	Peru*	40
Greece*	25	Poland	40
Austria*	20	France	35
Israel*	20	India*	30
Lebanon	20	Mexico	25
Spain*	20	Philippines	25
Jordan*	15	Belgium	20
Saudi Arabia*	15	Denmark*	20
		Haiti	20
Total votes	1,000	Hungary	20
		Netherlands	20
		South Africa	20
		Nicaragua*	15
		Yugoslavia*	15
		Total votes	1,000

* Countries failing to give appropriate notice of intention to adhere, as of May 7, 1954 (*26*, p. 149).

But the equivalence indicated in the tabulation is highly illusory. Four of the five largest blocs of votes were unlikely to be cast according to the direct consumer or producer interest they were presumed to represent. Nor did voting strength properly reflect the importance of smaller importing nations that obtain all or a major portion of their sugar supplies from the free market. The consumer interest was further diluted by the failure of several medium-sized importers to ratify the agreement they helped negotiate.

The interests of the United Kingdom, as already indicated, are extremely mixed. This country does pay the "world" price on sugar she imports from Cuba, the Dominican Republic, and comparable sources, whether to meet the small deficit in domestic requirements or for the export trade. But even in these respects her position was extremely well hedged in the first year under the new Agreement. In building up reserves against the hazards of derationing, the United Kingdom had entered into a contract with Cuba on April 13, 1953 for one million tons, on extremely favorable terms. Large, indeed excessive, stocks were thus accumulated at prices moderately below the Agreement minimum. Accordingly, even a higher Agreement floor would have penalized other free-market importers or re-exporters relatively more than the United Kingdom in the short run, while a lower minimum could have involved the Treasury in substantial inventory losses on government-held stocks. When one recalls also the function of a "reasonable" free-market price in the CSA, one certainly cannot infer the clear interest in a low world price that might be suggested both by her large net-import position in sugar and by her failure to participate in the International Wheat Agreement of 1953. The very delegation responsible for casting the United Kingdom's bloc of 245 votes on the importing side had, under the terms of the CSA (Art. 11), to include "as advisers" one representative from each of the sugar-exporting colonies.

The United States, though it divides 49 per cent of the importer votes equally with the United Kingdom, imports nothing on free-market terms and has for months on end maintained the domestic price of raw sugar (including the one-half cent tariff) at almost double the "world" level (27, p. 2). While so high a differential might be maintained indefinitely, the danger is that a quota premium judged excessive might bring Congressional correction. Besides, if Cuba ships sugar more vigorously into the United States during the key months of the marketing year for lack of other outlets, prices in New York can weaken though total imports are not increased. As for the State Department, it cannot ignore either the current well-being or reasonable expectations of the Cuban people, who happen to occupy a strategic island close to the Florida coast and in the main line of shipping via the Panama Canal. The United States will accordingly countenance, even if it does not actively promote, a higher free-market price.

Among other major importers, Japan comes closest to depending exclusively on free-market supplies. The one million or so tons that once came from Empire sources must now, as a result of the postwar territorial settlements, be purchased from foreign countries, if at all. While Canada now produces double its prewar volume of beet sugar, it continues to import some 500,000 tons, or about 80 per cent of its rising requirements. West Germany, a region with an estimated net deficit of more than 500,000 tons before the war, was by 1950 importing at a rate that entirely compensated for the loss of shipments from German territories now under Communist control. But expansion of domestic beet-sugar production has been so considerable that self-sufficiency is in the offing, and Germany's heavy voting strength remains somewhat anomalous. Mainland China (including Manchuria) imported one-third of a million tons before the war, but that commerce no longer is reflected in the trade statistics of Western countries, and Red China could not be represented in 1953 at a United Nations Sugar Conference.

The remainder of "world" sugar exports goes to a host of ultimate destinations. A group with net imports of 100,000 to 250,000 tons—including countries in Western Europe, Asia, and Latin America, both independent nations and also some dependent overseas territories—absorbed about 1.5 million tons in 1935–39 but close to 2.5 million in 1952 (*28*). Only the United States and the United Kingdom exceeded that collective prewar figure; the latter's postwar imports before 1953 had averaged only about 1.5 million, and even an unrationed consumption may not support imports in excess of the combined takings of the group. A still larger number of countries, none with net imports above 85,000 tons and most below 10,000 tons, provides a total market for one million tons. It is these voices that are subdued or still in the meetings of the International Sugar Council.

Among exporters generally, nations with an interest in high "world" prices can neutralize the votes of several that would prefer a larger volume of free-market trading. Australia and the Union of South Africa, which have no export quota at stake, can counterbalance the Dominican Republic, a nation capable of exporting more than 750,000 tons on the free market. Brazil, France, and the Philippines, which export a trivial portion of their crop on the free market, offset the scheduled voting strength of China (Formosa) and Peru, countries that enjoy no privileged outlets for the great bulk of their production. The position of the Soviet bloc is somewhat peculiar. The postwar direction of trade might have justified placing the USSR in the importer column. By definition, however, the Agreement excluded from the free market any sugar shipped to the USSR from the Soviet satellites. The USSR therefore remains in a position to strengthen the bargaining position of Eastern European exporters vis-à-vis the West.

The turns of the market are such, however, that major reversals of customary patterns can and do occur. Unexpected support for the free market came both in 1954 and in 1955 from heavy imports into countries classified as exporters by the ISA. In 1954, partly because domestic policy created an unwarranted price advantage in favor of gur at the expense of the refined product, India experienced a sugar shortage that required imports of some 900,000 tons (*29*, p. 23). Unfavorable weather compounded by difficulties in the internal organization of agriculture severely reduced sugar production in the Soviet satellite countries that same fall. As a result, net imports to Soviet countries were required in 1955, although the "basic export tonnage" of this group totaled over 800,000 tons.

Paradoxically, the major exporting nation stands as the chief defender of the consumer's interest. What Cuba seeks from an international agreement is not primarily a high price but rather some assurance against further shrinkage in the total free market. In selling outside the United States, Cuba's sole competitive advantage rests on an ability to offer sugar at prices that look quite unattractive to practically every other producing region. Even 2½-cent sugar need not be fatal for her, and might possibly bring redemption. At that price, expansion by competing exporters would run into strong resistance; occasional exports from normally self-sufficient countries would be discouraged; and the degree of sugar protectionism in the Commonwealth and Continental Europe might warrant reappraisal. A dent might conceivably be made on the sugar markets within South America, which have been expanding of late more rapidly than elsewhere, but for the most part without resort to imports. The higher the price target of a sugar agreement, the greater the restriction imposed on Cuban production, and the broader the umbrella held out to shield Cuba's competitors. She can accordingly be counted on to vote for a relatively high volume and a relatively low price, as she did also in the deliberations of the Sugar Conference (*30*, p. 56).

The distribution of benefits and burdens can hardly be appraised as precisely as the interests of participating countries can be pinpointed. That privileged sources of supply come out extremely well, there can be no doubt. This is most unfortunate. At the crux of any ISA are the restraints that major importers accept upon protected domestic output and upon supplies from specially privileged areas. Under the ISA of 1937, a serious attempt was made at least to maintain the *status quo* and to reserve for free-market exporters a share of increases in consumption. But in 1953 a further erosion of the free market was sanctioned. Consistent with the CSA, allowable exports of Commonwealth territories were 50 per cent higher than in 1937. Shipments from the Soviet satellites to the USSR were in no way inhibited. Up to 175,000 tons that five Western European countries[3] would formerly have imported from the free market were now reserved for exports originating within the group. Even the modest obligation to reduce protective margins as the free-market price rises (Art. 4, ISA of 1937), in order that domestic consumption might be encouraged and domestic production restrained, has now been dropped. In the view of the United States Department of Agriculture, neither increases in statutory marketing quotas of domestic producing areas nor occasional exports of over-quota domestic sugar would violate America's commitments as an importing nation (6, pp. 4, 52).

In other respects also, sheltered exporters stand on firm ground. Because they operate planned national economies, Eastern European exporters escape the duty of having ever to "restrict" production or even to report on subsidies. Since exports to the USSR are not chargeable against their ISA quotas, the satellite countries have in effect gained a considerable improvement over their prewar allowances. Australia and South Africa, though they vote as exporters, need not carry minimum stocks as most free-market exporters are required to do. Under perhaps the most peculiar clause of all (Art. 16(2)), British Commonwealth exporters can renounce all obligations, light though they are, if a Commonwealth importer enters into a "special trading arrangement" guaranteeing a specified portion of its market to any participating exporter. This clause tends to prevent the renewal of such trade agreements as Cuba and the Dominican Republic have had with Canada in some recent years, and Commonwealth exporters seem to be denying to relatively weak economies the kind of guaranteed markets that the British colonies and the Dominions have assured for themselves.

In the actual allocation of export quotas (Table 12), the countries naturally enjoyed uneven success. Of countries exporting exclusively to the free market, the Dominican Republic's basic export tonnage[4]

[3] Belgium, Luxembourg, Netherlands, France, and the Federal Republic of Germany.

[4] Quotas actually in effect at any one time will regularly differ from basic export tonnages. Adjustments are to be made, in accordance with prescribed rules, to take account of changed estimates of free-market requirements, reallocation of unused quotas, etc.

was 50 per cent higher, Peru's 15 per cent lower, than in 1937. Bargaining strength must have been a more important consideration than either performance or efficiency. Indeed Peru, claiming to be "possibly the cheapest producer in the world," insisted that "Neither the United Nations nor the present Conference could press an efficient producer to become an inefficient one" (30, p. 49), and accordingly withdrew its representation. Formosa and the Dominican Republic each enjoy a basic quota of 600,000 metric tons. Before 1953, the latter's production was rarely sufficient to fulfill five-sixths of its quota, while Formosa supplied fully a million tons as part of the Japanese sugar empire before the war and has since exceeded the quota figure under trying local conditions. Indonesia, which enjoyed the largest export quota of all in 1937, shipped out practically no sugar during 1945–53. But a 250,000-ton quota, and a contingent claim to favorable future treatment (Art. 14A(5)), were not enough to win its ratification. Brazil, Mexico, and France, while normally self-sufficient, look to export as a safety valve for occasional domestic surpluses, and have been provided for in various ways under the agreement.

TABLE 12.—FREE MARKET EXPORTS AND ISA BASIC EXPORT QUOTAS*

(*Thousand metric tons*)

Country	Basic quota 1937	Assigned "basic export tonnage" 1953	Requested quota 1953	Net exports 1951/52
Cuba[a]	940	2,250	2,500	2,362
Dominican Republic	400	600	700	524
Formosa	—[b]	600	750	398
Peru	330	280	360	304
Czechoslovakia	250	275[c]	450	315
Indonesia	1,050	250	500	8
Poland	120	220[c]	350	220
USSR	230	200	250	...
Brazil	60	175	400	2
Eastern Germany	120[d]	150[c]	250	340
Mexico	0	75	100	9
Denmark	0	70	75	104
Total[e]	3,623	5,390	7,005	4,905

* Data from F. O. Licht's *Sugar Information Service*, Sept. 2, 1953, p. 3.
[a] Excluding exports for consumption in the United States.
[b] Part of Japanese Empire.
[c] Excluding exports to the USSR.
[d] Prewar Germany.
[e] Including unlisted exporters.

The Cuban quota of 2.25 million tons can hardly be considered ungenerous. The figure is more than twice as large as in 1937, and represents fully 40 per cent of the total allotment to all exporters. But Cuba's potential for expansion in response to modest price stimulus is enormous. In the effort to compensate for supply deficiencies elsewhere, Cuba contributed almost half of all sugar that moved in international trade during 1945–49, protected and free-market supplies alike. Sugar production, the basis of her well-being, had by 1952 reached almost 8 million tons, as compared with a restricted crop of about 3 million tons in 1935–39 and a peak of just under 6 million in the late 1920's.

Though still the largest single element in the United States sugar supply, Cuba has consequently become increasingly dependent on the free market. As recently as the late 1920's, only one-quarter of her crop had to find an outlet outside the preferential American system. In the 1930's, conditions on the island were barely tolerable with a million-ton quota under the ISA and twice that volume of exports to the United States. The United States of late has been absorbing closer to 3 million tons a year, but Congress is under relentless pressure to further displace imports from Cuba by domestic production. In the crop years when Cuban exports were highest (1947/48 and 1950/51), more than half had to be sold outside the United States, many of the necessary dollars coming, however, from various United States aid programs. Sugar output in importing regions generally has now recovered to and in many cases surpassed prewar levels. Even without an ISA, poor market prospects induced Cuba to restrict her 1953 crop to 5.5 million tons by official decree. With a substantial quota under the Agreement, her best prospect was to limp along at less than two-thirds of productive capacity and a considerably lower fraction of obvious potential.

INSTABILITY OF THE FREE-MARKET PRICE

The "world" price of raw sugar fluctuates over a considerable range, though not so widely as the prices of many other primary commodities. The inflation that followed the outbreak of the Korean war jumped the quoted price of sugar, f.a.s. Cuba, from about 4.20 to almost 6.00 cents per pound, between June and August 1950. After a setback in February 1951 to about 4.80 cents, it climbed momentarily above 8.00 cents in June, only to fall back below 4.80 cents by the end of the year. Subsequently the "world" price sagged to a level little over 3.00 cents in October 1953 (31, p. 21). In earlier periods, price variations of comparable magnitude and speed were experienced from time to time.

In many respectable quarters, it is customary to argue that these price movements are due to the economic failings of unregulated marketing, and to look to an ISA as a price stabilizer. Thus a study by the Food and Agriculture Organization (FAO) states (*12*, pp. 3–4): "Sugar exhibits to an unusual degree the features that make the operation of an unregulated 'free' market undesirable." These undesirable features include "excessive price fluctuations" mainly because, "owing to the low short-term elasticities of demand and supply, even small changes in the balance of production and consumption tend to be associated with large variations in price." Moreover, in response to such price variations, "the cane-sugar industry has a chronic tendency to excess capacity." Both the FAO and the United Nations' Department of Economic Affairs, in a study of Latin-American trade (*32*, p. 70), support the further notion that market adjustments are in fact perverse, alleging that sugar producers can be expected to respond to low prices by increasing rather than by reducing output.

Such generalizations do not measure up well against the facts. There is clear evidence of some price instability and of severe "chronic surplus" in world sugar markets, but for reasons little associated with the characteristics of an "unregulated" market mechanism. The most important factors at work on world sugar prices have been the following:

1. Peacetime imbalances between productive capacity and sugar consumption have typically been huge, not "small." Cuba could turn out almost 6 million short tons in 1929, and Java about 3 million; within four years, sales had fallen to barely 2.5 million and one million respectively. Cuba could today produce some 8 million tons but has been fortunate to sell 5–6 million. Excessive capacity developed primarily because the heavy requirements for sugar imports that followed two world wars proved transitory. The second major factor contributing to imbalance in the 1930's was expansion in the Philippines, Puerto Rico, Hawaii, the Japanese Empire, India, and the British Empire, all of it to serve protected markets. Increased production in Europe and in the British Commonwealth is causing a repetition of the experience today. The natural counterpart of expansion in those regions is redundant capacity elsewhere.

2. If production and consumption of free market sugar *were* in close adjustment, then the narrowness of the free market, which supplies only the residual requirements of major importing countries, would be a prime cause of price instability. In actual practice,

"world" sugar prices can fall a long way because importing countries react to moderately cheaper foreign sugar not by increasing imports but by increasing their margins of protection to domestic producers.

3. While "relatively moderate changes in demand can have violent effects on prices" in a residual market (*33*, p. 583), "world" sugar is liable to perverse variability in free-market supplies quite as much as in free-market requirements. By definition, all imports into the 8-million-ton American market are excluded from the free market. Though the "world" price and the United States price largely go their separate ways, American action has serious external repercussions. The Secretary of Agriculture, by varying the total marketing quota, has had a considerable success in stabilizing the domestic price of sugar. Periods when commodity prices generally are low, here and abroad, find the domestic price of sugar relatively high. United States sugar consumption at such times is accordingly discouraged, while sugar crops are relatively attractive. These variations both in production—particularly of beet sugar—and in consumption lead the United States to import less sugar from Cuba during the very deflationary periods when that exporter is likely to be having greatest difficulty in marketing overseas; while more Cuban sugar tends to be absorbed in the United States when demand in the free market is also relatively strong, as it was for some months following the Suez crisis. Stabilization of the American price of sugar accordingly contributes to the destabilization of the free-market price, though it might be argued that the price to all comers would go higher in periods of scarcity if the United States were to bid for its requirements in open competition.

TABLE 13.—INTERNATIONAL DISPARITIES IN PRICES OF RAW SUGAR, 1948–54*

Market specifications	Unit (*lbs.*)	Annual average price						
		1948	1949	1950	1951	1952	1953	1954
"World" sugar, f.a.s. Cuba (sterling equivalent)	112	23s.6½d.	33s.3d.[a]	39s.9½d.	45s.8d.	33s.4d.	27s.4d.	26s.0d.
Commonwealth Sugar Agreement, c.i.f. U.K., at prewar freight	112	27s.3d.	27s.3d.	30s.6d.	32s.10½d.	38s.6d.	42s.4d.	41s.0d.
Cuba, f.o.b., sales to U.S.	100	$4.64	$4.86	$5.09	$5.07	$5.35	$5.42	$5.24
Cuba, f.o.b., sales to rest of world	100	$4.23	$4.08	$4.98	$5.67	$4.16	$3.41	$3.26

* Data from C. Czarnikow, Ltd., *Sugar Review* (London) and International Monetary Fund, *International Financial Statistics* (Washington, D.C.).
[a] Calculated at exchange rate of $2.80, which became effective Sept. 18, 1949.

4. Quite apart from *movements* in the world price, the *structure* of international sugar prices is complicated by importer policies. The high price that Cuba receives on its United States sales covers overhead costs of production and permits exports to the world market to be priced at lower rates appropriate to variable costs alone. It might be said that the United States "subsidizes" consumption abroad (or that Cuba "dumps" on the free market) under this arrangement. Only rarely is the pattern reversed. Out of obvious long-run interest, Cuba continued to honor its United States quota for some months during 1950 and 1951 even at prices considerably below those ruling in the free market. During the subsequent two-year period, prices in the American market edged upward in the face of a sustained sag in the world price (*34*, pp. 24–25).

Interesting price differentials are also created by British policy. In the one-million-ton United Kingdom–Cuban contract of 1953, at a moment when "world" sugar was quoted at about 3.40 cents per pound, a price of 2.75 cents was specified on shipments in 1953 and 3.04 cents on the portion remaining to be shipped in 1954. Exercise of British bargaining power forced the lowest price at which sugar had been traded in the postwar years. Cuba was not reluctant to concur because disposal of one million tons, even at a low price, made it easier to earn a higher return on the remainder of the crop. This, however, was an unusual transaction, preparatory to the termination of rationing in the United Kingdom. Over the longer haul, price features of the Commonwealth Sugar Agreement are more relevant. Throughout the late 1940's, the price at which the United Kingdom contracted to purchase Commonwealth sugar did not diverge widely from the price on world markets. During the Korean boom, the United Kingdom (like the United States) obtained its preferential supplies comparatively cheaply, for the contract price in 1951 was almost 30 per cent below the free level. World prices, however, began to decline in June 1951, whereas CSA prices subsequently climbed considerably more sharply than those in the United States and were reduced for the first time in 1954. That year, CSA prices stood more than 50 per cent higher than in 1949, whereas world sugar averaged a 20 per cent decline.

The widening disparity between the prices of Commonwealth and foreign sugar created a highly artificial internal price in the United Kingdom. So long as the Ministry of Agriculture, Fisheries, and Food remained responsible for purchasing on government account all sugar imported for domestic consumption, its policy remained one of reselling raw sugar at home at a rate that covered the expense of acquiring its sugar from all sources, and that resale price, as already indicated, became the pivotal price for beet sugar as well. The averaging process created something of a lag in the adjustment of internal prices to changes in external prices, while the high premium on CSA sugar kept the level of internal prices considerably above the world level. Indeed, so long as world prices remained low and relatively stable, variations in the proportion of foreign sugar imported for United Kingdom consumption were a more important influence on the internal price level than small changes in world price that entered into the average.

The White Paper (25) that looked forward to the complete return of private trading and to the entry of sugar on the basis of the world-market price sought also to establish a more rational domestic price structure. The particular device was to be a surcharge on all sugar (and molasses) for domestic consumption, whether home-produced or imported. The surcharge is to be calculated so as to cover deficits incurred in the reselling of CSA sugar below the negotiated contract price and to make deficiency payments to the British Sugar Corporation to cover losses incurred in purchasing home-grown beets at prices guaranteed by the government.

The official position is that "since the internal price level will be directly linked with movements in the world price, internal prices may be expected to move more freely than under present arrangements" (25, p. 4). To some extent, that is a correct interpretation. Alterations in the level of surcharge are to be made only infrequently, and accordingly all movement in world prices will in the short run be immediately reflected in the internal price level. But all changes in the surcharge will in fact be perverse. So long as only about 10 per cent of supplies in fact come from foreign sources, small changes in the guaranteed price for beets and in the negotiated CSA price tend to be more important than small changes in the level of world price in determining how large a surcharge is needed to put the trade on a self-financing basis. Large increases in the world price, particularly by reducing losses incurred on negotiated CSA sugar, make a smaller surcharge necessary. Indeed, once the world price climbs above the CSA rate, a rebate rather than a surcharge can become appropriate. Thus the margin between internal and external price declines the higher the external price, and United Kingdom consumption will in fact be promoted rather than restrained at such times as there are real shortages of world sugar. As in the United States, though the operational technique is very different, relative stability in domestic price tends to be at the cost of relative instability of the world price.

5. Sugar can hardly insulate itself against the effect of a general inflation or deflation of commodity prices, especially if open hostilities are an important causal factor (as in 1950–51). But the peculiar political environment of the day implies that during war scares precautionary stockpiling will be taking place on both sides of the Iron Curtain. The major areas of beet-sugar surplus before the war were Poland, Czechoslovakia, and eastern Germany, all now incorporated in the Soviet bloc. In the postwar period such surpluses as existed in these territories were mainly absorbed by the Soviet Union; any serious effort to improve consumption levels of the Russian people might be expected to continue the flow in an eastward direction. But even to the extent that these beet-sugar exporters recover their sugar trade with the West, exports can certainly not be counted on in a serious emergency. Here again supplies available to the free market seem likely to be cut back at the very moment when free-market requirements are heaviest. Regardless of the risk of open hostilities, international sugar markets now face the prospect of substantial, erratic fluctuations in nondollar beet-sugar supplies from Eastern Europe, in response as much to political considerations as to economic stimulus.

6. Europe's protected beet-sugar system, concentrated in a region liable to similar variation in weather conditions, introduces a further element of instability into the world market. European acreages planted to beets have tended to increase over time but do not show major fluctuations from year to year. Variable weather, by affecting the weight of harvested beets, does cause substantial changes in the annual crop. Since the direction of variation tends to be the same in all northern European countries, there is a counterpart in Europe's fluctuating imports of cane sugar from overseas. The United Kingdom, in promoting exports from Australia and South Africa at the expense of the free market, similarly accentuates the importance of natural hazards, for it is substituting sources which are drought-prone for those where yields are more reliable.

7. Mills typically expand processing facilities and farmers extend plantings when prices are profitable, not when they are depressed. For all commodities, expansion is a smoother process than contraction; as compared with other commodities, sugar production is particularly responsive to moderate price increases, but is no more unresponsive to price decreases. The notion that growers will respond

to low prices by perversely increasing output may be of some theoretical interest for primitive economies; it can have little practical application for commodities so closely tied to highly capitalistic processing facilities as are sugar beets and sugar cane. Cuban experience does not give evidence of a backward-sloping supply curve (*35*, p. 17).

To be sure, time series will demonstrate that, in some countries, "production actually tended to rise in periods when prices declined" (*12*, p. 5). More sugar was indeed supplied in the 1930's by "Mauritius, Trinidad, British Guiana, Jamaica, and Australia," but for three main reasons: (*a*) these producers sold sugar not at the very unsatisfactory "world" market price but at far more favorable absolute prices; (*b*) the price of sugar relative to alternative export staples was, in certain of these regions, reasonably attractive; and (*c*) the decline in real costs, resulting from technological improvement, would have made some expansion possible even at reduced relative prices. Java, which received only the "world" price, increased production between 1927 and 1930 in the face of declining realized prices, because advances in agricultural science were making a new price structure feasible.

PRICE STABILIZATION UNDER THE INTERNATIONAL SUGAR AGREEMENT

Standing cane available for harvest and stocks of raw sugar already accumulated in Cuba, excess factory capacity in various countries, and protected expansion in progress in others implied that low rather than high prices were likely to concern the present International Sugar Council for the early years of the Agreement, barring war or internal insurrection. The price-support features included adjustment of total export quotas to the estimated requirements of the free market; a ceiling on total stocks held in countries enjoying export quotas; and limitation of total production in quota countries to the sum of export quota, maximum stocks, and domestic consumption. Much was claimed (cf. *36*, p. 3) for the provision that export quotas had automatically and successively to be reduced by at least 5 per cent within 10 days after the "world" price had stood below the 3.25 cents floor for 15 consecutive market days.

This is by no means a powerful enforcement mechanism. The floor is not buttressed by guaranteed importer purchasers, as under the International Wheat Agreement. The one clear obligation upon importing countries (Art. 7(1)(i)) is that annual imports from non-participating countries do not exceed 1951–53 levels. Even that commitment can be circumvented by pleading monetary difficulties (Art. 25). A moderate clause included in the 1937 Agreement, which could be interpreted as requiring importers to provide some relief for exporters by maintaining "adequate reserve stocks" (Art. 26(a)), has been dropped. The weapon of lower export quotas gives out once cuts have reached 20 per cent of basic export tonnages. The floor price had to be maintained during the first quota year in the face of the high aggregate quota written into the Agreement, despite a heavy 1953/54 beet crop in Europe, and despite the provision that shipments under the United Kingdom contract were not deductible from Cuba's quota. The floor was held less by the International Sugar Council than by the Cuban Sugar Stabilization Institute, which intensified its pre-Agreement practice of regulating the flow of sugar into export. At that, prices stood below the Agreement for various periods after June 1954.

While the top of the "zone of stabilized prices" was initially of rather academic interest, the provisions of the new ISA were hardly likely to cope with the type of situation that could make the ceiling operative. Increases in quotas are provided for in the same fashion as decreases, though the automatic increment is 2.5 per cent larger. Quota countries must hold minimum stocks (amounting to 10 per cent of the basic export tonnage) for meeting increased free-market requirements, at the call of the International Sugar Council (Art. 13(3)). Since the Communist bloc and Commonwealth exporters are not bound by this article, this basic reserve cannot exceed 450,000 tons, which is considerably less than the year-to-year fluctuations in requirements that must be expected from variations in the European beet crop alone. By contrast with the Wheat Agreement, which specifically provided for allocating supplies among eager importers when there was pressure against the ceiling, the ISA merely assigned participating countries a priority over nonparticipating ones. Importers may withdraw from the Agreement in the event of an abnormal price rise (Art. 44(2)), or invoke the monetary escape-clause, but neither course of action can hold prices within the Agreement limit.

The main guarantee against the prospect of high prices, like the chief support of the market, was in fact unilateral: excess capacity already in existence in Cuba and the Cuban Stabilization Reserve. The latter, originally totaling some 1.75 million tons, was set aside from the record 1952 crop, for release against future United States quotas in five annual installments. Though these stocks were exempt from the maximum-stock provisions of the ISA, Cuba was expected to "consider" a Council request that they be made available to the free market (Art. 13(6)). As it happened, the Stabilization Reserve was largely exhausted by late 1956, when the Suez crisis together with a short beet crop in Europe suddenly raised free-market requirements. That the ISA was an inadequate instrument against dear sugar became obvious as the free-market price rapidly breached the Agreement ceiling. Once again, Cuba temporarily received a higher price on shipments to the world market than on those exported to the United States. As in the past, out of a sense of long-run interest, Cuba sought to fill any increase in its United States quota, regardless of incidental effects on the free market. She can be expected to act similarly in the future even without the penalties she would incur for doing otherwise, under the 1956 version of the United States Sugar Act.

Certain difficulties in the first year of operation distinguished the ISA of 1953 from its prewar predecessor. For one thing, stocks built up in the United Kingdom stood almost one million tons in excess of normal as of December 31, 1953 (*37*, p. 5), and the backlog had to be worked off during the first quota year. For another, Cuba's quota was substantially inflated by virtue of the slow restoration of Indonesian sugar production. Moreover, the portion of the free market not covered by the 1937 Agreement had been inconsequential, whereas the new ISA failed to enlist the membership of Indonesia, Peru, the Indian Union and Pakistan; and Brazil subsequently declined to ratify. Article 7(1)(i) of the Agreement, regulating imports by members from nonmember sources, accordingly became a more significant provision than originally anticipated, but the risk nevertheless remained that the sacrifices of member exporters would unduly operate to the advantage of nonmember exporters. Special generosity was shown by the International Sugar Council in the amount of sugar permitted to be imported into Japan from Indonesia (*38*, pp. 9–10), mainly because foreign exchange was available for purchases from that source and dollars were currently being used sparingly for sugar imports. The case of Brazil, which withdrew from the Council in the face of a particularly large crop and unusually large exportable surplus, raises the question of whether quotas for smaller exporters might not be cumulative for a second year, rather than rigid for each twelve-month period.

INTERNATIONAL PROBLEMS OF SUGAR

Participants have an official responsibility to initiate programs of economic adjustment that "ensure as much progress as practicable within the duration of this Agreement towards the solution of the commodity problem involved," and weak provisions do aim at lower subsidies and higher consumption (Arts. 3–5). But it is fair to say that toward the root difficulties of the world's sugar, the ISA makes no contribution whatsoever.

1. Two world wars disrupted former patterns of production and trade, and on both occasions the world temporarily (and rationally) turned to low-cost Cuban sugar to make up the deficiency. How is Cuba to adjust to a lower level of demand without social disaster? What is the peacetime function of an economy whose resources are called into full play only after the nations go to war? In a one-crop economy, Cuban sugar producers cannot, as do growers of many crops in the United States, look to the public treasury for supplementary income in foul weather.

2. The inclination of major importers to supply an increasing portion of national consumption from protected sources creates great disparities in the well-being of various exporting regions. In the Caribbean, Puerto Rico sells its entire output (though not its full potential) at the favorable United States price; the British West Indies and British Guiana currently sell under similarly favorable conditions in the United Kingdom. The Dominican Republic, however, must market practically its entire output at the world price, while Cuba straddles New York and points overseas. In the Orient, the Philippines enjoy the United States outlet, but Formosa (and Indonesia) take what they can get, especially from Japan. Political decisions in importing countries, not the efficiency of production methods, largely determine the prosperity of exporting regions. Cuba and the Dominican Republic, committed to international markets for 90 to 95 per cent of their sales, find the limits imposed on domestic policy extremely narrow and have little voice in external decisions that are crucial to their well-being. Several sugar-exporting islands enjoy independent nationhood without being viable economic units, whereas colonial sugar producers today have the ear and the aid of the metropolitan legislature. Is there perhaps a road out of the present difficulties through political federation of tiny economic units, not merely within but beyond the British West Indies (cf. *39*, p. 84)?

3. For several decades, markets have been further disturbed by the effects of a technological transformation of sugar-crop production. For beets, this process has been largely agricultural, notably mechanization of the harvest. It might also perhaps be described as "defensive": the new practices did not at first permit beet acreages in the

United States to hold their own in competition with other crops, but at least the basis was provided for resisting a previous downward trend. For sugar-cane technology, the changes have had far wider ramifications. Modernization of processing facilities, improvement in cane transport and agricultural practices, and introduction of higher-yielding cane varieties have led the way in transmitting modern science to tropical agricultural economies. In this process, free-market exporters frequently played a pivotal role. In Cuba, the advantages of rail transport and highly capitalistic milling have been most fully realized. Java set the pace in improving cane varieties. But the profits from scientific advance have come to accrue more to their specially favored rivals, who are in a better position to make long-term investments with full confidence of selling their final output in protected markets. Nationalistic sugar systems now provide the incentives that nurture the technological innovators, in field mechanization or bulk shipment, while free-market producers have become the laggards.

4. The incentive to promote domestic or Empire production has been reinforced in the postwar period because more of the sugar available from free-market sources requires dollars than formerly. However, while promotion of dollar-replacing production is in the customary case a measure taken to the advantage of relatively weak importing economies and to the disadvantage of relatively strong exporting ones (like the United States or Canada), dollar supplies of sugar originate in countries (Cuba, the Dominican Republic) that are distinctly less prosperous than the major importers.

If one asks why Cuba nevertheless prefers to restrict its crop severely rather than accept payment in soft currencies, the answer is not simple. To be sure, the economy has a high propensity to import American goods. The United States is a natural source of the rice, lard, cotton goods, and wheat flour that Cuban consumers require. Clearly the American manufacturer of machinery, vehicles, and metal products has an edge over rivals in the nearby Cuban market, especially when a respectable fraction of the orders for capital goods are placed by American-owned enterprises on the island. But part of the explanation lies in the fact that any alteration in Cuba's basic trade or monetary policies would jeopardize her delicate position in the American sugar-quota system. Despite marketing difficulties, the peso remains pegged to the dollar; the alternative of exchange control is not seriously considered; and reciprocal preferences with the United States inhibit trade with other countries. As a member of the dollar area, Cuba's ties are stronger, but her privileges perhaps less, than her counterpart within the sterling area.

Clearly, to criticize the ISA of 1953 for introducing a "restrictive quota system" is to exaggerate its powers. Under surplus conditions, Cuba has been inclined to restrict production and to hold stocks, Agreement or no. What is at fault with the present Agreement is that it accepts as its main objective the reconciliation of essentially irreconcilable national sugar policies, while failing to come to grips with identifiable underlying difficulties. In the twentieth century, as in the nineteenth, politics more than economics is at the root of the international sugar "problem."

CITATIONS

1 U.S., *Economic Report of the President, 1954.*

2 World Crops (London), November 1953.

3 F. O. Licht's *Sugar Information Service* (Ratzeburg), Sept. 2, 1953.

4 International Sugar Journal (London), January 1954.

5 United Nations, Dept. Econ. Affairs, *Commodity Trade and Economic Development* (New York, 1953).

6 U.S. Senate, Subcommittee of the Committee on Foreign Relations, *International Sugar Agreement, Hearing* 83d Cong., 2d Sess.

7 U.S. Staff Papers Presented to the Commission on Foreign Economic Policy (1954).

8 UN, *United Nations Sugar Conference 1953, Summary of Proceedings* (New York, 1953).

9 International Labour Office, *Intergovernmental Commodity Control Agreements* (Montreal, 1943).

10 UN, Econ. and Social Council, Resolution 30 (IV), Mar. 28, 1947.

11 UN, *General Agreement on Tariffs and Trade* (Lake Success, 1947), Vol. I, Art. XXIX (1).

12 Food and Agriculture Organization of the UN, *Observations on the Proposed International Sugar Agreement* (Commodity Policy Study No. 4, Rome, 1953).

13 L. B. Bacon and F. C. Schloemer, *World Trade in Agricultural Products: Its Growth, Its Crisis, and the New Trade Policies* (International Institute of Agriculture, Rome, 1940).

14 International Sugar Council, *Pocket Sugar Year Book, 1950* (London).

15 Supplement to *Statistical Bulletin of the International Sugar Council* (London), Vol. 13, No. 7.

16 Gr. Brit., Ministry of Food, *Bulletin,* June 17, 1950.

17 Weekly Statistical Sugar Trade Journal, Feb. 28, 1952.

18 Gr. Brit., Ministry of Food, *Bulletin,* Jan. 12, 1952.

19 F. O. Licht's *Sugar Information Service* (Hamburg), Dec. 29, 1954.

20 B. C. Swerling, "The International Sugar Agreement of 1953," *American Economic Review,* December 1954.

21 H. Frankel, "Controls and Subsidies on Agricultural Products and Requisites. Sugar-Beet and Sugar 1939–53," Supplement to the *Farm Economist* (Oxford), VII, 7, 1954.

22 Australian Sugar Year Book 1951 (Brisbane).

23 The Public Ledger (London), Sept. 19, 1953.

24 International Monetary Fund, *International Financial News Survey,* (Washington, D.C.), July 30, 1954.

25 Gr. Brit., Min. Agr., Fisheries, and Food, *Future Arrangements for the Marketing of Sugar* (Command Paper, series Cmd. 9519, H.M.S.O., 1955).

26 International Sugar Journal, June 1954.

27 U.S. Dept. Agr., Commodity Stab. Serv., Sugar Div., *Sugar Reports No. 25* (May 1954).

28 U.S. Dept. Agr., For. Agr., Serv., *Foreign Agriculture Circular,* July 7, 1953.

29 Statistical Bulletin of the International Sugar Council, Vol. 15, No. 4.

30 UN, Econ. and Social Council, *United Nations Sugar Council 1953, Executive Committee, Summary Record of Meetings,* E/CONF. 15/EX/S.R. 1–17, Oct. 21, 1953 (mimeo.).

31 U.S. Dept. Agr., Prod. and Mkt. Admin., Sugar Branch, *Sugar Reports No. 21* (October 1953).

32 UN, Dept. Econ. Affairs, *A Study of Trade Between Latin America and Europe* (Geneva, 1953).

33 "Shelter for Sugar's Cinderellas," *The Economist* (London), Aug. 29, 1953.

34 U.S. Dept. Agr., Prod. and Mkt. Admin., Sugar Branch, *Sugar Reports No. 20* (September 1953).

35 J. W. F. Rowe, *Studies in the Artificial Control of Raw Material Supplies, No. 1, Sugar* (London and Cambridge Economic Service, Special Memo. No. 31, September 1930).

36 UN, Interim Coordinating Committee for International Commodity Arrangements, *Review of International Commodity Problems 1953* (New York, 1954).

37 Public Ledger, Jan. 16, 1954.

38 International Sugar Council, *Annual Report 1954* (London, 1955).

39 D. H. Robertson, *Britain in the World Economy* (London, 1954).

40 Commonwealth Economic Committee [Gr. Brit.], *Plantation Crops* (1955).

41 Australia, Commonwealth Bur. Census and Stat., *Official Year Book of the Commonwealth of Australia 1954.*

42 Australia, *Official Year Book . . . 1951.*

INDEXES

NAME INDEX

Ahlfeld, Hugo, 277
Alpert, H., 199
Anderson, Clinton, 182, 184
Anderson, Edgar, 154
Andreev, A. A., 307, 308
Arkhimovich, A., 123
Armer, Austin, 122

Bacon, Lois B., 35, 278, 350
Bailey, W. R., 122
Bainer, Roy, 119, 121, 122
Ballinger, R. A., 198
Baran, P. A., 61
Baratte, J., 123
Barlowe, F. D., Jr., 15
Barnes, A. C., 35, 87, 152
Benediktov, I., 302, 308, 321
Benin, G. S., 320
Berdeshevsky, P. G., 16
Beresford, Hobart, 122
Bernhardt, J., 197
Beveridge, Sir William, 233
Bingham, J. B., 121
Black, J. D., 15, 62
Brandes, E. W., 152
Brandt, Karl, 277
Brigden, J. B., 87
Brown, Robert C., 121
Brykczynska, W., 278
Bulganin, N. K., 311
Burdick, R. T., 121
Butenko, G. P., 322

Chamberlain, Neville, 215
Chang, P'ei-kang, 61
Chrzanowski, L., 278
Ciriacy-Wantrup, S. von, 276
Coke, J. E., 122
Coller, F. H., 234
Conner, William, 122
Coons, G. H., 121
Corson, C. T., 15
Cottrell, R. H., 89, 122, 123
Cross, W. E., 152
Culpin, Claude, 124
Curtin, P. D., 35
Czarnikow, C., 323, 341

Davies, R. A., 123
Davis, J. S., 35

Debordes, J., 277
Decoux, Louis, 279
Deerr, Noel, 62, 88, 152, 276
Demidov, S., 322
Deming, G. W., 122
Deomano, F. V., 153
Dillewijn, C. van, 152, 153, 154
Dillner, G., 276
Dodds, H. H., 62
Dolgopolov, M., 320
Doyle, B. K., 16
Dubourg, J., 15
Duncan, R. A., 154
Dutt, N. L., 152

Earle, F. S., 152
Ebi, Saburi, 16, 88
Efferson, J. N., 154

Ferguson, E. F., 154
Fleher, F., 123
Frankel, H., 351
Freeland, E. C., 87, 153
Frejlich, J., 89
Fridman, S. E., 298, 303, 308, 321
Fuchs, W. H., 123

Galbraith, J. K., 61, 86, 88
Gazdarov, M., 123
Ghandi, M. P., 62, 69
Gilcreast, R. M., 121
Goswami, P. C., 62
Gourou, Pierre, 62
Grau, R. Ramos, 87, 88
Greene, Wilfrid, 203
Griffin, C. S., 87
Grist, D. H., 15
Guerra y Sánchez, Ramiro, 35, 88
Gurevich, S., 322

Hammond, R. J., 16, 197, 200, 233, 234
Hart, P. E., 234
Hassack, A. T., 154
Heinrich, K., 278
Holland, W. L., 87
Honig, Pieter, 61, 88, 153
Hook, Andrew van, 15, 198
Hrudka, G. E., 122
Hsieh, S. C., 15
Hughes, C. G., 152

355

SUBJECT INDEX

Agricultural Adjustment Administration (AAA), 160

Alcohol, industrial and beverage, 9–12, 28, 33, 177–78, 183, 246; motor fuel, 10; in Italy, 247; in France, 247–48, 264, 266, 272

All-Union Scientific Research Institute for the Sugar Industry (USSR), 297

American Sugar Refining Co., 70

Area Sugar Officers (U.K.), 208, 218, 222

Argentina, 6, 19

Armed Forces, 171, 178

Australia, 6; feeding practices, 5; trade with U.K., 22, 27, 28, 31; crop limitation, 24; use of tractors, 138; cane transportation, 143; bulk shipment, 147; prices, 330; see also Queensland

Austria-Hungary: subsidies, 18; loss of beet area, 18, 241; exports, 239

Bagasse, 5, 9, 33

Barbados, 5, 8, 173; as scientific breeding center, 127, 128, 129

Beet pulp, 66, 208, 215, 227, 229; nutritive value, 4, 5

Beet-sugar industry: development, 53–55; and tax policy, 64; farm size, 78–79; beet-purchase contracts, 99–100; social aspects, 106; price guarantees, 187; introduction in Europe, 235; see also individual countries

Belgium, 33, 241; beet acreage, 252, 258, 275; sugar supplies, 265

Booker Bros., Mc Connell, Ltd., 70

Boston and Preston, 70

Brazil, 6, 8, 19, 347; alcohol, 10; modernization of milling, 125; government tractor station, 138

British Guiana, 6, 143

British Sugar Corp., 206, 208, 209, 214, 217, 218, 219, 224–28 passim; wartime loss of efficiency, 227–28

Brussels Sugar Convention (1902), 18, 20, 65, 200, 238, 239, 243, 325

Bulk shipment, 144–48, 188; raw cane sugar, 34; liquid sugar, 145

Burlap, burlap bags, 34, 144, 146, 148

B.W.I. Central Sugar-Cane Breeding Station (Barbados), 130

By-products, 8–11, 228–29; beet, 4, 235, 249; cane, 5; see also Alcohol, Bagasse, Beet pulp, Feed, Molasses

Cake and Biscuit Manufacturers' Wartime Alliance, 218

California, 8, 188; crop competition, 6; irrigation, 48; refinery, 72; mechanization, 98, 102, 108; subsidy payments, 193

California Agricultural Experiment Station, 94

Canada, 24, 27; Beet Sugar Development Foundation, 94; imperial preference, 325; sugar requirements, 335

Cane-sugar industry: social aspects, 17, 70–71; interwar developments, 55–61; plantation system, 71–74; small farm system, 74–77

Cartels, 24, 82, 162, 238, 243

Central Asia (USSR): irrigation, 48, 283, 284, 286, 296–99; beet acreage, 286, 296–99 passim; beet yields, 286, 298, 299; collective farms, 298; harvesting, 299

Central black-soil region (USSR), 283, 287, 293, 294

Central Chaparra, 70

Central Cunagua, 70

Central Delicias, 66

Central Jaronú, 70

Central Research Institute of the Sugar Industry (USSR), 112

Central Rio Haina, 67

Chadbourne Agreement, 23, 25, 242, 244

Chancaca, 8

China, 8, 32, 335

Coimbatore Sugar Cane Breeding Station, 32, 49, 130

Colonial Sugar Refining Co., 70, 76; farming system in Fiji, 76, 138, 142–43

Colorado, 107, 188

Combined Food Board, 180, 181, 213

Commodity Credit Corporation (CCC), 85, 173, 175, 187; purchases, 169, 188; subsidy programs, 174

Commonwealth, British, 212; quotas to U.K., 328, 329, 332; sugar position, 332; see also individual countries